Biologically-Inspired Energy Harvesting through Wireless Sensor Technologies

Vasaki Ponnusamy
Universiti Tunku Abdul Rahman, Malaysia

Noor Zaman
King Faisal University, Saudi Arabia

Tang Jung Low
Universiti Teknologi Petronas, Malaysia

Anang Hudaya Muhamad Amin
Multimedia University, Malaysia

A volume in the Advances in Environmental Engineering and Green Technologies (AEEGT) Book Series

Information Science REFERENCE
An Imprint of IGI Global

Published in the United States of America by
Information Science Reference (an imprint of IGI Global)
701 E. Chocolate Avenue
Hershey PA, USA 17033
Tel: 717-533-8845
Fax: 717-533-8661
E-mail: cust@igi-global.com
Web site: http://www.igi-global.com

Library of Congress Cataloging-in-Publication Data

Names: Ponnusamy, Vasaki, 1974- editor. | Zaman, Noor, 1972- editor. | Low, Tang Jung, 1956- editor. | Muhamad Amin, Anang Hudaya, editor.
Title: Biologically-inspired energy harvesting through wireless sensor technologies / Vasaki Ponnusamy, Noor Zaman, Tang Jung Low, and Anang Hudaya Muhamad Amin, editors.
Description: Hershey, PA : Information Science Reference, [2016] | Includes bibliographical references and index.
Identifiers: LCCN 2015041966| ISBN 9781466697928 (hardcover) | ISBN 9781466697935 (ebook)
Subjects: LCSH: Energy harvesting. | Bionics. | Wireless sensor networks--Industrial applications.
Classification: LCC TK2896 .B56 2016 | DDC 621.31/2429--dc23 LC record available at http://lccn.loc.gov/2015041966

This book is published in the IGI Global book series Advances in Environmental Engineering and Green Technologies (AEEGT) (ISSN: 2326-9162; eISSN: 2326-9170)

British Cataloguing in Publication Data
A Cataloguing in Publication record for this book is available from the British Library.

Advances in Environmental Engineering and Green Technologies (AEEGT) Book Series

ISSN: 2326-9162
EISSN: 2326-9170

Mission

Growing awareness and an increased focus on environmental issues such as climate change, energy use, and loss of non-renewable resources have brought about a greater need for research that provides potential solutions to these problems. Research in environmental science and engineering continues to play a vital role in uncovering new opportunities for a "green" future.

The **Advances in Environmental Engineering and Green Technologies (AEEGT)** book series is a mouthpiece for research in all aspects of environmental science, earth science, and green initiatives. This series supports the ongoing research in this field through publishing books that discuss topics within environmental engineering or that deal with the interdisciplinary field of green technologies.

Coverage

- Industrial Waste Management and Minimization
- Alternative Power Sources
- Radioactive Waste Treatment
- Policies Involving Green Technologies and Environmental Engineering
- Electric Vehicles
- Biofilters and Biofiltration
- Water Supply and Treatment
- Contaminated Site Remediation
- Renewable Energy
- Sustainable Communities

IGI Global is currently accepting manuscripts for publication within this series. To submit a proposal for a volume in this series, please contact our Acquisition Editors at Acquisitions@igi-global.com or visit: http://www.igi-global.com/publish/.

The Advances in Environmental Engineering and Green Technologies (AEEGT) Book Series (ISSN 2326-9162) is published by IGI Global, 701 E. Chocolate Avenue, Hershey, PA 17033-1240, USA, www.igi-global.com. This series is composed of titles available for purchase individually; each title is edited to be contextually exclusive from any other title within the series. For pricing and ordering information please visit http://www.igi-global.com/book-series/advances-environmental-engineering-green-technologies/73679. Postmaster: Send all address changes to above address.

Titles in this Series

For a list of additional titles in this series, please visit: www.igi-global.com

Biologically-Inspired Energy Harvesting through Wireless Sensor Technologies
Vasaki Ponnusamy (Quest International University Perak, Malaysia) Noor Zaman (King Faisal University, Saudi Arabia) Tang Jung Low (Universiti Teknologi Petronas, Malaysia) and Anang Hudaya Muhamad Amin (Multimedia University, Malaysia)
Information Science Reference • copyright 2016 • 290pp • H/C (ISBN: 9781466697928) • US $200.00 (our price)

Handbook of Research on Waste Management Techniques for Sustainability
Ulas Akkucuk (Bogazici University, Turkey)
Information Science Reference • copyright 2016 • 438pp • H/C (ISBN: 9781466697232) • US $240.00 (our price)

Control and Treatment of Landfill Leachate for Sanitary Waste Disposal
Hamidi Abdul Aziz (Universiti Sains Malaysia, Malaysia) and Salem Abu Amr (Universiti Sains Malaysia, Malaysia)
Information Science Reference • copyright 2016 • 459pp • H/C (ISBN: 9781466696105) • US $220.00 (our price)

Toxicity and Waste Management Using Bioremediation
Ashok K. Rathoure (Vardan Environet Guargaon, India) and Vinod K. Dhatwalia (Uttaranchal University, India)
Engineering Science Reference • copyright 2016 • 421pp • H/C (ISBN: 9781466697348) • US $205.00 (our price)

Handbook of Research on Climate Change Impact on Health and Environmental Sustainability
Soumyananda Dinda (Sidho-Kanho-Birsha University, India)
Information Science Reference • copyright 2016 • 711pp • H/C (ISBN: 9781466688148) • US $325.00 (our price)

Computer Vision and Pattern Recognition in Environmental Informatics
Jun Zhou (School of Information and Communication Technology, Griffith University, Australia) Xiao Bai (School of Computer Science and Engineering, Beihang University, China) and Terry Caelli (Department of Electrical and Electronic Engineering, The University of Melbourne, Australia)
Information Science Reference • copyright 2016 • 407pp • H/C (ISBN: 9781466694354) • US $225.00 (our price)

Promoting Climate Change Awareness through Environmental Education
Lynn Wilson (SeaTrust Institute, USA) and Carolyn Stevenson (SeaTrust Institute/Open College@Kaplan University, USA)
Information Science Reference • copyright 2016 • 305pp • H/C (ISBN: 9781466687646) • US $195.00 (our price)

Handbook of Research on Uncovering New Methods for Ecosystem Management through Bioremediation
Shivom Singh (ITM University Gwalior, India) and Kajal Srivastava (G.B. Pant University of Agriculture and Technology, India)
Information Science Reference • copyright 2015 • 638pp • H/C (ISBN: 9781466686823) • US $325.00 (our price)

Editorial Advisory Board

Table of Contents

Detailed Table of Contents

the most immediate issues to be solved is the relatively low efficiency in a WPT system. The system's efficiency decreases gradually as the distance increases. To ensure highest efficiency, the path with the shortest distance and the lowest interference is to be selected. In this case, a few of the most popular biologically-inspired algorithms, i.e. Firefly Algorithm(FA), Simulated Annealing (SA), the Ant Colony Optimization (ACO) and Genetic Algorithms (GAs) could be employed for efficiently solving different optimization problems. In short, these algorithms mentioned involve the emulation of fireflies flash signals for mating, exploitation of the metallurgy process, nature of ants leaving pheromones along the trails traveled and also the mimic of gradual evolution of organisms.

Internet of Things (IoT) has becoming a central theme in current technology trend whereby objects, people or even animals and plants can exchange information over the Internet. IoT can be referred as a network of interconnected devices such as wearables, sensors and implantables, that has the ability to sense, interact and make collective decisions autonomously. In short, IoT enables a full spectrum of machine-to-machine communications equipped with distributed data collection capabilities and connected through the cloud to facilitate centralized data analysis. Despite its great potential, the reliability of IoT devices is impeded with limited energy supply if these devices were to deploy particularly in energy-scarced locations or where no human intervention is possible. The best possible deployment of IoT technology is directed to cater for unattended situations like structural or environmental health monitoring. This opens up a new research area in IoT energy efficiency domain. A possible alternative to address such energy constraint is to look into re-generating power of IoT devices or more precisely known as energy harvesting or energy scavenging. This chapter presents the review of various energy harvesting mechanisms, current application of energy harvesting in IoT domain and its future design challenges.

In this chapter, the characteristics of the propagation of waves in superconducting waveguides are investigated. To compute the propagation constant, the complex conductivity of the superconductor is incorporated into the set of characteristic equations which describes the propagation constant of waves in the waveguide. An important outcome from this analysis is that superconducting waveguides are shown to behave like a lossless waveguide, exhibiting literally lossless behaviour at frequencies above the cutoff and below the gap frequency. Above the gap frequency, however, the waveguide loses its superconductivity, giving attenuation which increases in correspond to frequencies. The result suggests strongly that superconducting waveguides can be applied in receiver systems to minimize the loss of propagating signals.

Section 2
Presenting Energy Harvesting in WSN and Its Perspectives

Electricity consumption will encompass a large converse about connected with international electricity demand while in the next 2 decades. Newly, this improving rate connected with fossil fuels and also issues about the environmentally friendly consequences connected with gas emissions get renewed the attention in the progress connected with alternative electricity resources. Renewable Energy Sources and Climate Change Modify Minimization offers a good estimation on the chapter for the technological, scientific, environmentally friendly, financial and also societal aspects of this factor connected with six renewable energy (RE) options for the minimization connected with weather adjust. This functioning chapter on environmentally friendly Energy Solutions and Local climate Change Minimization presents an assessment on the literature for the scientific chemical, technological, environment, economic in addition to social areas of the contribution connected with six environmentally friendly energy (RE) sources on the mitigation connected with climate alter. This chapter is definitely an overview of presentation of the Local climate Change Minimization expansion on the essential results. Considering this significant component of Renewable Energy Sources can be reduce carbon dioxide, there is an international relating to reducing carbon emissions. Due to the fact most of the United Nations wanted to greenhouse gas (GHG) emissions is carbon dioxide, there is a can be a global concern on minimizing carbon emissions. Emissions of greenhouse gases (GHGs) resulting from the provision of the services of one have contributed significantly to improve the historical concentrations of greenhouse gases to the atmosphere of MIT. The IPCC (AR4) concluded that "most of the observed global climate improving as it is very likely that as a result of the improvement observed in the concentrations of anthropogenic gases mit techniques this mid of 20th century confirms Recent Files the use of fossil power accounts for most of the international anthropogenic GHG emissions". Emissions always grow, in addition to CO2 concentrations of it had increased to more than 390 ppm, or perhaps 39% above pre-industrial levels, by holding from 2014-5. There are many options for reducing GHG emissions from energy system while satisfying the desire for global energy services. Some of these possible alternatives, such as energy conservation and competition, switching fossil fuel, RE, nuclear, plus carbon capture and hard drive (CCS) was evaluated from the AR4. A full assessment related to any profile minimization options will likely involve an evaluation of respective potential alongside minimization with his bargain with sustainable development as well as all associated risks, and costs. This phase will focus on the role that this display technology related to RE can participate in within the portfolio related to mitigation alternatives. In this sense, the only policies can be given to reduce emissions of carbon dioxide, to improve the implementation of green energy, and such encouraging technological innovation. At inclusion, supporting components, such as feed-in tariffs, rules Renewable side view in addition to tax insurance policies are used by governments to help develop green energy generation in addition to the implementation of the efficiency of energy use save energy. In this chapter, the various insurance policies could possibly be placed on reducing carbon emissions, for instance improving green energy deployment and also significant technologies. A pair of main clarifications may be realizing to scale back carbon emissions and also overcome the issue connected with weather adjust: exchange fossil fuel having green electricity options wherever possible and also enhancing energy proficiency. In this chapter, many of us discuss most up-to-date performance connected with technology intended for improving green electricity deployment and also electricity work with proficiency.

Advancement of technology brings advantages to human race to certain extent. Unfortunately, some of the technological inventions not only brings advantages, but also disadvantages. One of the disadvantages is the pollution of carbon emission in the environment from the non renewal energy resources. Therefore, to prevent pollution from getting worse, the environmental-friendly aspect is suggested to be included in new inventions nowadays. Thanks to the invention of solar panel, solar energy is able to become a renewable energy source that can be applied to various applications. Considering the increase of electricity consumption every year in Malaysia, solar powered street lighting system is proposed in order to reduce the burning of fossil fuels to generate electricity, hence reducing air pollution. The proposed street lighting is companied with automatic switching mechanism to further reduce the electricity consumption of street lights. The system will also be able to provide energy measurement to determine the efficiency of street light. Security features like motion detection and camera module interfaced to the street light for surveillance monitoring in a residential area to overcome intruders and crime such as theft. Theft is now major issue in Malaysia and in many developing and developed countries. The news reported that many people lost their belongings due snatch theft and averagely 16 people go missing every day. Overall design of the solar powered street lighting systems with the security features will help people live in green and protected environment. The implementations of client server connection and Remote Method Invocation (RMI) in wireless network system ease the monitoring process of the entire street lights in a residential area for giving continuous service to the community and environment.

Microbial Fuel Cells (MFC) are the main topic of this chapter. Different types of electrochemical devices are presented and their typical power output is compared with other energy sources, providing a framework for the uses and applications of MFC technology. Following an historical approach of how this technology came to be, a more detailed description of some aspects of a typical microbial fuel cell is then brought forward. The energy harvesting concept, its use on low power wireless systems and maximum power point tracking (MPPT) techniques are presented and described. Wastewater treatment plants are a kind of infrastructure where this technology could be applied with a major success to power wireless sensing networks. An experimental setup, develop to improve the use of MFC in waste water treatment plants is presented. This chapter also provides a review on research trends for microbial fuel cells and maximum power point tracking algorithms, therefore, pointing current researches on this technology.

Protection of environment implies the coordination of the local and regional authorities to implement actions for conservation of all natural resources in the actual context of globalisation. One of the most important direction for applying the above mentioned desiderates is a better planning of transportation, as we all know that globalisation implies an accelerated process of externalisation of ideas, know-how, products and labours. In this chapter, we consider discrete event systems (DES) modeled via a state space representation. The model objective is the avoidance of a given set of states, or equivalently the fact that certain predicates, specified in terms of states are always false. We address the state space controlled Petri Nets (PN) and a technique to reduce the complexity of these nets, by taking into account the fact that the complexity of the considered nets depends mainly on the representation of the control design, respectively on the forbidden sets of places. Our goal is to describe in an efficient way the forbidden sets of the modeled system, e.g. to describe in an efficient way the DES modeled with controlled Petri nets. An example from the railway systems illustrates this approach.

The impacts of climate changes are significantly influencing the approaches of organizations and governments to use resources, develop appropriate environment-friendly strategic frameworks and adopt a holistic approach to understand their operating environment. These impacts can be seen in the search for energy effective solutions, outsourcing processes and engaging into partnerships and alliances. At the same time, governments have been stepping up policy and legislative initiatives, assessment frameworks, and engagement in international conventions to cut carbon emissions and promote sustainability.

Biologically inspired computational solutions for protein coding regions identification are termed as optimized solutions that could enhance regions of interest in noisy DNA signals contrary to contemporary identification. Exponentially growing genomic data needs better protein translation. The solutions proposed so far rely on statistical, digital signal processing and Fourier transforms approaches lacking the reflection for optimal biologically inspired identification of coding regions. This paper presents a peculiar biologically inspired solution for coding regions identification based on wavelet transforms with notion of a peculiar indicator sequence. DNA signal noise has been reduced considerably and exon peaks can be discriminated from introns significantly. A comparative analysis performed over datasets commonly used for protein coding identification revealed the outperformance of proposed solution in power spectral density estimation graphs and numerical discrimination measure's calculations. The significant results achieved depict 75% reduction in computational complexity than Binary indicator sequence method and 32% to 266% improvement than other methods in literature (as a comparison with standard NCBI range). The significance in results has been achieved by efficiently denosing the target DNA signal employing wavelets and peculiar indicator sequence.

Wireless Sensor Networks (WSNs) are gaining popularity in many monitoring and event detection areas. But the limitation of low power sensors needs to be further researched to combat the self-healing aspect of sensor networks. Therefore this paper aims to present a survey on application of self-healing routing protocols in wireless sensor networks that have been designed by analyzing their operation, advantages and disadvantages. This presents an outline of each application of the protocol and an idea of how these features could be adapted into future research area in self-healing wireless sensor networks in terms of energy aspect. The self-healing characteristics are mainly derived from biologically-inspired mechanisms for autonomic behavior. Autonomic behavior systems are characterized into four functional systems as self-configuration, self-healing, self-optimization and self-protection. This survey mainly focuses on self-healing characteristics from the biologically-inspired perspectives as well as non-bio-inspired perspectives.

Green energy paradigm has been gaining popularity in the computing system from the software, hardware, infrastructure and application perspectives. Within that concept, data center greening is of utmost importance at the moment since data centers are one of the most energy conserving elements. Data centers are seen as the technology era's black energy-swallowing secret. Reducing energy consumption at data centers can reduce carbon footprint effect tremendously. Not addressing the issue immediately will lead to significant energy usage by data centers and will hinder the growth of data centers. The call for sustainable energy efficient data center leads to venturing into data center green computing. The green computing concept can be achieved by using several methods adopted by researchers including renewable energy, virtualization through cloud computing, proper cooling system, identifying suitable location to harvest energy whilst reducing the need for air-conditioning and employing suitable networking and information technology infrastructure. This paper focuses into several approaches used by researches to reduce energy consumption at data centers while deploying efficient database management system. This paper differs from others in the literature by giving some suitable solutions by looking into a hybrid model for green computing in data centers.

Section 3
Exploring and Illustrating the Current and Future Trends in Combining Biologically-Inspired Computing with Energy Harvesting in WSN and Energy Transmission

Wireless sensor networks (WSNs) has gain popularity due to their wide range of applications in almost all walks of life including industry controls, environmental monitoring, health, transportation, military, civil infrastructure, science, security and more. Wireless sensor nodes are cheap and tiny in size therefore its deployment is easier. They perform well in harsh environments where human intervention is almost difficult or not possible. However, wireless sensor networks are resource constraints and its power supply has been a big challenge to keep the sensor nodes functional for a longer period. Advancement in low power electronics helped a lot but the use and maintenance of conventional batteries with a limited life span cannot address the power supply problem effectively in a long run. Therefore, harvesting energy from ambient environment is an effective alternative both in terms of power and cost, which can help sensor networks to live longer. This chapter mainly focuses on different possible energy sources available in ambient environment and current technological mechanism to harvest energy for WSNs.

There are numerous forms of energy that can gratify the needs of computing, these energies include solar, thermal, mechanical, wind, acoustic, and wave for different devices and applications. With ubiquitous computing requirements for self-powered devices such as MEMS (Micro-electromechanical systems) and these devices have narrow capacity of finite power sources and it is needed that these systems are needed to be supplied with energy for the lifetime of the system. Mostly Batteries are the main source of energy for many of these mobile systems, embedded systems and remote system applications. There is a requirement for Energy harvesting which is a method of extracting energy from the nearby environment called Environmental energy which can be used as supernumerary to electrochemical battery. Studies are being conducted to harvest energy from solar, wind, thermal, tidal and other mechanical sources which have limited lifespan. This paper serves as a study to identify various sources of energy available for harvesting centered on various technical papers available and present the work carried out by investigators in identifying them.

Foreword

Wireless Sensor Network (WSN) technology is gaining popularity due to its use in the design of several natural systems such as environmental monitoring, natural disaster monitoring, human intrusion detection and more importantly in medical systems using implantable sensors. These systems require sustainable operation without power interruption or power failure. Therefore uninterrupted energy delivery has become an important objective in the design of new wireless sensor networks. This objective may be achieved through new and upcoming technologies in wireless computing such as energy harvesting mechanisms. Energy harvesting provides sustainable energy management features in wireless sensor networks. The first inspiration for energy harvesting comes from solar energy harvesting, which itself has roots in the study of biological objects looking towards the sun for their energy needs and the photosynthesis process by plants. This attracted researches to look for more biologically-inspired energy harvesting solutions ranging from human foot striking vibration energy and tree trunk swaying inspired wind energy.

Therefore, biologically-inspired computing can provide potential solutions for wireless sensor networks to extend sensor lifetime. Besides that, efficient energy harvesting mechanisms and energy transmissions techniques are also essential to address the stated problems in wireless sensor networks. An extensive literature on energy harvesting applications, energy harvesting mechanisms, energy transmission techniques, green information and green databases provided in this book can be a potential guide for researchers in this vast and emerging field. Besides that, bio-sensors and bio-fuel cells are also gaining popularity in the context of renewable energy.

This book assists the academic and industrial communities with up to date information about recent advances in sustainable energy with focus on energy harvesting, renewable energy and green computing. Further, it discusses various biologically-inspired solutions for autonomic computing such as self-healing and self-sustainability through energy-efficient computing. In addition, this book encourages more research in biologically-inspired solutions for sustainable energy management.

Muhammad Mehboob Yasin
King Faisal University, Saudi Arabia

Preface

Wireless networks are playing vital role in current emerging technological era by providing ease and access in multi daily life applications. Among the many challenges in wireless networks is the finite battery power in the nodes affecting the shorter lifetime of wireless nodes and thus the network life. Wireless nodes are normally powered by battery and the finite battery power shall impede the performance of wireless networks. Numerous techniques related to energy aware routing, clustering, scheduling, energy aware MAC routing protocols and data aggregation have been proposed to address energy efficient issue and focus to increase the lifetime of wireless networks. Though few of them provide better solution up to simulations and assumptions only, the stated energy efficiency issue still requires major attention. Possible solutions such as battery replacement, higher battery capacity, and low transmission range could mitigate such issues but they need to compromise on other factors such as cost, reliability and energy usage. In fact one of the promising alternatives to address such situations is to look into power recharging techniques for wireless nodes, in particular the energy harvesting approaches. Energy harvesting is the process of hunting energy from the environment such as solar, wind or other sources such as body heat, foot strike, frequency harvesting or vibration, and converting it into electrical energy. Proper harvesting rate can enhance the performance of the network and extend the lifetime of the entire networks.

Biological inspired algorithms are now becoming a new paradigm to look for naturally inspired solutions for self-healing mechanism to address energy issues in wireless networks. Biologically-inspired solutions use natural selection mechanisms to provide promising factors for sustainable energy management in energy harvesting. This proposal draws attention to the energy harvesting and transmission mechanism in wireless networks from the technological perspective with respect to the nature inspired solutions in energy harvesting. The advancement of wireless technology in the current trend is increasingly influenced and inspired by the biological norm.

The ultimate objective of this volume is to provide the academic and industrial communities with recent advances in technological development, application and impact of technologies for sustainable energy and environment. The objective of this book is two folds. The first objective is to encourage more research in energy harvesting and transmission technologies. The second objective is to publish the excellent research been conducted at this moment of time. Specifically, we seek to explore new possibilities based on biologically-inspired solutions for sustainability in the context of energy and environment.

ORGANIZATION OF THE BOOK

This book consists of 14 peer-reviewed invited chapters authored by several international researchers around the world. Book is divided into the following three sections:

Section 1: Introducing Biologically-Inspired Computing and Its Perspectives
Section 2: Presenting Energy Harvesting in WSN and Its Perspectives
Section 3: Exploring and Illustrating the Current and Future Trends in Combining Biologically-Inspired Computing with Energy Harvesting in WSN and Energy Transmission

Section 1 entitled "Introducing Biologically-Inspired Computing and Its Perspectives" is to specify is strictly about biologically-inspired computing. The second section focuses on energy harvesting and wireless transmission and leaving the third part open to general chapter submissions including merging of biologically-inspired computing and energy harvesting.

A brief description of each of the chapters follows:

SECTION 1: INTRODUCING BIOLOGICALLY-INSPIRED COMPUTING AND ITS PERSPECTIVES

Chapter 1: Applications of Vibration-Based Energy Harvesting (VEH) Devices

This chapter reviews present usage of vibration-based energy harvesting (VEH) devices and applications. The evolution of energy resources and advance in electronic technologies has resulting the need of self-sustainable wireless/portable electronic devices in current modern society. VEH mechanism is suggested in this chapter as the solution for the bottleneck. Various consideration of creating an optimal vibration energy harvesters are suggested through an analytical model of a mechanical transducer. Useful applications and usages of VEH are presented and some suggestions for improvement are also given.

Chapter 2: Biologically-Inspired Wireless Power Transmission System – A Review

This chapter gives an overview on Wireless Power Transfer (WPT) systems and the existing biologically-inspired architectures which could be used to optimize the efficiency of the systems. In general, wireless power transmission could be categorized into near-field and far-field mechanisms. At present, one of the most immediate issues to be solved is the relatively low efficiency in a WPT system.

Chapter 3: Energy Harvesting Methods for Internet of Things

This chapter presents the review of various energy harvesting mechanisms, current application of energy harvesting in IoT domain and its future design challenges. Internet of Things (IoT) has becoming a central theme in current technology trend whereby objects, people or even animals and plants can exchange information over the Internet. IoT can be referred as a network of interconnected devices such as wearables, sensors and implantables, that has the ability to sense, interact and make collective decisions autonomously.

Chapter 4: Analysis of Energy Loss in Superconducting Waveguides

In this chapter, the characteristics of the propagation of waves in superconducting waveguides are investigated. To compute the propagation constant, the complex conductivity of the superconductor is incorporated into the set of characteristic equations which describes the propagation constant of waves in the waveguide. The result suggests strongly that superconducting waveguides can be applied in receiver systems to minimize the loss of propagating signals

SECTION 2: PRESENTING ENERGY HARVESTING IN WSN AND ITS PERSPECTIVES

Chapter 5: Different Resources Consumption of Renewable Energy

Renewable Energy Sources and Climate Change Modify Minimization offers a good estimation on the chapter for the technological, scientific, environmentally friendly, financial and also societal aspects of this factor connected with six renewable energy (RE) options for the minimization connected with weather adjust. This functioning chapter on environmentally friendly Energy Solutions and Local climate Change Minimization presents an assessment on the literature for the scientific chemical, technological, environment, economic in addition to social areas of the contribution connected with six environmentally friendly energy (RE) sources on the mitigation connected with climate alter.

Chapter 6: Solar Powered Smart Street Light with Maintenance Service System

Advancement of technology brings advantages to human race to certain extent. Thanks to the invention of solar panel, solar energy is able to become a renewable energy source that can be applied to various applications. Considering the increase of electricity consumption every year in Malaysia, solar powered street lighting system is proposed in order to reduce the burning of fossil fuels to generate electricity, hence reducing air pollution.

Chapter 7: Harvesting Energy from Microbial Fuel Cells – Powering Wireless Sensor Networks Operating in Wastewater Treatment Plants

Microbial Fuel Cells (MFC) are the main topic of this chapter. Different types of electrochemical devices are presented and their typical power output is compared with other energy sources, providing a framework for the uses and applications of MFC technology. Following an historical approach of how this technology came to be, a more detailed description of some aspects of a typical microbial fuel cell is then brought forward.

Chapter 8: Modelling Formalisms for Green Transportation Systems

In this chapter, we consider discrete event systems modeled via a state space representation. The model objective is the avoidance of a given set of states, or equivalently the fact that certain predicates, specified in terms of states are always false. We address the state space controlled Petri Nets and a technique to reduce the complexity of these nets, by taking into account the fact that the complexity of the considered nets depends mainly on the representation of the control design, respectively on the forbidden sets of places.

Chapter 9: On the Decision Criteria for "Greening" Information Systems

The impacts of climate changes are significantly influencing the approaches of organizations and governments to use resources, develop appropriate environment-friendly strategic frameworks and adopt a holistic approach to understand their operating environment. These impacts can be seen in the search for energy effective solutions, outsourcing processes and engaging into partnerships and alliances. At the same time, governments have been stepping up policy and legislative initiatives, assessment frameworks, and engagement in international conventions to cut carbon emissions and promote sustainability

Chapter 10: A Biologically-Inspired Computational Solution for Protein Coding Regions Identification in Noisy DNA Sequences

This paper presents a peculiar biologically inspired solution for coding regions identification based on wavelet transforms with notion of a peculiar indicator sequence. DNA signal noise has been reduced considerably and exon peaks can be discriminated from introns significantly. A comparative analysis performed over datasets commonly used for protein coding identification revealed the outperformance of proposed solution in power spectral density estimation graphs and numerical discrimination measure's calculations.

Chapter 11: Application of Self-Healing in Wireless Sensor Network – A Survey

This presents an outline of each application of the protocol and an idea of how these features could be adapted into future research area in self-healing wireless sensor networks in terms of energy aspect. The self-healing characteristics are mainly derived from biologically-inspired mechanisms for autonomic behavior. Autonomic behavior systems are characterized into four functional systems as self-configuration, self-healing, self-optimization and self-protection. This survey mainly focuses on self-healing characteristics from the biologically-inspired perspectives as well as non-bio-inspired perspectives.

SECTION 3: EXPLORING AND ILLUSTRATING THE CURRENT AND FUTURE TRENDS IN COMBINING BIOLOGICALLY-INSPIRED COMPUTING WITH ENERGY HARVESTING IN WSN AND ENERGY TRANSMISSION

Chapter 12: A Critical Review on Energy Harvesting Techniques

Advancement in low power electronics helped a lot but the use and maintenance of conventional batteries with a limited life span cannot address the power supply problem effectively in a long run. Harvesting energy from ambient environment is an effective alternative both in terms of power and cost, which can helps sensor networks to live longer. This chapter mainly focuses on different possible energy sources available in ambient environment and current technological mechanism to harvest energy for WSNs.

Chapter 13: Green Energy in Data Centers

The green computing concept can be achieved by using several methods adopted by researchers including renewable energy, virtualization through cloud computing, proper cooling system, identifying suitable location to harvest energy whilst reducing the need for air-conditioning and employing suitable networking and information technology infrastructure. This paper focuses into several approaches used by researches to reduce energy consumption at data centers while deploying efficient database management system. This paper differs from others in the literature by giving some suitable solutions by looking into a hybrid model for green computing in data centers.

Chapter 14: Exploring Current Trends of Energy Harvesting

Mostly Batteries are the main source of energy for many of these mobile systems, embedded systems and remote system applications. There is a requirement for Energy harvesting which is a method of extracting energy from the nearby environment called Environmental energy which can be used as supernumerary to electrochemical battery. Studies are being conducted to harvest energy from solar, wind, thermal, tidal and other mechanical sources which have limited lifespan. This paper serves as a study to identify various sources of energy available for harvesting centered on various technical papers available and present the work carried out by investigators in identifying them.

Vasaki Ponnusamy
Universiti Tunku Abdul Rahman, Malaysia

Acknowledgment

I would like to acknowledge the help of all the people and close reference research friends involved in this project and, more specifically, to the authors and reviewers that took part in the review process. Without their support, this book would not have become a reality.

First, the editors would like to thank each one of the authors for their contributions. Our sincere gratitude goes to the chapter's authors who contributed their time and expertise to this book.

Second, the editors wish to acknowledge the valuable contributions of the reviewers regarding the improvement of quality, coherence, and content presentation of chapters. Most of the authors also served as referees; we highly appreciate their double task.

Vasaki Ponnusamy
Universiti Tunku Abdul Rahman, Malaysia

Section 1

Introducing Biologically-Inspired Computing and Its Perspectives

Chapter 1
Applications of Vibration-Based Energy Harvesting (VEH) Devices

Ooi Beng Lee
Quest International University Perak, Malaysia

Thein Chung Ket
Taylor's University, Malaysia

Yew Chun Keat
University of Hull, UK

A. Rashid A. Aziz
Universiti Teknologi Petronas, Malaysia

ABSTRACT

This chapter reviews present usage of vibration-based energy harvesting (VEH) devices and applications. The evolution of energy resources and advance in electronic technologies has resulting the need of self-sustainable wireless/portable electronic devices in current modern society. Batteries are non-beneficial in the miniaturization process of electronic designing and alternative power supplies are desperately needed to fill in the falling behind technologies gap to drive the advance of the wireless/portable development further. VEH mechanism is suggested in this chapter as the solution for the bottleneck. Various consideration of creating an optimal vibration energy harvester are suggested through an analytical model of a mechanical transducer. Useful applications and usages of VEH are presented and some suggestion for improvement are also given. Lastly, the trend of energy harvesting is annotated and commented in-line with the demand of electronic sensors market.

ENERGY RESOURCES

Since the last century, energy has been essential in building civilization revolution for modern society. In the 19th century, most of the energy resources were obtained from coal and timber. In the 20th century, crude oil and natural gas were added to the list and followed by hydroelectric and nuclear power. For the past decades, with the increasing demand and raising of global climate discussion, energy transi-

DOI: 10.4018/978-1-4666-9792-8.ch001

Figure 1. Global Mobile Sensing Health & Fitness Sensor Shipments (Mareca, Darryl, & Chi, Q2 2013)

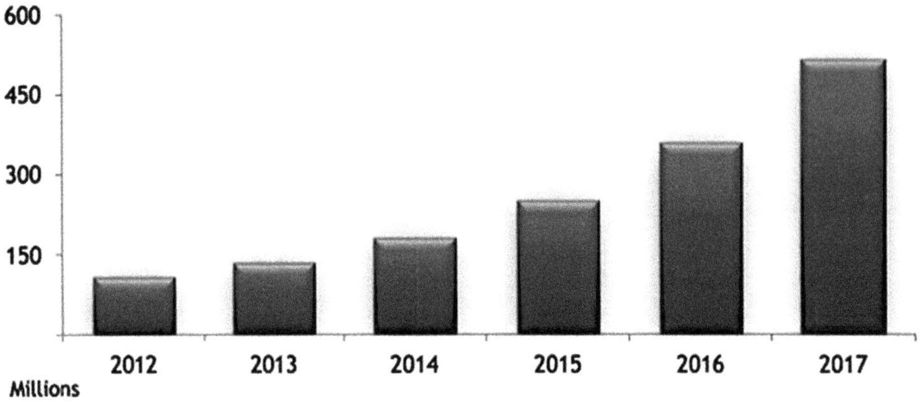

tion at the global scale takes place by adding new green energy resources to the existing list, such as solar, wind, thermal, tidal and vibration. The European Commission has committed to reduce Europe's greenhouse gas emissions by 80% before year 2050 compared to 1990 levels (European Commission, 2014). To achieve the low-carbon economy worldwide, all other developed and developing countries are expecting to endorse the objective by a similar degree.

In electronic technology, the advancement in silicon size and power consumption of CMOS technology has led to a wide research activities on wireless/portable devices. By removing the wires from the devices, there is a potential for embedding sensors in previously inaccessible locations, such as roof tops, underneath the raised-access floor panel, implant into the concrete wall and etc. At present, most of the remote electronic devices are still taken the energy from batteries with finite lifespan, which not only limit the miniaturization process of the device size, but also create environmental pollution caused by the improper disposal of used batteries. Not to mention the cost of labors and parts involved during the batteries replacement process, especially in the large-scale Wireless Sensor Network (WSN) and portable electronic device systems. Due to the increase in wireless sensing devices (Figure 1) that led to a large research effort to seek for alternative long lasting energy sources, especially in applications where the replacement of batteries is practically unfeasible or costly. For instance wireless sensors that use for health monitoring, animals tracking, security system and some military applications. With a self-powered energy resources, large amounts of saving can be achieved for the ownership of wireless sensor systems and portable electronic devices in long run.

According to the Law of conservation of energy, energy can be neither created nor destroyed. Heat, electricity, dynamic, chemical and biomass forms of energy are all stored differently, but can be converted from one form to the other. One of the promising solutions to a self-powered wireless device is by using energy scavenging from ambient sources such as solar, thermal, wind, vibration and etc. These energy resources are redundant and have almost infinite lifetime. Some of these energy sources have some limitations like in case of wind energy as the wind turbine cannot be used in lower altitude or urban areas, at where there is an inconsistent and weak wind density. In the case of solar energy, the effectiveness of the harvester will drops at location where there is absence of sunlight. Among these energy sources, vibration sources claimed to be one of the most attractive approach due to their abundance availability in many targeted wireless sensor or portable electronic device applications (Roundy, Wright, & Rabaey, 2003).

There are many proposed vibration-based energy harvesting (VEH) devices produced for various meaningful applications. However to date, most of the review articles and the majorities of the research on vibration energy harvesting are focus on energy conversion mechanism and the effectiveness. Hence, this chapter is mainly reviewing on the state-of-the-art of VEH products and applications. Firstly, some general overviews of energy resources evolution are given. Followed by the discussion of various types of energy resources that can be utilized for energy harvesting. A general vibration-based energy converter is also presented and the design restriction toward an optimize energy harvester is described. Last but not least, some VEH on various applications are presented and reviewed. Lastly, the future prospect of energy harvesting is also commented.

ENERGY SOURCES SUITABLE FOR SCAVENGING

Electrochemical batteries have been the dominant power source for mobile electronic devices over the past century. However, they provide a finite amount of energy and need to be replaced for the lifetime of the device. Although batteries have improved over the years (Jose, Antonio, & Francesc, 2003), the improvement is fairly gradual compared with other areas of electronics (Thad, 2003) and cannot satisfy the simultaneous demands for long life, low volume, low weight and limited environmental impact. There are many sources of energy which have been considered for energy harvesting and their lifetime is potentially infinite, such as solar power, electromagnetic radiation, thermal, human power, wind energy, ambient vibration and the like. Energies can be found in various form at different locations, but in order to capture them, a proper conversion technology must be applied. Although not all solutions listed here are fit for wireless sensors and portable electronic devices, it is worthwhile to study the scavenging techniques and the possible amount of energy available from the proposed sources.

SOLAR ENERGY (PHOTOVOLTAICS-PV)

Solar energy is abundant outdoors during the daytime. It is one of the common and the main clean energy source which is harvested directly from the sunlight. Recently, a four-junction solar cell compound semiconductor materials developed by Soitec and CEA-Leti and Fraunhofer Institute for Solar Energy Systems ISE has proven to be able to achieve 46% of cell efficiency. Meaning almost half of the solar light can be converted into electrical energy by their PV modules which characterized at 800W/m^2 at 47°C (Dimroth, 2014). Certainly, during the sunny day, solar energy is considered one of the best resources if the harvesters are placed directly under the sun light, yet the disadvantage is solar energy is only available during the day time and it is climate dependent. For instance, indoor lighting has far lower power density than outdoor sun light. There was a comparison of solar power density measured at varying distances from an indoor 60 watt incandescent bulb by (Roundy, Steingart, Frechette, Wright, & Rabaey, 2004) in Table 1 which has shown that the indoors power density with ambient office lighting is not sufficient for self-powered electronic devices.

Table 1. Solar power measurements taken under various lighting condition (Roundy, Steingart, Frechette, Wright, & Rabaey, 2004)

Distance	8 Inches	12 Inches	18 Inches	Office Lighting
Power (μW/cm^2)	503	236	111	7.2

Radio Frequency (RF) Electromagnetic Radiation

RF harvesting circuits have been investigated for more than 50 years, but due to the availability of the sufficient radio frequencies, the potential of harvesting the energy from radio frequency wave are only widely explored in the recent decade. In modern society, RF energy is readily available in most inhabited areas from sources such as wireless transmitter, cellular base stations, radio station, TV broadcasting and even portable communications devices. The RF energy can be harvested via an antenna that is designed to pick up a wide range of radio frequencies and convert the electromagnetic waves into an electrical current. A prototype has been developed by Nokia Research Laboratory and proven that the harvester was able to scavenge between 3 to 5mW from the ambient radio waves, emitting from mobile antennas, satellite communication, radio, TV masts and other sources (EE Times, 2009). However, the antenna and the receiver circuit need to be designed to pick up a very wide range of frequencies, typically from 500MHz to 10GHz. Most recently, (Xia, Cheng, Glover, & Chiang, 2014) have demonstrated to perform a 3m distance data transmission with a battery-less temperature monitoring system that constituted by multiple 0.56V RF-powered wireless sensor nodes at sensitivity level better than -20 dBm at 904.5MHz. This shows RF energy harvesting is a promising technology for future self-powered miniature wireless devices. Apart from that, RF signals can also be used in powering many passive electronic devices, such as smart card, radio frequency identification (RFID tags) and etc. In reality energy received at the receiving end antenna falls rapidly with distance from the energy sources. Therefore, in these ambient conditions, to obtain a sufficient level of energy to drive electronic devices, many strong signal sources from a wide range of frequencies are required for the harvester. Not to mention that the antennas and receiver circuits are required to be designed accordingly to the existing RF signals.

Thermal (Temperature Gradients)

Naturally, energy can be harvested during the variation of environment temperature. Harvesting electrical energy from thermal energy sources can be achieved by using pyroelectric (PE) or thermoelectric (TE) conversion techniques. The efficiency of conversion for a typical thermal generator from thermal source is limited to the Carnot efficiency, $\eta = \left(T_{high} - T_{low}\right) / T_{high}$., where T_{high} and T_{low} are the absolute higher and lower temperate on the each side of the device, respectively. Hence the greater the temperature variation, the greater the efficiency of the energy conversion that can be achieved. It has been proven that the harvested power of a standalone pyroelectric harvester with an auxiliary circuitry that had been attached to a micro heat engines generated 3.3 and 15.7μW from a temperature difference of 45 and 85K, respectively (Ravindran, Kroener, & Woias, 2012). A thermal energy harvester was also demonstrated to powering autonomous wireless sensor nodes in an aircraft by (Elefsiniotis, Kokorakis, Becker, & Schmid, 2014). In this study, one side of the thermoelectric generator (TEG) was attached to the thermal mass inside the cabin while the other side of the generator was fixed to the heat source, in

Figure 2. Schematics of the thermoelectric harvesting device when attached to the fuselage of an aircraft (Elefsiniotis, Kokorakis, Becker, & Schmid, 2014)

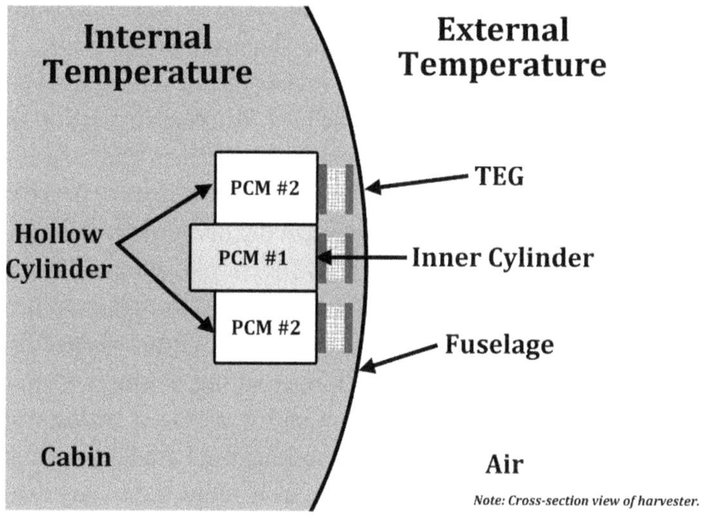

this case the fuselage of the aircraft (see Figure 2). The experiment ansimulation show that the amount of energy of ~30.99J (worst condition) and ~69.78J (optimum condition) that produced by the TEG is sufficient to power up a conventional low-powered wireless sensor node from the temperature difference across two sides of the TEG during the aircraft take-off and landing operations. The draw back to the TEG is that, to maintain the device at higher efficiency, it is advisable to operate the device at the temperate range of 150 to 250 °C and gradient of at least 100 °C (TECTEG MFR, n.d.), however miniaturization of TEGs operating at temperature above 200 °C are rarely available. While recently, it is true that (Borghino, 2014) has found a new way to transform heat energy into electricity, particularly in situations where the temperature gradient is below 100 °C and large amounts of wasted heat produced in the industrial processes and power plants can be recycled.

Human Power

From various research, it is known that the human body is a tremendous storehouse of energy. (Starner, 1996) noted that an average human body burns about 10.5 MJ of energy per day. This is corresponds to an average power dissipation of 121W, which showing the great potential to harvest energy from human body. A distinction was also made between active human power and passive human power by (Roundy, 2003). For active human power, it normally required the user to perform a specific power generating motion. On the contrary, passive human power scavenging occurs during normal activities. Therefore, active human power is not very appropriate for most of the applications since it is not a wise method to generate power by performing a specific activity. A significant amount of work has been done on the possibility of scavenging power from the human body for use by wearable electronic devices, such as generating electricity from the compression of the shoe sole. For this device, a generator is mounted in the shoe sole and energy can be captured during normal walking. (Paradiso & Starner, 2005) developed a piezoelectric generator in shoes which produced an average power of 8.3mW at the heel and 1.3mW at the toes during a standard walk. The shoe inserts method offers a good solution for energy scavenging, but the application space for this generation method is extremely limited. (Donelan, et al., 2008) developed the biomechanical energy harvester which was mounted at the knee and the tested subjects (see Figure 3), walking with one device on each leg and produced an average of 5 watts of electricity, which was significantly more than what can obtained from the shoe-mounted devices. This method seems to be well-suited for charging some portable electronic devices, however the drawback is the harvester is too bulky in size and it must be custom made to fit the user. Another restriction to harvest energy from human motions is that normal human movements occur in three dimensions and involve a high degree of rotational rather than oscillatory motions. Yet, most of the harvesters were designed for only rectilinear degree of freedom, so reducing the possibility to optimize the harvester's efficiency. To overcome this, (Rao, McEachern, & Arnold, 2013) has presented a fully functional, compact and self-sufficient human-powered energy harvesting system for passively capturing energy from human activities (see Figure 4). Charging rates ranging from 33μW to 234μW was recorded from this system that came at volume size of $70cm^3$ under normal human activities (such as walking, jogging, cycling) when was worn on different parts of the test subject body. In this device, a magnet ball was placed within a spherical housing that wrapped with 1400 turns of copper wire to form the coil for the Faraday's Law induction effect when the internal ball moved chaotically during the operation as it subjected to any external human motions.

Figure 3. Biomechanical energy harvester device by (Donelan, et al., 2008)

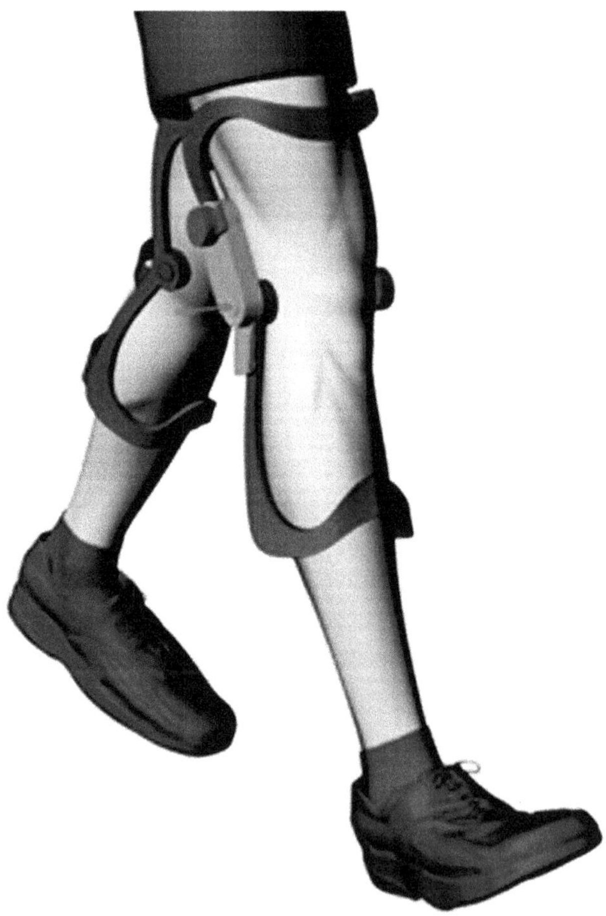

Figure 4. Photograph (left) and 3-D schematic (right) of the spherical housing energy harvester presented by (Rao, McEachern, & Arnold, 2013)

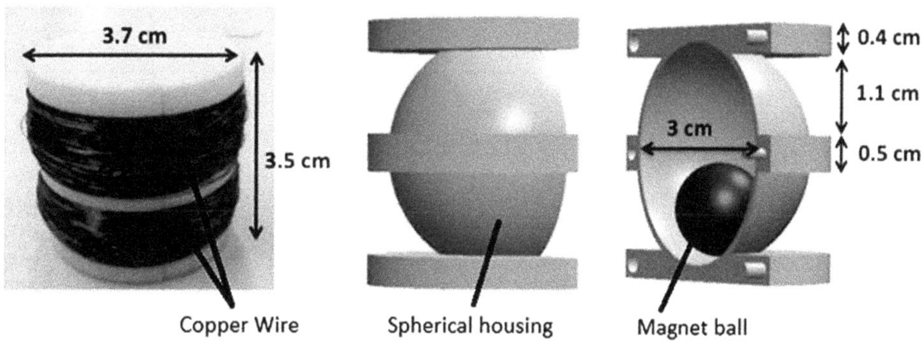

Ambient Vibrations

Plenty of low-amplitude mechanical vibrations are existing in our daily environment and vibration resources are found in abundance in many targeted electronic devices locations, such as the vibration from the building next to the busy street, fly-over bridges, road surfaces, home appliances, industry machineries and so forth. For this reason, the potential of converting these mechanical vibrations energy into a useful electrical form is feasible and redundant to be used. The utility of this type of energy depends on the amplitude of the vibration and its dominant frequency. Vibration sources may vary considerably in amplitude and dominant frequency on different objects. In

Table 2 gives several vibration source measurements that recorded by various authors. It is shown that the fundamental frequencies of the measured resources are fall between range from 26 to 200 Hz and acceleration amplitudes range from 0.059 to 12 ms^{-2}. Yet, the vibration present in most environments is not made up of a single frequency but is typically made up of a number of fundamental frequencies and their harmonics. For example the vibration data that shown in Figure 5, was taken by (Ooi, 2010) from a desktop computer case when a CD room running, indicates a fundamental frequency of 120 Hz with an acceleration amplitude of 0.54 ms^{-2} with the 2nd and higher harmonics present at lower amplitudes. Since most of the vibration-based conversion devices have a relatively narrow range of operating frequencies (high Q-factor), it is important that the nature of the source be understood first so that the generator can be designed to resonate at an appropriate frequency peak in the vibration spectrum, thus optimal power can be harvested. Based on the aforementioned potential energy resources survey in this section, notice that many energy resources look promising to be used as an alternative power source for portable/wireless electronic devices, however this chapter will only focus on energy harvesting from a vibration source. This does not suggest that vibration-based power sources is the best energy scavenging solution, but only recommend as one of the most appropriate energy resources for the self-powered electronic systems.

Modeling of General Vibration Conversion

A simple model based on the schematic in Figure 6 has been proposed by (Williams & Yates, 1996) to model common inertia vibration-based converter devices. The model consists of a seismic mass, *m*,

Table 2. Acceleration amplitude and frequency of fundamental vibration mode for various resources

Vibration Resources	Acceleration A (m/s²)	Resonant frequency F_r (Hz)	Source of Information
Windows next to busy road	0.7	120	(Roundy, 2003)
Bread Maker	1.03	121	(Roundy, 2003)
Washing Machine	0.5	109	(Roundy, 2003)
Car engine compartment	12	200	(Roundy, Wright, & Rabaey, 2003)
Clothes dryer	3.5	121	(Roundy, Wright, & Rabaey, 2003)
Small microwave oven	2.5	121	(Roundy, Wright, & Rabaey, 2003)
Desktop computer casing [On top of the casing]	0.5	120	(Ooi, 2010)
Desktop computer casing (with CD running) [On top of the casing]	0.54	120	(Ooi, 2010)
Domestic Freezer [On top of the Freezer]	0.1	50	(Ooi, 2010)
Vacuum Cleaner	1.55	100	(Mohd Fauzi Bin Ab Rahman & Leong, 2011)
Treadmill (Reebok)-Without loading	0.059	26	(Mohd Fauzi Bin Ab Rahman & Leong, 2011)
Grinding Machine (Mc D&D)-Without load	3.51	49	(Mohd Fauzi Bin Ab Rahman & Leong, 2011)
CNC Machine –Without load	0.216	83	(Mohd Fauzi Bin Ab Rahman & Leong, 2011)

Figure 5. Vibration spectra for desktop computer casing with CD is running (Ooi, 2010)

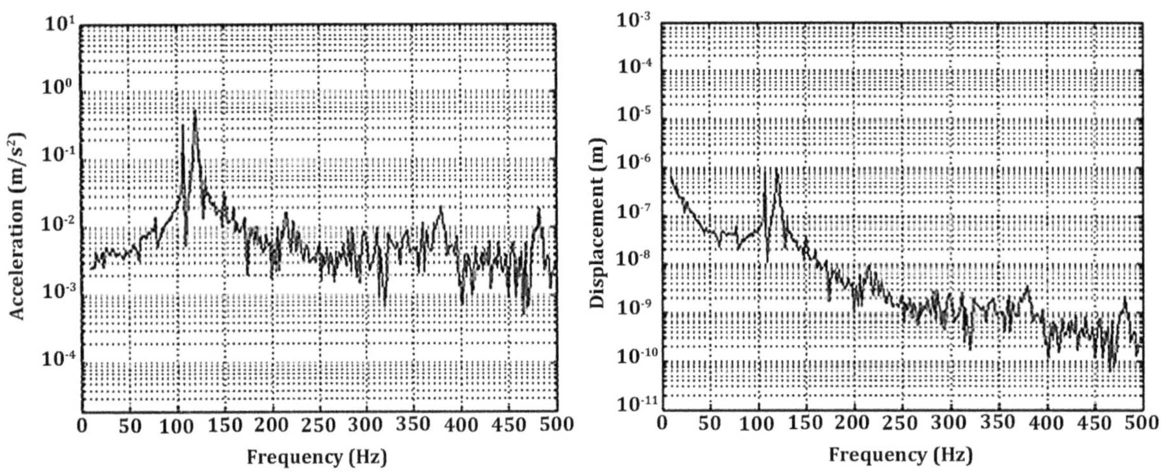

Figure 6. Model of common inertia vibration-based converter

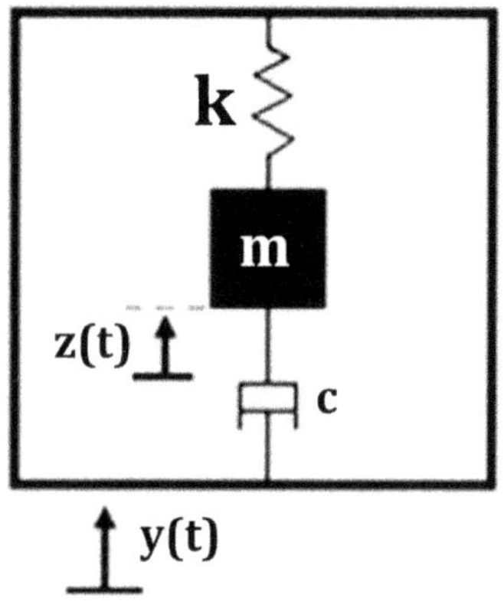

a spring of stiffness, *k*, the damping coefficient, *c*, the time varying position, *z(t)*, of the seismic mass relative to the support casing that is driven by an external time varying vibration force, *y(t)*. When the converter model is subjected to an external vibration force, the mass moves out of phase with the converter housing, so that there is a net movement between the mass and the housing, which cause loss of mechanical energy that is represented transformation of energy into electrical form. The motion equation for this system may be described as

$$m\ddot{z}(t)+c\dot{z}(t)+kz(t)=-m\ddot{y}(t) \tag{1}$$

For a sinusoidal external excitation vibration force where, $y(t)=Y\sin(\omega t)$, the steady-state solution for *z(t)* is

$$z(t)=\left[\frac{Y\omega^2}{\sqrt{\left(\omega_o^{\,2}-\omega^2\right)^2+\left(2\omega_o\omega\zeta_F\right)^2}}\right]\sin(\omega t-\phi) \tag{2}$$

and the phase angle ϕ is

$$\phi=\tan^{-1}\left(\frac{2\omega_o\omega\zeta_F}{\omega_o^{\,2}-\omega^2}\right) \tag{3}$$

where $\omega_o=\sqrt{k/m}$ is the natural frequency of the converter system, $\zeta_F=c/(2m\omega_o)$ is the full damping factor and *Y* is the amplitude of external excitation force. If the damping factor of the model is the sum of electrical induced (ζ_e) and mechanical loss (ζ_m) damping, $\zeta_F=\zeta_e+\zeta_m$, the average power delivered to the electrical domain can be written as

$$P_{avg} = \frac{m\zeta_e Y^2 \left(\frac{\omega}{\omega_o}\right)^3 \omega^3}{\left[1-\left(\frac{\omega}{\omega_o}\right)^2\right]^2 + \left[2\zeta_F \frac{\omega}{\omega_o}\right]^2} \quad (4)$$

Under an ideal condition where there is no power loss due to the converter's internal impedance, then the theoretical maximum average power delivered occur when the external vibrating force frequency is matched with the resonant frequency, $\omega = \omega_o$, and when $\zeta_e = \zeta_m$ which yield:

$$P_{avg(\max)} = \frac{mA^2}{16\omega\zeta_m} \quad (5)$$

where $A = Y\omega^2$ is the amplitude of external excitation force. From Equation (5), is can be seen that the power is inversely proportional to frequency. Therefore, if the acceleration amplitude of the vibrations is constant or decreasing with frequency, then the vibration-based converter should be designed to resonate at the possible lowest fundamental frequency and highest acceleration.

The relationship of the full damping ratio, external driving frequency and average power in Equation (4) is plotted in Figure 7 by assuming that the proof mass, m and amplitude, Y, are constant with unity gain that the resonant frequency is 100Hz and input vibration frequencies vary from 10 to 1000Hz. By assuming that $\zeta_m = \zeta_e$, three different damping ratios (ζ_F) output response are plotted. It can be seen that a converter will operate at its optimum condition when the vibration frequency is matched with the system resonant frequency ($\omega = \omega_o$). It may also be noted that increasing the damping, ζ_F has the effect of reducing the peak output power and Q-factor but increasing the operational bandwidth. To create a highly efficient converter, the converter should be designed to resonate at the lowest frequency peak in the vibration spectrum provided that higher frequency peaks do not have higher acceleration amplitude. Or more specifically, higher value of A^2 / ω. If the case in Figure 5 is used as an example, for a system with proof mass (m) of 1grams and full damping factor (ζ_F) of 2% as suggested in (Kumar, Sharma, Kumar, Vaish, & Chauhan, 2014), according to Equation (5), the theoretical maximum average power of ~2.42μW could be delivered from the vibration-based energy harvester, which is sufficient to give core power to drive most of the ultra-low-power sensors (European Editors, 2013) and this raise the potential for vibration-based energy harvesting in ultra-low-powered electronic systems.

To further enhance the performance, the converter needs to be designed at the minimal value of damping factor to provide a greater operational output at the resonant frequency. Equation (5) indicates the generated power will tends to infinity as ζ_m tends to zero, but in practice it is not possible to reduce the ζ_m to an arbitrarily small value. This implies that the generated power will always be a finite value and depends on the design of the converter. The amplitude is, $Z = Y / 2\zeta_F$, and so reducing the damping factor implies increasing the displacement range of the mass. The maximum movement range of the mass can move is limited by the size and the design of the resonator. Therefore, the damping ratio must be designed to be large enough to keep the range of movement within the limits which the resonator is capable of making without damage. Thus, if the vibration source has a fixed frequency then a low level

Figure 7. Output response for systems with ζ_F = 0.01, 0.05, and 0.1 (Ooi, 2010)

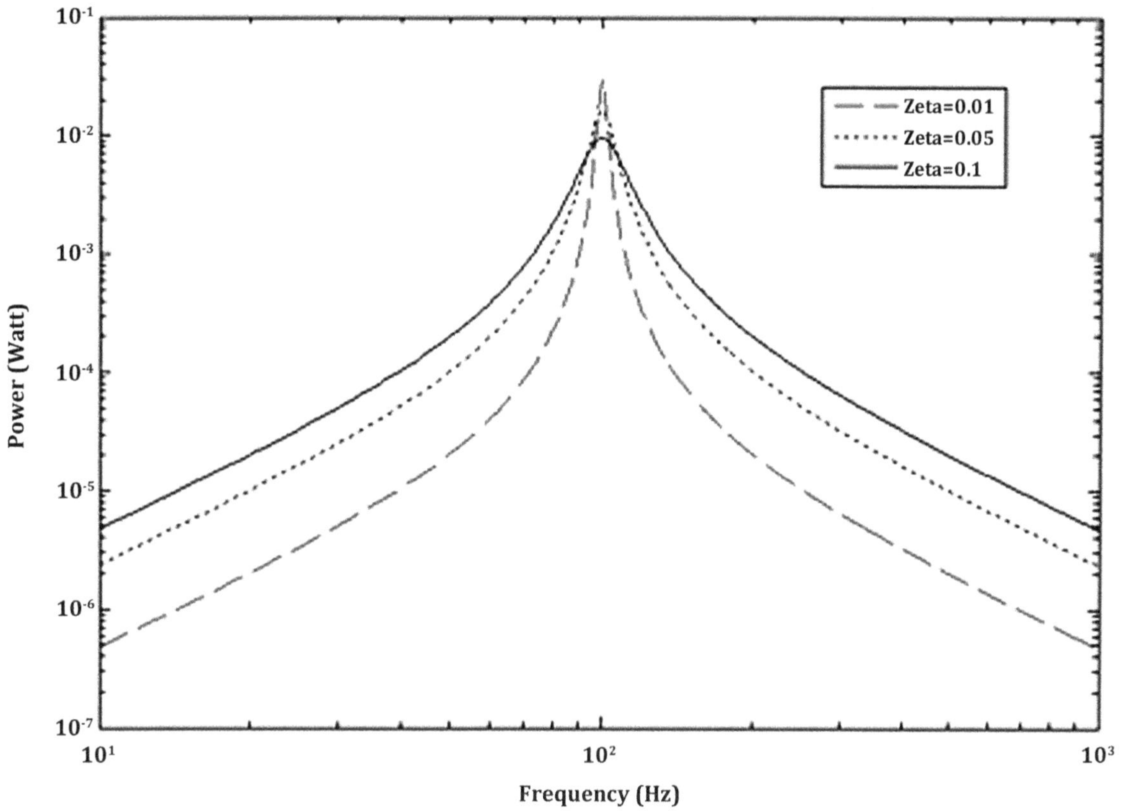

of damping (high Q-factor) is preferable but if the source frequency varies, it may give better average power to use a higher damping factor (low Q-factor) resonator. On the contrary, one of the main penalty for this solution is the power output is still less than what it would be if the resonance frequency matched the driving frequency. To date, frequency matching is a common issue for all the linear vibration-to-electricity converters, in order to resolve this, many alternative solutions have been proposed in the literature. These approaches are non-linear oscillation (Cottone, Vocca, & Gammaitoni, 2009), frequency-up-conversion (Zorlu, Topal, & Kulah, 2011), dual-resonator cantilevers (Ooi & Gilbert, 2014) (Ooi, A. Aziz, & Gilbert, 2014), multiple harvesters (Liu, et al., 2008) (Sari, Balkan, & Kulah, 2008) and resonance frequency tuning (Wischke, Masur, Goldschmidtboeing, & Woias, 2010). Every aforementioned approach has its own advantages and drawbacks in term of device size, output density, design cost and so on. Many alternative strategies and approaches were also very well documented by (Zhu, Tudor, & Beeby, 2010) and (Tang, Yang, & Soh, 2010), hence the wide-band devices are set beyond the scope of this chapter.

Applications of Vibration-Based Energy Harvesting (VEH) Conversion

Recall that the biggest challenge in portable systems is the batteries limited lifespan. Typical battery-powered vibration sensor lifetime is around 3 years (Weddell, 2010), due to environmental uncertainty the lifetime period generally is shorter than what is expected. Many portable systems are installed at physically inaccessible location or implant into the human body (for medical micro-sensors), considering the cost of ownership and quality of live, changing the system batteries periodically is not an option, especially for a ubiquitous systems. Not to mention that batteries are not practically applicable in many applications, such as automotive engine sensors that operate in high thermal range. Hence, this raise the interest of research activities in energy harvesting to seek for an alternative solution to replace the battery or the concept of "Integrated Power unit" – combination of battery and harvester as a primary and secondary energy sources to prolong the battery lifespan. There are many energy harvesting resources that are potentially feasible to be used as a replacement, however each resource has their own limitation, one resource may be good for certain application/place but not the other. For example solar energy is not available in an indoor environment, such as the monitoring sensors in the tunnel, whereas vibration and air-flow energy harvesters may be suitable. Consequently, this chapter only focus on the applications of VEH and any other resources are set beyond the scope for exploration. Added to the above, the following section in rest of the chapter will not fully cover all the applications of VEH, but to the best knowledge of the author of this chapter, only those that are impactful and insightful will be further analyzed.

Energy Harvesting from Traffic Flow

VEH has been extended to many applications for different purposes. One of the most common application VEH applied on is to harvest the traffic-induced vibration from the road surface. (Lucy, 2010) in association with Technion Israel Institute of Technology has developed a VEH converter which was based on piezoelectric materials that exerted mechanical energy by the weight of passing vehicles into electrical energy. The piezoelectric harvester was embedded along two wheel footprints as in Figure 8. In this work, the electrical storage system was also integrated in the roads and did not take up any public space and able to function in all weather conditions. This demonstration claims that provided there are 600 heavy vehicles travelling at speed of 72km/h through the interval per hour on average, an average of 200kWh/h of electrical power can be harvested from generators along a single traffic lane for a distance of one kilometer, which is sufficient to supply for the average electricity consumption in 200-300 households. Apart from supplying the harvested electricity to various roadside applications, the system can also be used to monitor the vehicle's speed and weight effectively. For example it can

Figure 8. Parallel Roadway Harvesters

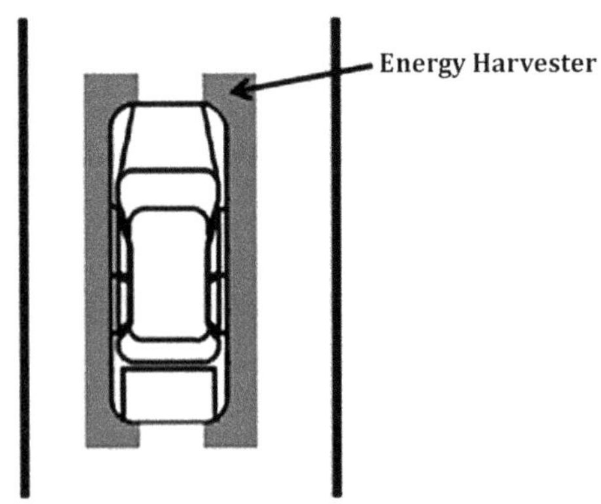

Figure 9. (a) Schematic of roadway harveter implementation on the road and (b) Schematic diagram of piezoelectric roadway harvester (Jiang, Li, Li, Wang, & Yao, 2014)

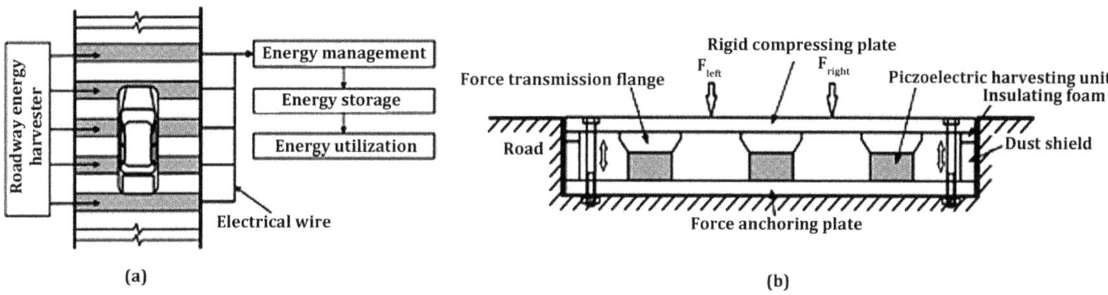

be used as a weigh bridge to measure the load of the vehicle by monitoring the average output from the generator. This will not reduce the turnaround time for a vehicle to entering or leaving the place, but also diversify the cost of ownership of road surface embedded generator. However, the drawback for this parallel roadway harvesters is the alignment issue, since it is very difficult to ensure that the vehicle wheels are always perpendicularly on top of the harvester footprints. Not to mention the inconsistent driving attitudes such as overtaking, right/left turning and etc., which will further reduce the available electrical power that can be harvested.

There was an alternative roadway harvester proposed by (Jiang, Li, Li, Wang, & Yao, 2014). Similarly, the piezoelectric energy harvester was installed along the road width but perpendicular to the vehicle direction (Figure 9a). A compression-based roadway energy harvester was embedded into pavement to scavenge electrical energy from the traffic flow vibrations. Group of three piezoelectric multilayer stacks harvesting units had been embedded underneath the road asphalt (Figure 9b) and electrical energy produced when vehicle passing through the road surface. By implementing the perpendicular form of harvesters on the roadway, the alignment issue can be resolved with the penalty of lower maximum generator output. The harvester was demonstrated capable of producing 2000W/h with 2000 vehicle/h traffic flow or 1W DC power per passing of a two-axle vehicle at 100km/h speed. Although lower output was recorded, the conversion implies that the electrical power could still be used for powering common lower-power-needed wireless sensors systems or common roadside electrical equipment such as lighting system, CCTV, speed camera and advertisement board. The technology can also be used for road data collection purposes through ubiquitous wireless sensors network system (Smart Road).

Energy Harvesting from Automobile Vibration

In automotive industry, aware that the function of vehicle suspension system is to support the weight of vehicle chassis, to isolate the body from road disturbances and to ensure comfort of the passenger in the vehicle. From the energy flow study done by (Bandivadekar, et al., 2008) on a 2.5L year 2005 model Toyota Camry as depicted in Figure 10, where 20% of the fuel energy is converted into driving force and only 25% from the driving force is transferred to the wheel rolling (or 5% from the total fuel energy). Typically, most of the available fuel energy is used to overcome the aerodynamic resistance, road friction, engine loss and etc. One of the important loss comes from dissipation of vibration energy by shock absorbers in the vehicle suspension that is caused by road irregularity during acceleration/

Figure 10. Vehicle energy flows of a 2.5L 2005 Camry in urban driving cycle (Bandivadekar, et al., 2008)

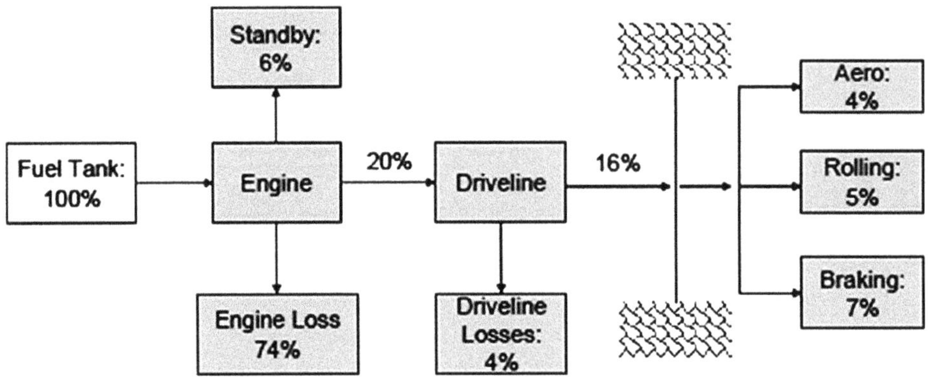

deceleration of the vehicle. So instead of having the vibration energy dissipating into heat wastes, vibration energy harvesting (VEH) mechanism can be applied to turn the kinetic energy into electricity or other potential energy and store it for later use. Due to the vast demand to improve the fuel efficiency, the topic of kinetic energy harvesting from the vehicle suspensions has attract many great interest from various research clusters to investigate on it.

According to the vibration conversion model in Equation (5), recall that larger seismic mass, m, provide a greater possible maximum average power. The other two main elements in the suspension are spring and damper. Thus in this case, the vehicle chassis weight become the seismic mass, the rebounding force and losses in the suspension become the spring and dissipative damper respectively in the model. Typically, the vehicle chassis provides a very sufficient weight to the suspension, this implies that it is very promising to extract energy out from the vehicle suspension systems when vibration is transmitted from road excitation. There are two types of regenerative suspension configuration as suggested by (Jin-qiu, Zhi-zhao, Lei, & Yu, 2013) according to the working principle, i.e., mechanical regenerative suspensions which are reformed from the traditional hydraulic/pneumatic suspension and electromagnetic regenerative suspensions which are constituted by permanent magnets and inductive coil. However, the latter type was claimed to be more promising structure due to its greater performance, better efficiency and compactness. (Zuo & Tang, 2013) reported that for a middle-sized passenger car driving on an average road condition, the power generated from each set of VEH suspensions was approximately 100W (100W×4 = 400W for 4 sets) at a speed of 96.56 km/h. According to (Rowe, 2012), the electrical usage of a vehicle with all optional system turned off is 250-300W, hence the 400W that harvested from the suspension system is considered sufficient and should not be underestimated.

(Patil & Gawade, 2012) have demonstrated on an electromagnetic suspension system consists of grade N32 NdFeB ring shaped permanent magnets, ring shaped spaces and center rod for stacking of magnets and spacers as shown in Figure 11. The coil assembly and the chamber body was fixed to vehicle frame whereas the center rod was fixed to the wheel. The output terminal wires were connected to a DC rectifier to convert the AC into DC voltage. Voltage is induced in the coils according to the Faraday Law of induction when the vehicle moved over an irregular road surface. The energy generation is also highly dependent upon the velocity of vehicle travel on road, the greater the velocity and vibration excitation, eventually greater energy can be generated. The proposed suspension system was claimed to be able to produce 8W of energy at 0.25-0.5m/s RMS suspension velocity for one complete scale of regenerative

Figure 11. The electromagnetic regenerative shock absorber by (Patil & Gawade, 2012)

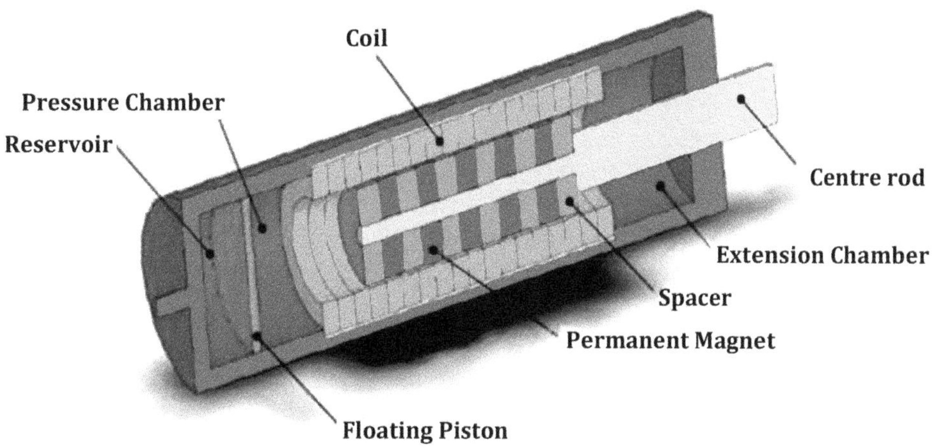

shock. The performance of the suspension system may also be improved by using a stronger permanent magnet, such as grade N48 NdFeB permanent magnets that gives greater residual flux density to the magnetic circuit.

The energy scavenge from the suspension system can be used to power the internal usage of the vehicle and if there is any excessive power left, it can be used to charge the nickel-metal hydride battery (NiMH) for a hybrid vehicle to run the motor. The motor can typically convert more than 90% of energy stored in the battery cells into motive force, whereas the efficiency of a gasoline internal-combustion engine can barely hit 30%. Thus, the fuel efficiency can be further improved for hybrid vehicles. On the other hand, the drawback of having a regenerative suspension system is that it is sometime difficult to handle the vibration control performance, which may cost unsmooth shock to the vehicle chassis, invalidating the intended purpose of implementing a suspension to vehicles for better ride comfort and handling stability.

Vibration Energy Harvesting (VEH) from Civil Structures

Another promising area of vibration energy is harvest from the civil structures. Civil structures like tall buildings, fry-over bridges, tunnels, communication base stations, are very susceptible to external vibration. In the modern society, to provide more space on a fixed amount of land, building structures expansion are lean toward the vertical direction. Especially in the urban area, where most of the sky-scrapers are built to accommodate more and more population onto an acre of land. Yet, vibration that cause by huge dynamic loadings from winds, earthquakes, traffics or even human motions has been a serious concern since the early days. Large deflection amplitude of the building structure not only cause damage to the civil component but also cause discomfort to its human occupants (Kareem, Kijewski, & Tamura, 1999). Hence, vibrations of a tall building are serious concern to engineers and many research has been carried out to mitigate the harmful vibration into a useful electrical energy (Ni, Zuo, & Kareem, 2011) (Tang & Zuo, 2011). Instead of dissipating the vibration energy into heat waste in the conventional vibration suppression devices, the vibration energy can be converted into electricity through dual functional tuned mass damper (TMD) energy transducers (Zuo & Cui, 2013). Practically, TMD is implemented in many tall buildings, such as Taipei 101 building in Taipei (See Figure 12). To

Figure 12. (a)location of where the TMD system in Taipei 101 building (b) Illustration of TMD system (c) Schematic of TMD model in two-DOF system (Deicon Dynamics & Controls, nd)

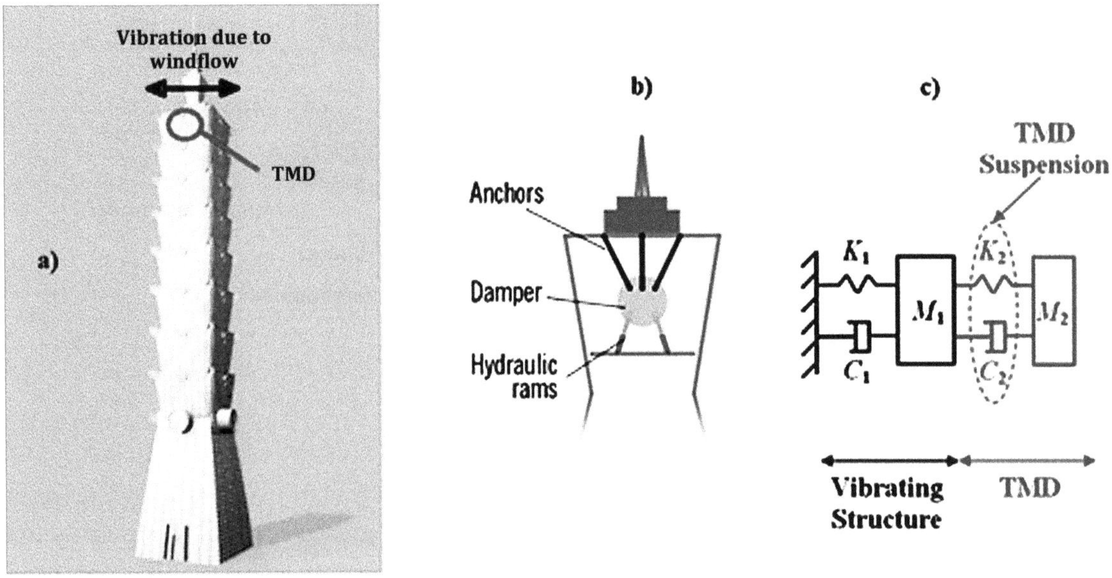

make the 508m tall building endurable to earthquakes, typhoons and fierce winds atmospheric assaults, a 730-ton giant pendulum TMD (See Figure 12b) has been included to counteract the building movement that due to external force. The TMD uses a computer controlled counter moving weight to adaptively move against the building sway, it is push in the opposite direction to the building sway to dampen the oscillation, reducing sway due to wind flow up to 36cm or 40% from 60cm initially (WordPress.com, 2011). The schematic of TMD model in two degree of freedom (DOF) is shown in Figure 12c, where the terms with subscript '1' representing the model of the vibrating structure (the building in this case) and the terms with subscript '2' representing the model of TMD suspension. This model is very useful for use in the modern mechanical analysis and simulation.

(Ni, Zuo, & Kareem, 2011) investigated the power available to be harvested from the TMD system when a 76-storey building was subjected to high wind event of 13.5m/s at height of 10m. The result shows that more than 85kW of power is available for harvesting from the TMD dynamics. In this study, similar performance level of energy harvesting can also be achieved by using multiple TMDs system, constructing in series configuration with significantly smaller mass than the single TMD system, but a larger stroke or building sway was required. A year later, for the first time (Tang & Zuo, 2012) demonstrated that both vibration mitigation and energy harvesting can be achieved simultaneously. In the three-story building prototype experiment, a conventional energy-dissipating mechanical suspension was replaced with an electricity-generating tuned mass damper (TMD), in which the damping force can be controlled by the duty cycle of pulse width modulation (PWM) signal that was generated according to the building sway condition. The TMD composed of a high energy density rotational motor with gears and rack-pinion mechanism. From the experiment, energy of 60mW was harvested from the proposed regenerative electromagnetic TMD (simultaneously with vibration control) under a harmonic force with amplitude of 15N at 3.3Hz. From the result, it can be seen that the harvested output is sufficient to supply to most of the ultra-low-powered microsensors that is ubiquitous in most of the smart building, such as fire alarm sensors, temperature & humidity sensors, security sensors and etc. It may also be suggested that to obtain a true output for a real physical three-story building, one may just upscale the output according to the up-scaling size of the prototype as well as the input force and frequency.

Apart from building structures, traffic tunnels are also one of the sensitive civil infrastructure. Traffic surveillance and structure condition are mandatory requirements for the safety operation of the traffic tunnels. Hence, a consistent monitoring system is required to provide a systematic overview of the tunnels during regular operation to minimize human losses especially during the event of landslide, traffic accidents or even terrorist attacks. This surveillance can be achieved by implementing a highly efficient sensor network that consistently providing the real- time update information in the tunnel. However, due to the narrow civil structure of the tunnel, installation of sensor within the tunnel pose a great challenge, not to mention that the sensor node might be embedded completely into the concrete. Thus wired connection between nodes is impractical and wireless communication is highly preferable under these circumstances. For this application, a sustainable and low maintenance power sources for the sensor nodes are needed, such as vibration energy harvesting. Traffic tunnels are frequently the bottleneck for traffic and working at full capacity most of the time, so sufficient vibration energy can be harvested from the vibration excitation caused by the traffic flow in the tunnel. (Wischke, Masur, Kroner, & Woias, 2011) studied the vibration available to be harvested from two different types of tunnel: a road traffic and a fixed railway tunnel. The sensors were mounted on the tunnel concrete walls, asphalt road surface and railway track for each of the tunnel, respectively. The results show that the vibration amplitude levels at the tunnel wall from both types of tunnel are below $0.01 ms^{-2}$, which is way too weak for a transducer, not to mention that the different resonant frequency that created by various vehicle/train sizes, further preventing the employment of a high-Q harvester. Added to that, the vibration amplitude obtained from the road surface were insignificant too, only maximum peak of $0.015 ms^{-2}$ amplitude was recorded when big truck passing by the road, which was not yield sufficient energy to power any wireless sensor devices. The weak vibration transferred from the vehicle was caused by the suspension and pneumatic tires that were implemented for passenger comfort. Not to mention that to harvest the vibration, a harvester has to be embedded into the road deck in order to protect them against the impact from the passing by vehicles, which further make the application complex and costly. In the railway tunnel, vibration amplitudes of 3.5 and $1 ms^{-2}$ were recorded at the railway and at the railway sleeper, respectively. To avoid any unwanted disturbance to the train, the railway sleeper was chosen as the vibration energy harvesting position. A four cantilever-based piezoelectric energy harvester which electrically connected in parallel was mounted to the railway sleeper and average energy of 395μJ was recorded per passing by train, which proved to be sufficient to power for one radio frequency (RF) protocol cycle. Other than a RF module, the harvested energy may also be used for the monitoring and analyzing purposes in the inclinometer and crackmeter to prevent the aging problems in the civil structures other than tunnels, such as bridges and embankment (Soga, 2011).

Implantable Devices

Due to the significant advances in extremely low power and miniaturization in electronics devices, implantable biomedical devices have considerable potential for energy harvesting from the human body itself. At present, most of the implantable biomedical devices are powered through wires (see Figure 13). This may cause discomfort, skin allergy or infections and many unwanted problem to the patients. To avoid the wires, implanted batteries may be used to power the devices, but due to its finite lifetime,

Figure 13. A chest X-ray image of implanted pacemaker with wires/leads (Deterre, 2013)

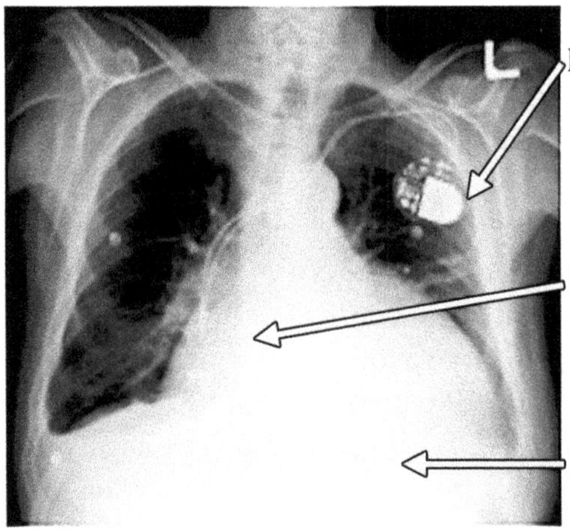

this may increase the patients' surgical risk and burden on maintenance operations, especially to the elders. Hence, a battery-less and a standalone implantable device designs are highly desired. By eliminating the batteries and wires, this can further extending the devices' lifespan and resulting in a smaller, longer lasting, safer implantable biomedical devices, which provide a greater comfort to the patients as well as reducing the surgical risks and costs. Currently, one of the most frequently used bioengineering technologies is pacemakers. Due to its relationship to human health, this particular implantable device has significantly grown as one of the lucrative business. For example, 3.6 million of pacemakers were in used in year 2003 (Hanna, Mutashar, Samad, & Hussain, 2014) and the market of pacemakers is forecasted to reach 5.1 billion USD in 2018 with annual growth rate of 11% by then (Martin, 2012).

As the global population continues to grow exponentially yet advances in healthcare, this increases in the number of elderly population. There are increasing incidences of cardiac disorders health concern even in a developed country like United States, this creating the need of cardiac pacemaker to maintain proper heart rhythm of patients. A very extensive review on current issues, challenges, advantages and disadvantages of various methods to harvest the energy for implantable biomedical devices have been documented by . Without straying from the context of this chapter, only the methods and devices that are about to commercialization will be further discussed. The human heart circulatory system, cardiac cycle, cardiac rhythm management and the evolution of pacemaker technologies have been studied in details by in the effort to seek for the solution of leadless pacemakers. Three different designs of piezoelectric spiral transducers (see Figure 14) including their complex electrodes patterns have been investigated. Average power density of 2.23 $\mu J/cm^3$/cycle and maximum power density of 2.92 $\mu J/cm^3$/cycle (#3 design) among three types of the proposed spiral transducers were recorded. The result shows that the outputs are too low for the current state-of-art, however it was argued that the output power generated by the proposed transducers should be at least ten times higher with further addressed developments suggested in the report, thus it should be sufficient to power the next generation of pacemakers perpetually after the improvement.

Dagdeviren, et al., (2014) have demonstrated to use a flexible energy-harvesting devices that can convert the movement of body organs like the heart into electricity that could be used by implantable devices like pacemakers. To avoid any irregular heartbeat reactions that is caused by an energy har-

Figure 14. Three Optical micrographs of 3 piezoelectric micromachined spirals transducers

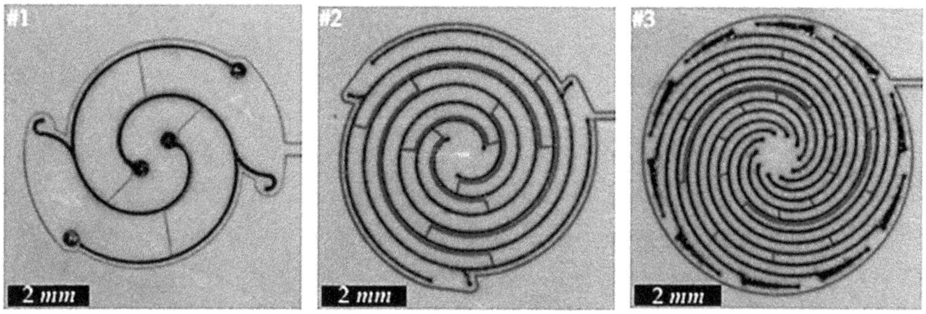

vester attached to the external heart tissues, the device need to be engineered to be ultra-flexible and 'weightless' so that any action of the heart is unaffected by the integration of the device onto the organ. Figure 15 shown the piezoelectric heart beating film that made of 500nm thick ribbons of lead zirconate titanate (PZT) surrounded by gold and platinum electrodes. The entire patches was encased in a layer of polyimide (biocompatible plastic) so that the organ tissues will not disrupted by the device's materials. Subsequently, the devices was attached in vivo to the organs of cows, sheep, and pigs using sutures. Power of 0.2μW/cm^2 was recorded from the devices, which demonstrates that the system could be used to power an implantable device like the pacemaker without the batteries, which pose risks to other organs with its hazardous chemicals. Although at present the experiment has not involved any long term and actual human implantation studies, from the power density obtained through the device, it is fairly to envisage that a self-powered, leadless and indefinite powering of pacemakers system is very feasible and can be accomplished.

Figure 15. Piezoelectric heart beating patches that stitched onto the hearts of cows (Simedics, 2014)

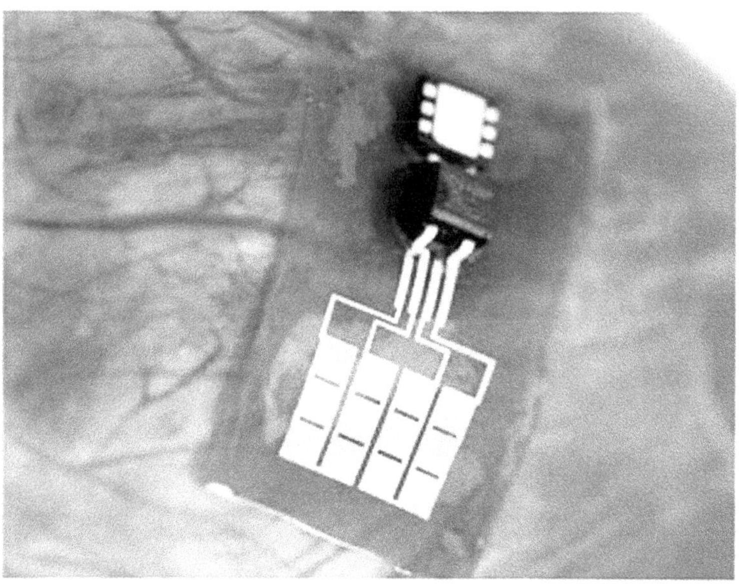

Portable Devices

Apart from using the vibration energy harvesting (VEH) in very specific applications such as those discussed in previous sections, VEH may also be used in many portable devices in our daily life. (Arveni, nd) has developed a high yield and high efficient piezoelectric mechanical to electrical energy harvester, which claimed to be able to achieve >80% and up to 40% efficiency in laboratory and real product conditions, respectively. Some innovative self-powered products were commercialized using the conversion mechanism as shown in Figure 16, such as pushing buttons wireless wallswitch and battery-less infrared remote control, see Figure 17. These devices utilize the motion while the button is pressed by the finger, converting the wasted energy kinetic energy into electrical form to power the wireless devices.

The wireless wallswitch system as depicted in Figure 18 claimed to be able to save up to 75% installation cost as compared to a wired solutions (Arveni, nd). Without the wiring restriction, the wireless wallswitch exhibits many advantages. Firstly, the switch is retrofittable in any existing system. A wallswitch together with the radio receiver installation can be easily completed in just a few minutes without any disturbance to the occupants. Secondly, the wallswitch is also very flexible to be fixed at many locations that was not suitable using the wired solution previously, such as fixing on the bedside table, wardrobe, glass or metal material surfaces and etc. Unlike the conventional wall switch, at where

Figure 16. Mechanical to electrical energy conversion mechanism between finger of user and the microgenerator (Arveni, nd)

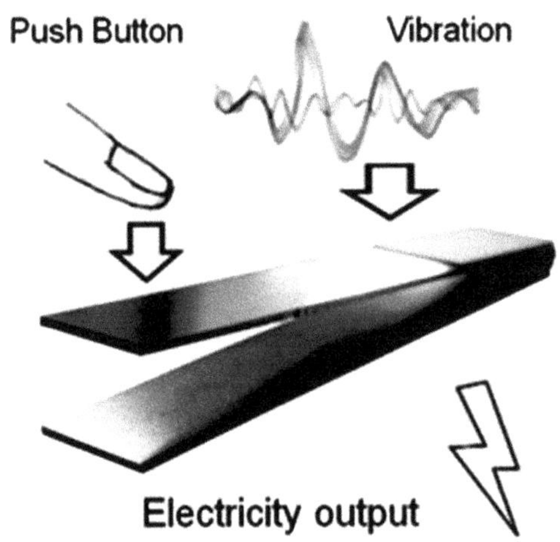

Figure 17. a) Wireless wall switch button and b) Battery-less infrared remote control (Arveni, nd)

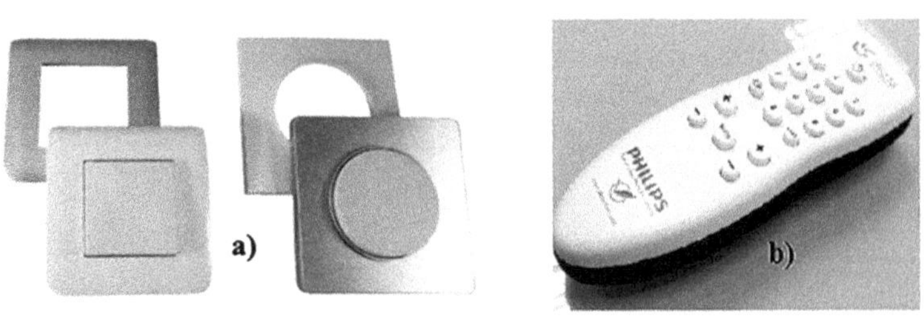

Figure 18. Wireless Wallswitch lighting Management Solutions by (Arveni, nd)

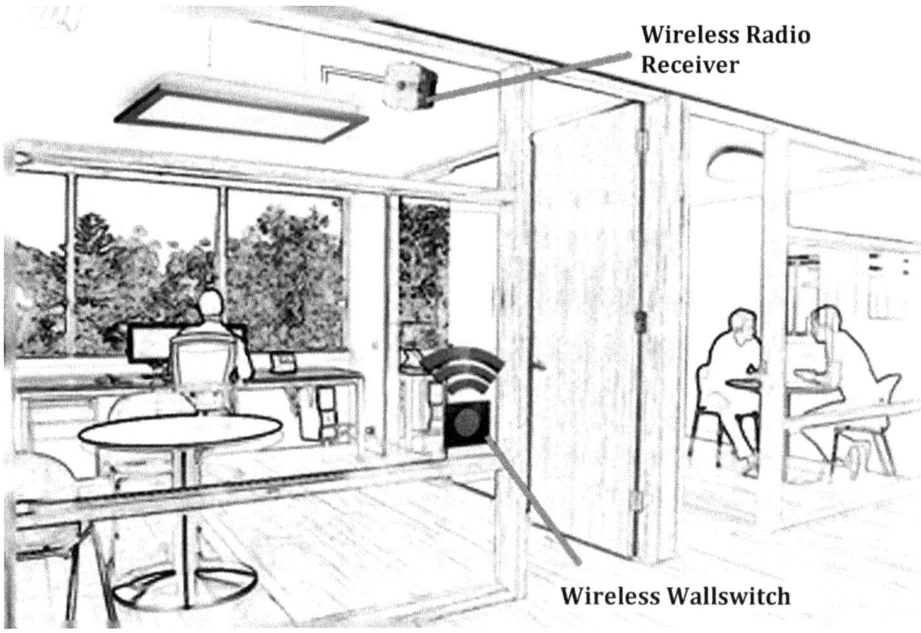

the location of the switch is fixed after installation, the wireless wallswitch can still be moved to other locations in within the radio frequency range easily. Lastly, not to mention the battery-less solution from the self-powered wallswitch can also further reduces the maintenance cost of batteries replacement program periodically. World first multikey battery-less remote control was first demonstrated by Arveni with Philips as shown in Figure 17b). The infrared remote control units has an integrated piezoelectric energy harvester in the unit. When the buttons on the remote controller unit was pressed, roughly 2.1mJ of energy could be delivered to the load, which representing efficiency of 11% (Das, 2009). The use of RF signal instead of infrared is also under exploring to make the remote control compliant to more electronic products in the future.

CONCLUSION AND FUTURE PROSPECTS

This chapter presents the current overview of the trend of energy resources used by sensors and portable devices. The environmental impact caused by batteries is identified and various types of energy available to be harvested from the surrounding environment are also reviewed. The arguments suggested that vibration energy harvesting has the potential to be one of the most promising alternative energy source due to its redundancy at most of the famous application locations. A general vibration-based energy harvesting (VEH) transducers model is also presented and design consideration toward an optimal VEH devices are discussed and analyzed. From the modelling, it also shows that the output power generated by the excitation vibration created from home appliances is sufficient to be used for powering the sensor and portable devices. Lastly, some insightful and useful applications of VEH are also presented and reviewed.

A typical sensor uses Lithium Thionyl Chloride ($LiSOCI_2$) batteries that come with lifespan of roughly 4 years (Weddell, 2010). However, most of the batteries start to show uncertainty of output after 2 years, thus maintaining a ubiquitous sensor systems is problematic and costly. Not to mention that $LiSOCI_2$ batteries are strictly regulated and cannot be transported on passenger planes in large packages (IATA, nd), hence this may lead to weeks of lead time for batteries replacement. Energy harvesting system is always being compared with batteries of its higher up-front cost. However, many incidents have shown that the higher up-front cost can be offset by the reduction in maintenance when batteries in conventional devices fail. On top of that, wireless technologies are also eliminating the need of wires within the system and hence saving installation and wiring costs. Nonetheless, it seem that technologists are still struggling to find some killer applications for energy harvesting systems (Special Interest Group, 2013), which would boost the scale of manufacturing and allow wider adoption of energy harvester as a replacement of batteries, thus significantly reducing the up-front cost of ownership of energy harvesting systems. While is it true that the application and demand of electronic sensors is increasing over the years as they play a vital role in this technology era (Technavio, 2015), so it is believed that the higher up-front cost of an energy harvesting system will further reduced when the demand for a self-power solution increases.

ACKNOWLEDGMENT

This research is supported by the Fundamental Research Grant Scheme (FRGS) MoE Grant FRGS/1/2014/TK03/QUEST/03/1 from the Ministry of Education (MoE) Malaysia. The authors would also like to thank to all the involved higher learning institutions for the facility supports.

REFERENCES

Arveni. (n.d.). *Arveni*. Retrieved from http://www.arveni.fr/en/

Bandivadekar, A., Bodek, K., Cheah, L., Evans, C., Groode, T., Heywood, J., & Weiss, M. (2008). *On the road in 2035: Reducing transportation's petroleum consumption and ghg emissions*. MIT Laboratory for Energy and the Environment.

Borghino, D. (2014, May 23). *Gizmag*. Retrieved Feb 3, 2015, from MIT finds new way to harvest energy from heat: http://www.gizmag.com/thermogalvanic-harvesting-waste-heat/32212/

CDC. (2013, July 23). *Centers for Disease Control and Prevention*. Retrieved March 9, 2015, from Division for Heart Disease and Stroke Prevention: http://www.cdc.gov/dhdsp/atlas/index.htm

Cottone, F., Vocca, H., & Gammaitoni, L. (2009, February). Nonlinear energy harvesting. *Physical Review Letters*, *102*(8), 806–811. doi:10.1103/PhysRevLett.102.080601 PMID:19257728

Dagdeviren, C., Yang, B. D., Su, Y., Tran, P. L., Joe, P., Anderson, E., & Rogers, J. A. et al. (2014). Conformal piezoelectric energy harvesting and storage from motions of the heart, lung, and diaphragm. *Proceedings of the National Academy of Sciences of the United States of America*, *111*(5), 1927–1932. doi:10.1073/pnas.1317233111 PMID:24449853

Das, R. (2009, November). *Energy Harvesting Journal*. Retrieved from http://www.energyharvesting-journal.com/articles/1842/batteryless-infrared-remote-control-from-arveni

Deicon Dynamics & Controls. (n.d.). *Tuned Mass Dampers*. Retrieved March 5, 2015, from http://www.deicon.com/tuned-mass-dampers/

Deterre, M. (2013). *Toward an energy harvester for leadless pacemakers*. (PhD Thesis). Universite Paris Sud - Paris XI. Retrieved from https://tel.archives-ouvertes.fr/tel-00868838

Dimroth, F. (2014, Dec 1). *Fraunhofer Institute for Solar Energy Systems ISE*. Retrieved January 14, 2015, from Fraunhofer ISE: Press Releases 2014: http://www.ise.fraunhofer.de/en/press-and-media/press-releases/press-releases-2014/new-world-record-for-solar-cell-efficiency-at-46-percent

Donelan, J. M., Li, Q., Naing, V., Hoffer, J. A., Weber, D. J., & Kuo, A. D. (2008, February 8). Biomechanical Energy Harvesting: Generating Electricity During Walking with Minimal User Effort. *Science*, *319*(5864), 807–810. doi:10.1126/science.1149860 PMID:18258914

EE Times. (2009, November 6). Retrieved Jan 20, 2015, from Connecting the Global Electronics Community: http://www.eetimes.com/document.asp?doc_id=1253967

Elefsiniotis, A., Kokorakis, N., Becker, T., & Schmid, U. (2014, February 1). A thermoelectric-based energy harvesting module with extended operational temperature range for powering autonomous wireless sensor nodes in aircraft. *Sensors and Actuators. A, Physical*, *206*, 159–164. doi:10.1016/j.sna.2013.11.036

European Commission. (2014, Dec 11). Retrieved January 6, 2015, from European Commission: Climate Action: http://ec.europa.eu/clima/policies/brief/eu/

European Editors. (2013). *Digi-Key Corporation*. Retrieved 2 24, 2015, from Article Library-Smaller, Smarter, Ultra-Low-Power Sensors Raise Potential for Energy Harvesting in Medical Implants: http://www.digikey.com/en/articles/techzone/2013/sep/smaller-smarter-ultralowpower-sensors-raise-potential-for-energy-harvesting-in-medical-implants

Hanna, M. A., Mutashar, S., Samad, S. A., & Hussain, A. (2014). Energy harvesting for the implantable biomedical devices: Issues and challenges. *Biomedical Engineering Online*, *13*(1), 79. doi:10.1186/1475-925X-13-79 PMID:24950601

IATA. (n.d.). *iata.org*. Retrieved from http://www.iata.org/html_email/car1001654/lithium_batteries.pdf

Jiang, X., Li, Y., Li, J., Wang, J., & Yao, J. (2014). Piezoelectric energy harvesting from traffic-induced pavement vibrations. *Journal of Renewable and Sustainable Energy*, *6*(4). Retrieved from http://scitation.aip.org/content/aip/journal/jrse/6/4/10.1063/1.4891169

Jin-qiu, Z., Zhi-zhao, P., Lei, Z., & Yu, Z. (2013). A Review on Energy-Regenerative Suspension. *Proceedings of the World Congress on Engineering 2013*.

Jose, G. L., Antonio, R., & Francesc, M. (2003, March). Human Powered Piezoelectric Batteries to Supply Power to Wearable Electronic Devices. *International Journal of Society of Materials Engineering for Resources*, *10*, 34–40.

Kareem, A., Kijewski, T., & Tamura, Y. (1999). Mitigation of motions of tall buildings with specific examples of recent applications. *Wind and Structures*, *2*(3), 201–251. doi:10.12989/was.1999.2.3.201

Kumar, A., Sharma, A., Kumar, R., Vaish, R., & Chauhan, V. S. (2014). Finite element analysis of vibration energy harvesting using lead-free piezoelectric materials: A comparative study. *Journal of Asian Ceramic Societies*, *2*(2), 138–143. doi:10.1016/j.jascer.2014.02.001

Liu, J.-Q., Fang, H.-B., Xu, Z.-Y., Mao, X.-H., Shen, X.-C., Chen, D., & Cai, B.-C. et al. (2008). A MEMS-based piezoelectric power generator array for vibration energy harvesting. *Microelectronics Journal*, *39*(5), 802–806. doi:10.1016/j.mejo.2007.12.017

Lucy, E.-A. (2010). *Innowattech: Harvesting Energy and Data A stand alone technology*. Retrieved March 2015, 2, from Innowattech: http://www.iroads.co.il/sites/default/files/mtsgt_1_innowattech_presentation_-lucy_edery-azulay.pdf

Mareca, H., Darryl, G., & Chi, C. (2013). *Mobile Sensing Health & Wellness*. ON World Survey Report.

Martin, M. (2012). *Cardiac Pacemakers Market: Global industry size, share, trends, analysis and forecasts 2012-2018*. Retrieved March 9, 2015, from http://www.allvoices.com/contributed-news/12404227-cardiac-pacemakers-market-global-industry-size-share-treds-analysis-and-forecasts-2012-2018

Ni, T., Zuo, L., & Kareem, A. (2011). Assessment of Energy Potential and Vibration Mitigation of Regenerative Tuned Mass Dampers on Wind Excited Tall Buildings. In Proceedings of 2011 ASME design engineering technical conferences, (vol. 1, pp. 333-342). doi:10.1115/DETC2011-48728

Ooi, B. L. (2010). *Optimisation and Frequency Tuning Concepts for a Vibration Energy Harvester*. (PhD Thesis). University of Hull.

Ooi, B. L., & Gilbert, J. M. (2014). Design of wideband vibration-based electromagnetic generator by means of dual-resonator. *Sensors and Actuators. A, Physical*, *213*(0), 9–18. doi:10.1016/j.sna.2014.03.037

Ooi, B. L. A., Aziz, R. A., & Gilbert, J. M. (2014). Analytical and numerical modelling for dual-resonator vibration-based electromagnetic generator. *5th International Conference on Intelligent and Advanced Systems (ICIAS)* (pp. 1-5). Kuala Lumpur: IEEE. doi:10.1109/ICIAS.2014.6869443

Paradiso, J. A., & Starner, T. (2005). Energy Scanvenging for Mobile and Wireless Electronics. *IEEE Pervasive Computing / IEEE Computer Society [and] IEEE Communications Society*, *4*(1), 18–27. doi:10.1109/MPRV.2005.9

Patil, R. U., & Gawade, S. S. (2012). Design and Static Magnetic Analysis of Electromagnetic Regenerative Shock Absorber. *International Journal of Advanced Engineering Technology*, *III*(II), 54–59.

Rahman, & Leong, K. (2011). Investigation of useful ambient vibration sources for the application of energy harvesting. *IEEE Student Conference on Research and Development (SCORed)* (pp. 391-396). Cyberjaya, Malaysia: IEEE.

Rao, Y., McEachern, K. M., & Arnold, P. D. (2013). A compact human-powered energy harvesting system. *Journal of Physics: Conference Series*, *476*(1). doi:10.1088/1742-6596/476/1/012011

Ravindran, S., Kroener, M., & Woias, P. (2012). *A standalone Pyroelectric Harvester for Thermal Energy Harvesting* (pp. 80–83). Atlanta, GA: PowerMEMS.

Roundy, S. (2003). *Energy Scavenging fro wireless sensor nodes with a focus on Vibration to electricity conversion.* (PhD Thesis). University of California Berkeley.

Roundy, S., Steingart, D., Frechette, L., Wright, P. K., & Rabaey, J. (2004). Power Sources for Wireless Sensor Networks. In H. Karl, A. Wolisz, & A. Willig (Eds.), Wireless Sensor Networks (Vol. 2920, pp. 1-17). Springer Berlin Heidelberg. doi:10.1007/978-3-540-24606-0_1

Roundy, S., Wright, P. K., & Rabaey, J. (2003, July 1). A study of low level vibrations as a power source for wireless sensor nodes. *Computer Communications*, *26*(11), 1131–1144. doi:10.1016/S0140-3664(02)00248-7

Rowe, D. (2012). *Modules, Systems, and Applications in Thermoelectrics.* CRC Press. Retrieved from https://books.google.com.my/books?id=CVLSBQAAQBAJ

Sari, I., Balkan, T., & Kulah, H. (2008). An electromagnetic micro power generator for wideband environmental vibrations. *Sensors and Actuators A: Physical, 145-146*(0), 405-413. Retrieved from http://www.sciencedirect.com/science/article/pii/S0924424707008564

Simedics. (2014, Aug). Harvesting Energy to Power Medical Devices. *The Journal of mHealth, 1*(4), 38-39.

Soga, K. (2011, May 4). *eh-network.org*. Retrieved from 2nd Energy Harvesting Research Theme Workshop - Structural Monitoring: http://eh-network.org/events/workshop2/slides1.pdf

Special Interest Group. (2013). *Report on the UK Energy Harvesting Mission to Japan*. Energy Harvesting.

Starner, T. (1996). Human-powered wearable. *IBM Systems Journal, 35*(3-4), 618-629. doi:10.1147/sj.353.0618

Tang, L., Yang, Y., & Soh, C. (2010, December). Toward Broadband Vibration-based Energy Harvesting. *Journal of Intelligent Material Systems and Structures*, *21*(18), 1867–1897. doi:10.1177/1045389X10390249

Tang, X., & Zuo, L. (2011). Enhanced vibration energy harvesting using dual-mass systems. *Journal of Sound and Vibration*, *330*(21), 5199–5209. doi:10.1016/j.jsv.2011.05.019

Tang, X., & Zuo, L. (2012). Simultaneous energy harvesting and vibration control of structures with tuned mass dampers. *Journal of Intelligent Material Systems and Structures*, *23*(18), 2117–2127. doi:10.1177/1045389X12462644

Technavio. (2015, January 09). *Technavio*. Retrieved March 13, 2015, from Discover market opportunities: http://www.technavio.com/pressrelease/increased-demand-from-consumer-electronics-segment-driving-growth-in-the-global-mems

TECTEG MFR. (n.d.). Retrieved February 3, 2015, from TECTEG MFR Home: http://thermoelectric-generator.com/

Thad, S. E. (2003). Powerful Change Part 1: Batteries and possible alternatives for the mobile market. *IEEE Pervasive Computing / IEEE Computer Society [and] IEEE Communications Society*, *2*(4), 86–88. doi:10.1109/MPRV.2003.1251172

Weddell, A. (2010). *Notes from "Energy Harvesting for Wireless Automation" Conference*. Munich: Hanson Wade.

Weddell, A. (2010). *Notes from "Energy Harvesting for Wireless Automation"*. Munich: Conference.

Williams, C., & Yates, R. (1996). Analysis of a micro-electric generator for microsystems. *Sensors and Actuators A: Physical, 52*(1-3), 8-11. Retrieved from http://www.sciencedirect.com/science/article/pii/092442479680118X

Wischke, M., Masur, M., Goldschmidtboeing, F., & Woias, P. (2010). Electromagnetic vibration harvester with piezoelectrically tunable resonance frequency. *Journal of Micromechanics and Microengineering, 20*(3), 035025. doi:10.1088/0960-1317/20/3/035025

Wischke, M., Masur, M., Kroner, M., & Woias, P. (2011). Vibration harvesting in traffic tunnels to power wireless sensor nodes. *Smart Materials and Structures, 20*(8), 085014. doi:10.1088/0964-1726/20/8/085014

WordPress.com. (2011, March 21). *Cool Formula For Calculating Skyscraper Sway*. Retrieved March 5, 2015, from https://mathspig.wordpress.com/tag/taipei-101-tower/

Xia, L., Cheng, J., Glover, N. E., & Chiang, P. (2014, June 6). 0.56 V, –20 dBm RF-Powered, Multi-Node Wireless Body Area Network System-on-a-Chip With Harvesting-Efficiency Tracking Loop. *IEEE Journal of Solid-State Circuits, 49*(6), 1345–1355. doi:10.1109/JSSC.2014.2305074

Zhu, D., Tudor, M. J., & Beeby, S. P. (2010, February). Strategies for increasing the operating frequency range of vibration energy harvester: A review. *Measurement Science & Technology, 21*(2), 1–29. doi:10.1088/0957-0233/21/2/022001

Zorlu, O., Topal, E. T., & Kulah, H. (2011, February). A vibration-based electromagnetic energy harvester using mechnical frequency up-conversion method. *Sensors Journal, IEEE, 11*(2), 481–488. doi:10.1109/JSEN.2010.2059007

Zuo, L., & Cui, W. (2013). Dual-Functional Energy-Harvesting and Vibration Control: Electromagnetic Resonant Shunt Series Tuned Mass Dampers. *Journal of Vibration and Acoustics, 135*(5), 510181–510189. doi:10.1115/1.4024095 PMID:23918165

Zuo, L., & Tang, X. (2013). Large-scale vibration energy harvesting. *Journal of Intelligent Material Systems and Structures, 24*(11), 1405–1430. doi:10.1177/1045389X13486707

Chapter 2
Biologically-Inspired Wireless Power Transmission System: A Review

Agnes Ruey Chyi Cheah
UTAR, Malaysia

Kim Ho Yeap
UTAR, Malaysia

Kee Choon Yeong
UTAR, Malaysia

Kazuhiro Hirasawa
University of Tsukuba, Japan

ABSTRACT

This chapter gives an overview on Wireless Power Transfer (WPT) systems and the existing biologically-inspired architectures which could be used to optimize the efficiency of the systems. In general, wireless power transmission could be categorized into near-field and far-field mechanisms. At present, one of the most immediate issues to be solved is the relatively low efficiency in a WPT system. The system's efficiency decreases gradually as the distance increases. To ensure highest efficiency, the path with the shortest distance and the lowest interference is to be selected. In this case, a few of the most popular biologically-inspired algorithms, i.e. Firefly Algorithm(FA), Simulated Annealing (SA), the Ant Colony Optimization (ACO) and Genetic Algorithms (GAs) could be employed for efficiently solving different optimization problems. In short, these algorithms mentioned involve the emulation of fireflies flash signals for mating, exploitation of the metallurgy process, nature of ants leaving pheromones along the trails traveled and also the mimic of gradual evolution of organisms.

DOI: 10.4018/978-1-4666-9792-8.ch002

INTRODUCTION

Of all the current emerging technologies, wireless signals transmission has turned out to be one of the most prominent on-going development leaders. It acts as a catalyst to promote the growth of technology in the society. Wireless technology is ubiquitous, providing great contribution to the society, particularly, in health care, education, and community empowerment. Its remote capability allows anyone to connect or interact with each others at anywhere anytime. Such convenience has, therefore, encouraged a vast accessibility of information, allowing it to be provided and collected anywhere. Transparency in the information flow has also, in many ways, helped to foster socially positive impact. Take for instance the situation where one is required to travel to an unfamiliar or completely foreign destination. In the past, cumbersome homework has to be done before the start of the journey – one may have to refer to maps, writing down the signboards or landmarks that one might encounter en route, and memorizing the path shown in the map. With the advent of the wireless technology, however, all these have become very much redundant. With a simple click on the Global Positioning System (GPS), the user would then be guided easily to the destination he or she wishes to arrive at. In fact, besides its route-guiding capability, some Global Positioning Systems may even be equipped with additional features – such as, updating the drivers on the conditions of road blocks or traffic jams in real time. Also, with the application of wireless technology in health care environment, patients are ensured to be better connected to the medical personnel (Shobha, Chittal & Kumar, 2007). Medical experts could be alerted of emergency cases via wireless communication. This allows almost immediate treatment to be provided. Some of the hospitals in the United States have implanted Radio Frequency Identification or RFID tags into the patients for identification purposes. Since the tags make use of the propagation of electromagnetic fields for data transfer, doctors and medical experts are able to track the location or to obtain the physiological status of the patients on spot, during emergency cases.

There is no doubt that the portable cordless convenience found in wireless systems has remarkably enhanced the living standards of mankind. Despite the prevailing advancement in wireless technology, however, it seems that there still comes a time where most, if not all, of these devices are to be plugged to the power grid for batteries re-charging. As a result of this, the need for physical wire connections, at a certain instant of time, is still inevitable in wireless devices. Hence, wireless systems are, strictly speaking, not completely wireless after all. Indeed, the literal meaning of "wireless", in most cases, is referring to telecommunicating wirelessly; and not necessarily the device or system has to be functioning wirelessly at all time (Chen, Kawasaki, & Carvalho, 2013). The hope to realize a truly wireless device, i.e. to eliminate the very last wire which is necessary for recharge-ability, has prompted the scientific community to devote vigorous research in the field of wireless power transmission (WPT). With the availability of wireless power transfer, energy can be collected without the physical need of connecting a load to a power source. Such technology certainly provides great convenience, particularly, to the residents in desolated areas, where electricity is a scarce luxury to them (Kurs et al., 2007; Buonanno, D'Urso, & Pavone, 2011; Khripkov, Hong, & Pavlov, 2012).

The technology for wireless power transmission (WPT) has been evolving rapidly throughout the years. Although workable systems have already been developed, there still lie open issues and challenges ahead, yet to be overcome. One of the more critical issues faced by current WPT system is the restriction of distance – the amount of energy transferred wirelessly is highly constrained by the distance between the antennas. Beyond certain distances, a significant amount of energy would be lost in the transmission process. According to the findings by Kurs et al. (2007), at a distance above 2 m, the efficiency of a WPT system would degrade below 50%. Since the efficiency of the entire system is highly dependent on the amount of energy detected by the receiver system, researchers have been working on ways to overcome the limitation of the distance, as well as to improve the coupling efficiency of the receiver circuit. One way to improve the efficiency of the WPT system is to introduce biologically inspired topology management model to optimize the route for wireless energy transfer.

In this chapter, a brief review on different wireless power transfer methods is first discussed. Indeed, it could be seen in the subsequent section that, some of the mechanisms implemented in wireless power transmission – particularly those applied in far-field – are actually inspired from the wireless communication system, where a pair of antennas are required to transmit and detect the wireless signal traveling in free space. An elaboration on techniques which could be applied to enhance the efficiency of the WPT systems is subsequently performed. These techniques include some of the most popularly implemented biologically inspired architectures in wireless sensor networks. By incorporating the appropriate energy-efficient biologically inspired topologies into the system, the efficiency of the WPT system would certainly be optimized.

BACKGROUND

Nikola Tesla was best known to be the first person who conceived the brilliant idea of wireless power transmission. Due to his bold idea on wireless power transmission at his time and his discovery on wireless communication, he was also referred to as the "Father of Wireless Technology". Over the years, considerable development based on the concept introduced by Tesla has been achieved in wireless technology. Today, besides being able to communicate wirelessly via mobile handsets and wireless internet systems, the world has been able to realize what Tesla had yearned to build decades ago – energy can now be harvested via wave signals propagating in free space.

Based on the distance between the transmitting source and receiving device, as well as the wave characteristics, the mechanisms applied for wireless power transfer can, in general, bifurcate into (i) the near-field and (ii) far-field categories (Balasubramaniam, 2013).

As graphically depicted in Figure 1, when the distance from the transmitting antenna to the detector circuit is within the size of a wavelength (λ), the WPT system is classified as operating based on the near-field mechanism. The field patterns tend to change rapidly when the distance of the receiving antenna is close to the transmitting antenna. Within the near-field region, the energy that is coupled to the receiver includes both reactive and radiating energy. Hence, the load could be regarded mainly as a reactance which mostly stores energy (Tuovinen, Berg, Yazdandoost, Salonen, & Iinatti, 2012). On the other hand, when the distance between both the transmitter and receiver antennas is greater than 2λ, the system operates, mainly, based on the far-field mechanism. This region is dominated by the radiated fields, with the electric (**E**) and magnetic (**H**) fields supporting and regenerating one another perpendicularly. Like a plane wave, the **E** and **H** fields are transverse to the direction of propagation. Hence, far-field mechanism dissipates energy as a result of the alterations in magnetic and electric fields. In between the near- and far-fields, there is the transition zone. This is where both field characteristics found in the near- and far-fields overlap each other.

WIRELESS POWER TRANSFER MECHANISMS

The Near-Field Mechanism

The near-field wireless power transfer mechanism delivers energy based on the change in the current or charge distribution. This mechanism constitutes primarily two zones, i.e. the reactive and radiating regions. Within the reactive near-field region, the energy transfer process is strongly dependent on the distance – in which, energy drops dramatically with the increment in distance. On the other hand, the radiative area represents the beginning of the far field region. Unlike the reactive region, however, energy in the radiating region remains relatively constant and it is not affected by the variation of distance from the transmitting antenna.

The near-field mechanism comprises both inductive coupling and resonant inductive coupling. Inductive or magnetic coupling is the induction of voltage by the magnetic field generated due to the alteration of current flow in the transmitter coil. Faraday's law succinctly states that, any changes made to the magnetic flux in a coil of wire will induce a voltage (i.e. emf) in it:

$$\varepsilon = -\frac{d\phi_B}{dt} \quad (1)$$

where ε is the induced EMF and ϕ_B represents the magnetic flux.

As shown in Figure 2 (Bou, Alarcon, & Gutierrez, 2012), the part of the circuit at the transmitter and the receiver are mutually affected by the same magnetic flux. The total flux can be separated into two parts: (i) ϕ_{11} and ϕ_{22} which represent the individual magnetic flux yielded in the transmitter and receiver coils, respectively; whereas (ii) ϕ_{12} and ϕ_{21} are the mutual flux produced by the transmitter coil that interlinks to the receiver coil, and vice versa. Energy is coupled from the transmitter coil to the receiver coil through a magnetic field, without the need of physical contact between the transmitter and receiver circuits. The measurement for the mutual inductance is based on the total inductive coupling between the two circuits. It is to be noted that, the system loses part of the energy from the primary coil when the energy is being transmitted across free space inductively, into the secondary coil. Hence, the

Figure 1. Field types with respect to distance source, x

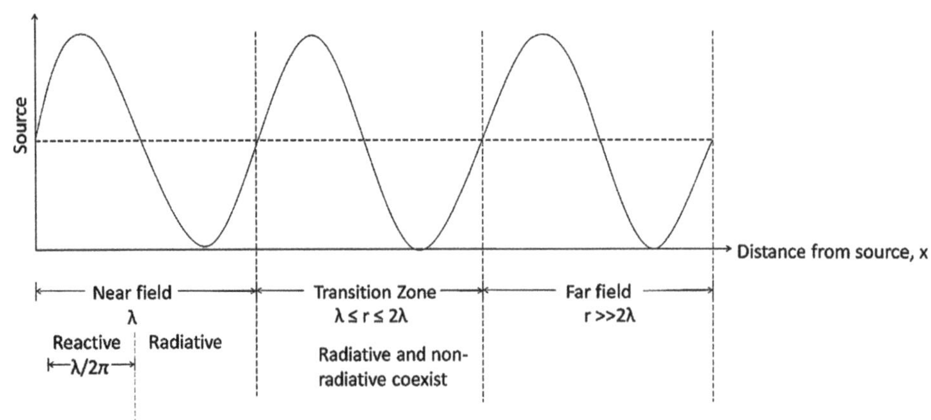

Figure 2. Magnetic coupling

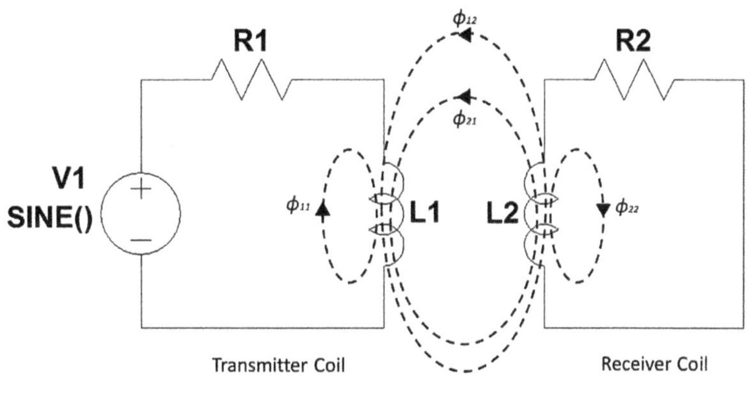

coupling between the circuits can be improved by placing the circuits on the same plane and decreasing the distance between the circuits. Since the intensity of the magnetic field produced by the coil is in direct proportion with the number of turns of the coil, by increasing the number of turns, the efficiency of the system could be improved accordingly, as well.

Electric toothbrush happens to be one of the most common applications of inductive coupling that may be encountered in one's daily life. A complete set of electric toothbrush consists of both the toothbrush and a base for recharging purpose. There are turns of coils at the handle of the toothbrush, as well as at the recharging unit in the base. When there is a change in current flow across the primary winding in the charger, a magnetic field is produced. By placing the toothbrush into the base, the magnetic field will then induce current in the secondary winding in the toothbrush. The power transmission principle in recharging the toothbrush is indeed similar to the inductive coupling mechanism used for wireless power transfer. Inductive coupling also finds wide applications in the biomedical field, such as medical implants, etc. This is mainly because at frequencies which fall within the near-field range, the field absorption rate in biological tissues is quite low (Flynn & Fotopoulou, 2013). Hence, it hardly imposes any detrimental effect to the patient's body.

Resonant inductive coupling (RIC) is another technique for near-field wireless power transmission. RIC is modified from the inductive coupling system mainly to minimize power loss due to the distance between the coils. For power transmission to occur, first of all, the coils in the devices have to be tuned to resonate at the same frequency. This is important because if it happens that the coils are resonating at different frequencies, then power transmission will fail to take place. When current passes through the coil, the coil will start to resonate. Resonant frequency f_{res} can be obtained by multiplying the capacitance values of the plates C with the inductance values of the coils L, i.e.

$$f_{res} = \frac{1}{2\pi\sqrt{LC}} \tag{2}$$

When current oscillates in a coil ring, an oscillating magnetic field will be produced, causing the coil to be capacitively loaded and becomes highly resonant. Energy flowing in the coil takes some time to fade after many cycles. When a second coil is then brought near to certain distance, most of the energy from the primary coil can be transferred to the second coil. Maximum power transfer occurs when the system is coupled at resonant frequency. This is because, at resonant frequency, the capacitive and inductive reactances cancel out each other, making the system to be purely resistive. The quality factor (Q factor) describes the ratio of energy stored in the resonator to the energy supplied by generator. Hence, Q factor is normally used to measure power loss at resonance.

The Far-Field Mechanism

As mentioned in the earlier section, the signal in the far-field zone loses its energy during its propagation. In fact, the power decreases with the inverse square of the distance r from the transmitting antenna, shown in Equation (3) below (Cheng, 1989)

$$P_r \propto \frac{P_t}{r^2} \tag{3}$$

where P_t and P_r are the transmitted power of the transmitting antenna and the received power of the receiving antenna, respectively.

A far-field wireless power transmission WPT system operates in a similar manner with that of a wireless telecommunication device. Like a wireless communication system, signals in a WPT system are transferred from the transmitter to the receiver via free space propagation. Unlike a communication system which transfers useful information, however, a WPT system detects the signal and harvest energy from it wirelessly. Due to their similarity in signal transmission, the underlying theory of a WPT system is actually developed based on that of the transceiver circuits in the telecommunication system.

As schematically depicted in Figure 3, the dc signal from the rectifier, at the transmitter module, is modulated by an oscillating signal to radio frequencies (RF). The RF signal is then transmitted via free space to a rectenna. A rectenna is invariably the key component in the far-field WPT technology. It is the merging of an antenna and a rectifier circuit that is used for converting RF signals into direct current or dc power. The output of the rectenna is connected to a resistive load. Upon coupling the electromagnetic field to the receiving antenna, the RF signal is fed to a non-linear device. When the signal goes through the non-linear device, it produces various spectral components at its output. These include the dc component, the ac fundamental frequency and its respective harmonics, as well as, the intermodulation mixing products (Boaventura, Collado, Carvalho, & Georgiadis, 2013). The non-linear device is connected in cascade with a low pass filter (LPF) which helps to filter the unwanted components. Although all ac components are to be ideally removed, leaving only the dc component; in practicality, some with lower frequencies would still remain at the output of the LPF. Hence, the signal is required to go through another round of rectification process to convert it to a mere dc signal. In order to allow maximum wave coupling between the signal and the receiver circuit, the standing wave induced by the reactive energy must be suppressed (Yeap, Tham, Nisar, & Loh, 2013). This can be done by connecting a matching network between the LPF and the rectifier circuit. The function of the matching network is to adjust the input impedance Z_{in} of the load so as to match the characteristic impedance of the transmission line Z_c. Due to their fast switching action and low forward voltage drop which reduces the loss caused by conduction and the switching process, Schottky diodes are commonly used for the rectification purpose. Schottky diodes are applied to convert the signal to dc voltage and current before sending it to the load.

Figure 3. General WPT system

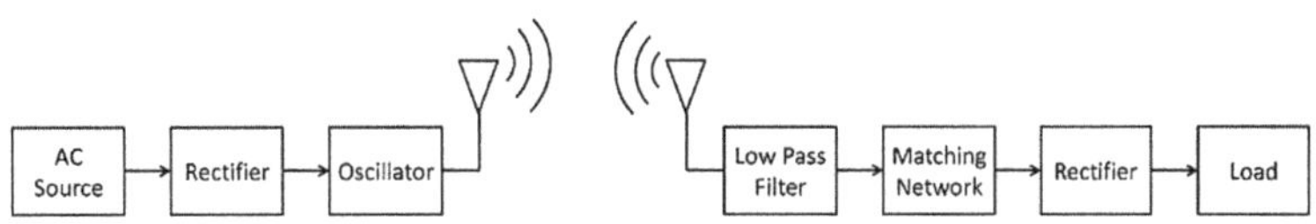

Envelope detector is the typical topology used for rectifier circuits. It provides an output which is in a smooth curve, outlining the extremes of the points locating on the original signal amplitude by taking in high frequency signals as the input.

Microwave power transmission (MPT) is a subset of the far-field WPT. MPT refers to the technology of transmitting a large amount of power at microwave frequencies. In MPT, the signals oscillate at relatively high frequencies, i.e., around 10^9 to 10^{11} Hz. The bandwidth in the microwave region is about 30 times greater than all the other electromagnetic spectrum below it. Such broad bandwidth allows a huge amount of information to be transferred wirelessly. This technology is used on a broad scale in point-to-point communications. With their small wavelength, microwave signals can be directed sharply into narrow beams at the receiving antenna better than any other radio wave signals. Because of this reason, equipments which transfer signals at the microwave system are, usually, able to communicate without any obstruction. Besides using it for communication in spacecrafts and sending power from space to the earth using satellites, MPT is also applied, based on the concept of space solar power (SSP), to power unmanned aerial vehicles (UAVs) remotely (Strassner & Chang, 2013). It is worthwhile noting that, one disadvantage found in microwave signals is that it may be attenuated significantly due to the high absorption rate of water vapor and oxygen found in the atmosphere.

BIO-INSPIRED ALGORITHMS FOR OPTIMIZATION

The universe has been around for approximately 13.7 billion years; while biological livings have already gone through millions of years of evolution. It is apparent that any living organism which exists until today has managed to develop a unique way of its own to ensure its survival. Plants like trees and flowers, for instance, would naturally grow towards the direction where they could obtain maximum sunshine, in order for them to undergo photosynthesis. Similarly, birds would take turns leading the flocks while migrating from one place to another during winter seasons. Clearly, nature has its own way to cope with tough situations. Living organisms have the capability to decide for themselves the optimum choice which benefits them the most. Therefore, besides installing impedance matching network to reduce the noise and improving the efficiency of the rectenna circuits, employing an appropriate biologically inspired algorithm could be an effective alternative to ensure optimum efficiency obtained during the transmission process. The signal transmitted via free space may take different paths to reach the receiving antenna. In fact, there could be infinite possible paths the signal could take. However, there is only a handful in which the highest efficiency could be obtained. Therefore, a biologically inspired algorithm can be implemented to identify the exact transmission path which has been consistently giving the highest efficiency to the receiver circuit. Once the path is identified, the transmission process would then be locked on to that particular path, using directors and reflectors, for example, to ensure optimum wireless power transfer.

Multiple types of nature inspired algorithms have been developed, hitherto. Some of which include the Firefly Algorithm (FA), the Cuckoo Search, the Ant Colony Optimization (ACO), the Genetic Algorithms (GA), Simulated Annealing (SA), Harmony Search, Bat Algorithm and so on. Each of these algorithms has their own strengths and drawbacks. In this section, discussion will be given on four of these algorithms most commonly applied in wireless power transfer – the FA, SA, ACO and the GA.

Firefly Algorithm (FA)

The flash signals found in fireflies are observed to play various significant functions. Besides acting as the primary courtship signals for mating, they are also found to serve as a self-protection mechanism – particularly, for warning off potential predators. According to the study performed by Yang (2013), three significant behaviors of fireflies are found to be useful in developing algorithms for wireless power transfer. Firstly, it is observed that (i) fireflies are all unisexual and that any individual firefly can be attracted by any other fireflies. Secondly, (ii) the attractiveness of a firefly is determined by its brightness which, according to the physical formula for light intensity I, is inversely proportional to the square of the distance between the fireflies r. Thirdly, (iii) the firefly moves arbitrarily if no fireflies are brighter than the given firefly. These three behaviors have become the fundamental building block for the firefly algorithm (FA), as depicted in the flowchart in Figure 4.

As presented in Figure 4, a random initialization of fireflies population, x_i (i=1,2,…, n) is generated. The light intensity I_i at x_i is obtained by $f(x_i)$. The maximum number of fitness function evaluations in the algorithm is associated with MaxGeneration. The firefly search process is performed within the while loop, controlled by the MaxGeneration. The population of the fireflies will be sorted out according to their fitness values, thus the best individual within the population will be selected. Based on distance between both fireflies r_{ij}, if firefly j has higher light intensity (brighter) than firefly i, firefly i is attracted and moves its position in the search space towards firefly j. New solutions will then be evaluated and current global best is determined.

There are 3 crucially important parameters that determine the behavior and performance of FA. They are the (i) randomization parameter α, (ii) attractiveness β, and (iii) the absorption coefficient γ. Light intensity $I(r)$ can be determined using the equation as stated below (Yang, 2013):

$$I(r) = I_0 e^{-\gamma r^2} \tag{4}$$

where I_0 is the light intensity from the source and γ denotes the light absorption coefficient.

As mentioned, the firefly is attracted to another firefly which has higher light intensity than itself. Therefore, the attractiveness β of a particular firefly is proportional to the light intensity $I(r)$, thus the attractiveness β can be defined by

$$\beta = \beta_0 e^{-\gamma r^2} \tag{5}$$

Equation (6) defines how a firefly i moves towards another brighter firefly j, with respect to the attraction β, the randomization parameter α, and the random numbers drawn from a Gaussian distribution or uniform distribution $\in_i$.

Figure 4. General flow of Firefly Algorithm (FA)

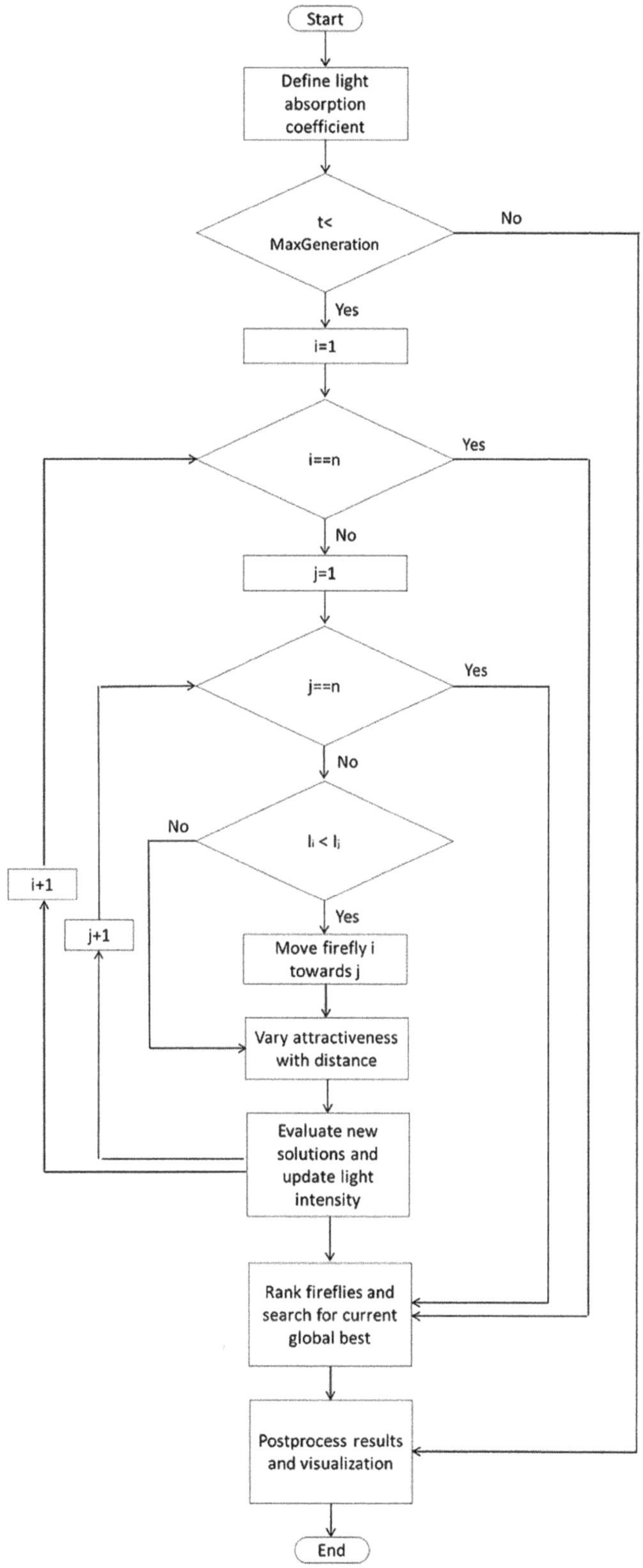

$$x_i = x_i + \beta_0 e^{-\gamma r_{ij}^2}(x_j - x_i) + \alpha \in_i \quad (6)$$

Studies performed by researchers show that FA outperforms many other optimization algorithms, in terms of speed and efficiency. This algorithm, has therefore, been expanding rapidly. The applications of FA have, in fact, been rather widespread, especially in the engineering field like wireless sensor networks, antenna designs, robotics, image processing etc.

A few of the FA's major strengths are worth highlighting so as to illustrate how actually this algorithm works better than other algorithms. Firstly, an automatic subdivision algorithm is present in FA. As mentioned in the beginning, FA works in the way where fireflies flash with the intention of attracting other fireflies for mating. However, since the intensity of light fades when they are farther apart, the attractiveness decreases along with distance as well. In this case, the entire population of the fireflies can eventually subdivide into smaller groups (subgroups) as the local attraction turns out to be stronger than the long-distance attraction. Therefore, FA is particularly efficient in solving highly non-linear, multimodal optimization problems. Secondly, FA possesses the ability to search for all or most of the multiple solutions for a particular problem. In other words, FA is capable of performing multimodal optimization. Continuous problems, clustering and classifications or even combinatorial optimization are the areas of applications of FA where the algorithm can give excellent performance. This is because FA serves as a swarm-intelligence-based algorithm and behaves as a global problem solver. Therefore, it is able to generate desired solutions for several continuous optimization problems. Besides, its flexibility can be seen by varying the scaling parameters. Tuning the scaling parameters enables FA to manipulate the randomness, as well as, getting adapted to the problem landscape for speeding up the convergence.

It is worthwhile noting that, despite the robustness found in FA, a few deficiencies are actually discovered when this method is applied alone in solving optimization problems. One particular drawback of FA is that it is only found to be effective in computing problems which are highly dimensional and non-linear. To circumvent these short comings, enhanced version of this algorithm such as the modified firefly algorithm and hybrid firefly algorithm are developed. As the topic is very broad, detailed studies of the enhanced firefly algorithms do not fall into the scope of this chapter. The interested reader should refer to refer to the papers written by Bidar & Rashidy (2013) and Luthra & Pal (2011).

Simulated Annealing (SA)

Simulated annealing (SA) is a probabilistic method proposed by Kirkpartick et al. in 1983, inspired by the process of annealing solids (glass or metal) in metallurgy. This technique is basically used for solving global optimization problems. Annealing is a heat treatment process (metallurgy) which involves heating a material and let it cool for increasing the material's ductility. Simulated annealing algorithm exploits the analogy of the physical process whereby a material is heated up, and undergoes slow cooling process by using a set of controlled cooling operations, while this slow cooling process permits higher probability for finding configurations with lower internal energy (global solution) than the initial one (Dowhan, Wymyslowski, & Urbanski, 2009).

The diagram depicted in Figure 5 basically sums up the structure of Simulated Annealing algorithm.

To start simulated annealing (SA), an arbitrary initial temperature is first defined and this is followed by evaluating the temperature with a random initial solution. As long as the thermal equilibrium (ending criteria in general) is not met, it proceeds in searching for neighbors. This stage is fundamental as it is

Figure 5. Structure of simulated annealing (SA) algorithm

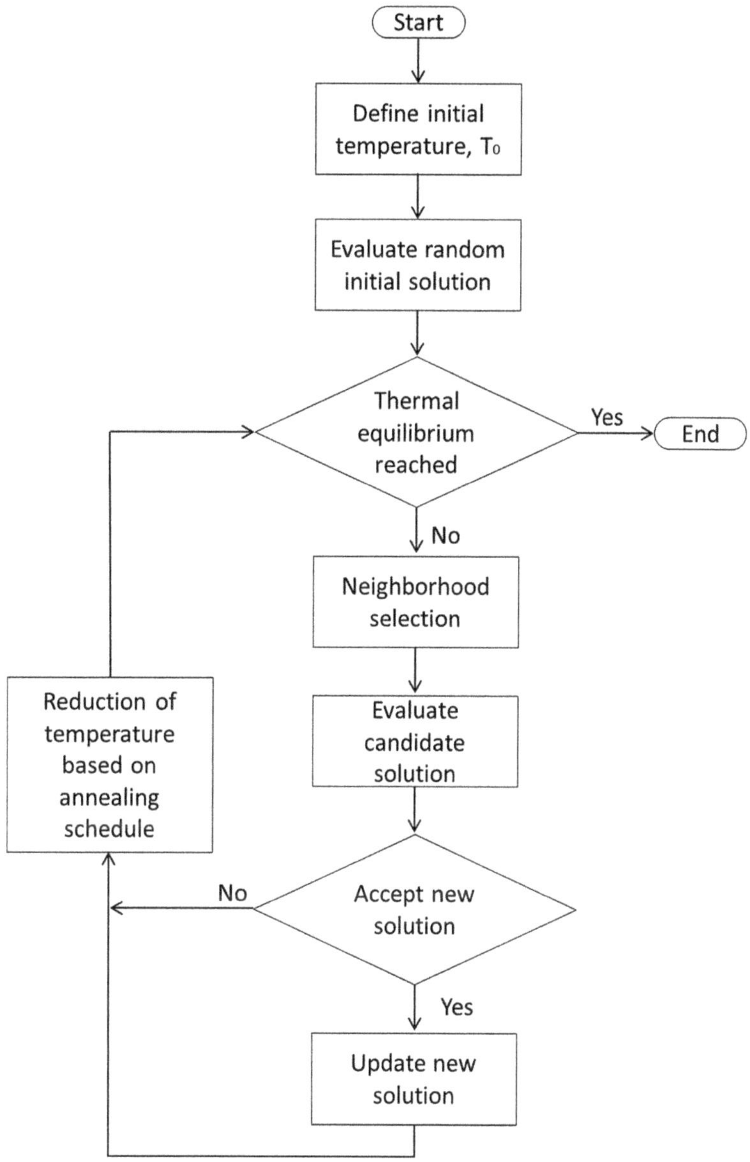

the key for determining the final solution after moving through all the successive neighbors. When a candidate solution is opted, evaluation of that solution based on the quality of the solution is performed. The algorithm compares between the existing and the candidate solutions. In general cases, a better solution will be accepted and updated as the new solution. Otherwise if a worse solution is found, the existing assignment remains unchanged. However, it will occasionally accept worse solutions under certain circumstances. The acceptance probability function will be further discussed later. The temperature is reduced in accordance with the annealing schedule (cooling schedule) after each execution.

There are two important criteria in Simulated Annealing: the neighborhood selection and the cooling schemes. With reference to the acceptance probability function mentioned previously, it can be seen that the starting temperature is the main cause for the algorithm to accept worse solution. The algorithm operates initially with sufficiently high temperature, allowing the process to accept an inferior solution. But as the simulation proceeds, it helps to narrow down the search results and improves the quality of the solution since there will be a gradual reduction in temperature according to the scheme of the annealing schedule. This characteristic of SA avoids having the solution being trapped in local minima, having more freedom in exploring global minima, and therefore, resulting in a better search for better solutions. SA accepts better neighbor solutions but when no any better neighbor solutions can be found, it stops its execution. Even though SA does not ensure that global optimum can be reached at every iteration, it produces solutions that are significantly better compared to some other methods. It is almost certain that, high quality solutions can be obtained although sometimes this may come at the cost of excessive computational time.

Annealing schedule is a scheme to estimate and control the parameters in the algorithm. It helps in determining the degree of uphill movement allowed throughout the search process. Adopting appropriate schedule is critical in defining algorithm's performance as it influences the quality of the solution obtained in the execution. A set of parameters (cooling schedule) has to be specified with a finite sequence of values in order to conduct the convergence of SA. The cooling schedule precisely comprises 3 major components, which are the (i) initial temperature, (ii) frozen value, and (iii) ending condition

First of all, the initial temperature has to be high enough in order to have enough freedom to make sufficient move to go along different local minima or else the solution obtained will be having a value which is almost equal to the random initial solution defined. On the other hand, if the initial temperature is set too high, it transforms the algorithm into a random search. This means that the search is free to move to any neighborhood state and it may need a long time to reach for the results. So either criterion needs to be taken into account in order to opt for an appropriate initial temperature.

The temperature decrement parameter is another important parameter in determining the performance of the algorithm. In order to achieve good results, the frozen value can be obtained by the equation (Prajapati, Agravat, & Hasan, 2009):

$$Frozen\_value = \frac{T}{\Delta T} \tag{7}$$

where $\Delta T=0.95$.

Since the temperature reduction value differs based on the types of annealing schedules being selected, it is common in SA to allow the temperature to decrease until it reaches zero. Doing so, however, might result in a longer time for the algorithm to complete its execution. It is a waste of time to continue excessive iterations if the worst or best solution is almost similar to the solution found when the temperature is equal to zero. Thus, a termination (ending) condition can be set to stop the process when there is no better or worse solution being accepted.

If the algorithm can no longer search for better solutions and if the ending criteria is met, SA algorithm will stop the searching process accordingly. Then the existing best solution found will be recorded and updated.

Simulated annealing can be applied on broad fields for global optimization problems, such as the Travelling Salesman Problem (TSP). The algorithm starts by selecting a random initial tour from all the possible selected cities (solution space). Going along the initial tour helps to search for a neighbor tour and this neighbor tour could be shorter or longer in distance than the initial tour. If the randomly selected candidate tour is shorter, it is updated as the new tour. Otherwise if the candidate tour is longer than the existing tour, that candidate tour can still be accepted under specific circumstances, like those being mentioned in the previous text. The chances of accepting a worse tour is based on the function of how much longer the candidate is compared to the current tour, as well the temperature of the annealing process. At higher temperature, the algorithm is more likely to accept an inferior tour. Then, the temperature is slightly reduced and the step of selecting a candidate tour from the neighbors is repeated, until the termination condition is met (low temperature is obtained).

Simulated annealing is diffusing broadly due to its excellent performance in the convergence of the best solution. Even though this algorithm requires a longer period of time when dealing with massive search executions, it has been a commonly used tool by many researchers in the development of optimal experimental design problems.

Ant Colony Optimization

Ant colony optimization (ACO) technique is a heuristic method for designing metaheuristic algorithms that helps in solving difficult combinatorial optimization issues. It was proposed by Marco Dorigo in 1992, at his initial attempt in determining the optimal path from a graph inspired by the behavior of the ant. In view of this, it would be better to explain the inherent behavior of ants first. At the beginning, the ants randomly pick a path to search for their food. For every path that they have traveled, pheromones are deposited on it. The pheromones will stay on the path for some time. But they are not permanent. They will evaporate and disappear over time. Hence, the pheromones would only stay longer at a path if the number of ants which passes by the same path increases. In other words, the path turns out to be more appealing to the future ants if the pheromones are reinforced by more ants (Putha, Quadrifoglio, & Zechman, 2012). It could be observed that a similar concept like this can be adopted to ensure that optimum efficiency can be consistently obtained at the rectenna of the receiver circuit. In this case, the path with the strongest pheromones represents the highest efficiency to the traveling signal.

The construction of the solution starts with an ant traveling from node i to node j. Mathematically, the probability of an ant traveling the path can be expressed as (Wei & Zhi, 2010)

$$Ant\_moving\_probability, P(A = (_{i,j})) = \frac{\tau(A)^{\alpha} \bullet n(A)^{\beta}}{\sum \tau(A')^{\alpha} \bullet n(A')^{\beta}} \quad (8)$$

where τ is the attractiveness of the pheromone trails, n the reciprocal of distance, A the path visited by the ant, A' the path which has not been visited by the ant, and α and β are the parameters for controlling $\tau(A)$ and $n(A)$, respectively. The amount of pheromones on the trail is updated using the following equation

$$\tau(A) = (1-\rho)\tau(A) + \Delta\tau(A) \quad (9)$$

where ρ is the pheromone evaporation rate and $\Delta\tau(A)$ represents new pheromone. If ant k travels on path A, then

$$\Delta\tau(A)^k = \frac{1}{L^k} \quad (10)$$

where L^k is the length of the path traveled by ant k.

Over the years, different ACO metaheuristic cases have been proposed. The three most successful ones are (i) the ant system (AS), (ii) the ant colony system (ACS), and (iii) the MAX-MIN ant system (MMAS). The ant system was the first ACO algorithm proposed. The algorithm for searching for the optimum solution could be better illustrated using the traveling salesman problem (TSP). As shown in the flow chart in Figure 6, after the pheromone trails are initialized, the ants will be located randomly on the cities. Then the ants will be moved to other cities according to the probability computed, until the ants completed the tour. Upon each tour completion, the new pheromone will be updated. Comparing the local solution with the global solution, the solution that carries a greater value will be stored.

Figure 6. Ant system algorithm for TSP

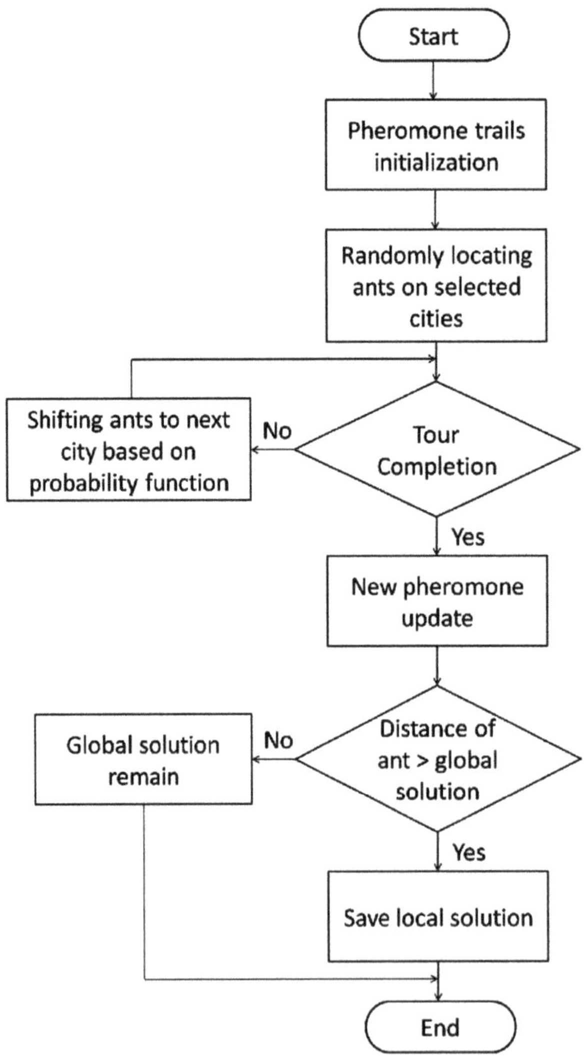

Based on the AS, algorithmic improvements are made to achieve a better performance. Thus, another system known as the ant colony system (ACS) was developed, which can be regarded as the improved version of the ant system (AS). What distinguishes the ACS from its predecessor, i.e. the AS, is the decision applied on the ants in the construction stage. A new transition rule, known as the pseudorandom proportional rule is introduced in ACS. Pseudorandom proportional rule helps the ants to select, among the possible ones, the components that are able to give the maximum product $\tau(A)n(A)$ (Nanda & Das, 2011). The third Ant Colony System, i.e. the MAX-MIN Ant System (MMAS), sums the pheromones of only the best solutions to achieve strong exploitation of the search history. Also, by limiting the MAX and MIN values of the pheromone, this system works very effectively in preventing premature convergence of the search. Indeed, MMAS is found to be one of the best performing methods for solving TSP and Quadratic Assignment Problem (QAP) (Stützle & Hoos, 2000).

Genetic Algorithms

During the 1960s and 1970s, a randomized search and optimization technique known as the Genetic Algorithms (GAs) was pioneered by John Henry Holland together with his students at the University of Michigan. GA is an adaptive heuristic search algorithm developed based on the ideas of natural process of genetic recombination and evolution. Genetic algorithms simulate processes in natural system which is necessary for evolution (Kumar & Jyotishree, 2014). The underlying principle of GAs, to put it in layman's term, can be best described using Herbert Spencer's famous quote – the survival of the fittest. This is to say that, evolution will take place until, eventually, the best species prevails, while the weaker ones are eliminated. This technique is widely applied in the engineering fields to solve problems in alternative ways. Although GAs are computationally pricey and time consuming, they strike a perfect balance when there is difficulty in finding the solution for a problem with a global optimum and many local optima. It can simultaneously manipulate a large number of parameters and also take into consideration every feasible search pathway. Since they are not instantaneous, genetic algorithms perform well on complex landscape and are well-fitted for solving non-linear problems. Therefore, when it comes to optimizing large numbers of discrete parameters, genetic algorithms consistently outperform other traditional techniques (Oyama, Obayashi, Nakahashi, & Nakamura, 1997).

The performance of GAs is influenced by the encoding scheme and also the genetic operators. In genetic algorithms, search space parameters are defined for all those feasible solutions for a given problem. The search space parameters are encoded in the form of genes, known as chromosomes. A set of these chromosomes forms a population. To put it simply, a population is referring to a set of possible solutions for a particular problem. After an initial population is randomly selected, the fitness is then evaluated for each individual chromosome. Each of the fitness function represents the degree of fitness of the string. Selection of the strings is carried out, on the basis of the principle of survival of the fittest. In other words, the individual chromosome that possesses a greater fitness will be chosen for reproduction. Once 2 parents are chosen in the selection process, crossover takes place. The mating of the two parents allows parts of the parents' chromosomes to be passed down to their offspring. As graphically depicted in Figure 7, new generation of strings are yielded through the crossover process.

Crossover is the recombination of individual genetic material to produce better new strings. Should there be no crossover, the chromosomes of the offspring would end up identical with those of their parents. Hence, crossover is one of the most critical stages in genetic algorithms. Crossover probability P_x dictates how often the crossover process would occur. This is to say that, P_x would be exactly 100% if

Figure 7. Single point crossover

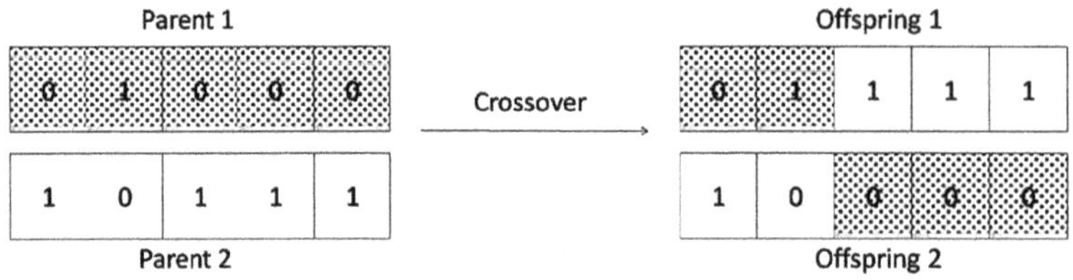

Figure 8. Single point mutation

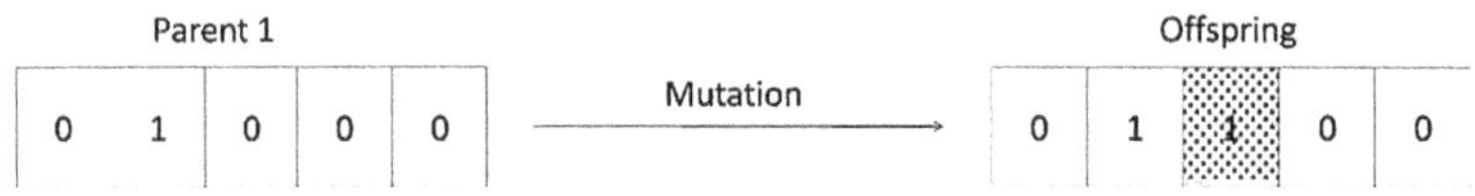

the crossover process is perfect. On the contrary, if P_x is 0%, this implies that the crossover process has failed and the exact chromosomes from the old population are copied to the offspring. The main objective of the crossover process is to pass the parents' good chromosomes to the offspring, so that the good part has a continuation to the next generation. It is worthwhile noting, however, that if the crossover rate appears to be too high, the selection is often not possible to be in the same speed with the elimination of high performance strings. On the other hand, if the crossover rate is too low due to low exploration speed, then stagnation will occur. Hence, a balance is to be taken between the two extreme cases. For effective crossover, the rate should fall in the range of 80% to 95%.

After the crossover process is complete, probabilistic mutation is applied on the new chromosomes. At this stage, some of the offspring is randomly modified, as depicted in Figure 8. The randomness introduced by the mutation process acts as an insurance policy which helps to prevent the algorithm from getting stuck in local optimal solutions; while, at the same time, allows the algorithm to search for new search space. This process is actually important to avoid the falling of genetic algorithms into local extreme and to establish ergodicity. Much like the crossover process in which the effectiveness is measured by the crossover probability P_x, the effectiveness of mutation is measured by the mutation probability P_m. This means that, the chromosome is completely changed if P_m is 100% and the offspring remains constant if P_m is equivalent to 0%. The optimum speed for the mutation process is between 0.5% and 1.0%.

The process of genetic algorithms is summarized in Figure 9. In short, the fittest parents' chromosomes from the old population will undergo crossover and mutation to produce offspring. A new population is generated from these offspring. The new population will then undergo a subsequent round of crossover and mutation to produce yet another newer generation of offspring. This process iterates for several generations until the most optimum condition is attained or until a termination condition is met (Maulik & Bandyopadhyay, 2000). Basically, the successive new generation produced contains more good genes, or in simple words, a better partial solution than those in the previous generations. Based on this concept,

Figure 9. General flow of GA computation

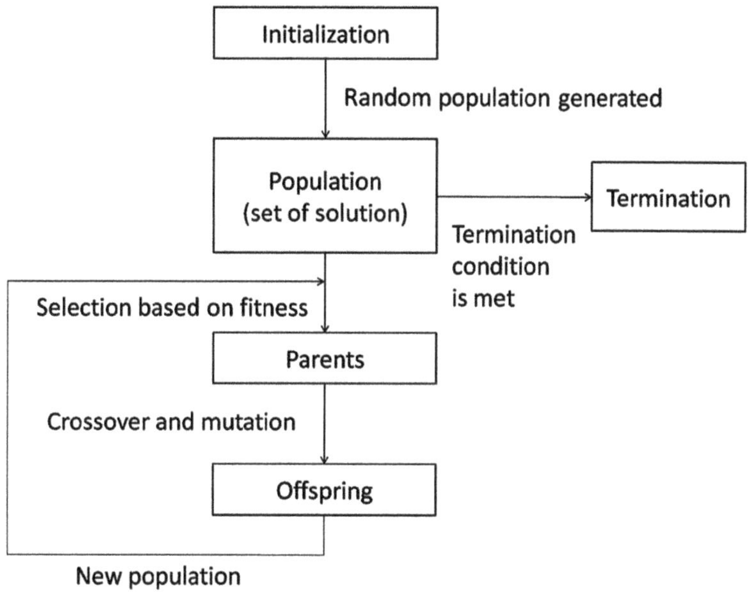

a wireless power transmission system which applies the genetic algorithms (GAs) would be able to find the optimum path which ensures the highest efficiency for the system. Starting with a random transmission path, GAs allow the system to evolve on its own, so that the most optimum path could be obtained after the process undergoes multiple generations of crossover and mutation.

Ant Colony Optimization vs. Genetic Algorithms

Both the Ant Colony Optimization (ACO) and Genetic Algorithms (GAs) have been shown to be feasible to be implemented in optimizing the efficiency of a wireless power transfer system. Both approaches, in fact, have their own strengths and weaknesses. Putha et al. (2012) have carried out an experiment to investigate the performance of the ACO and GAs approaches. In the experiment, the two approaches have been employed to solve traffic signal congestion. The effectiveness of both approaches were compared and assessed. The model network used for testing the ACO and GAs is graphically depicted in Figure 10 (Girianna & Benekohal, 2002). The model network shows a system consisting of 20 signals in total, which operates only in a single direction (one-way) and allows no turning movements. Each of the signals operates on a two-phase plan and the flow rate is defined as vehicle per hour per lane (vphpl) which is shown in Figure 10. The signals located on the arterials with thicker lines are coordinated (signal 20 to signal 5 and east-west running arterials). Initially, 20 vehicles are located in a queue on each lane and there are 2 arterial lanes. Maximum green time limits the maximum time for the green interval, whereas minimum green time defines the minimum duration of the green interval for each movement. The lost green time tells that, regardless of whether the traffic signal displays a green signal, no vehicles are able to pass through an intersection. Shockwave is the movement of point which demarcates the two stream conditions, for example, the transition from a flowing and speedy state to a congested and standstill state or the other way round. The parameters are adopted from the experiment demonstrated by Girianna & Benekohal (2002).

Figure 10. One way traffic with 20 signals

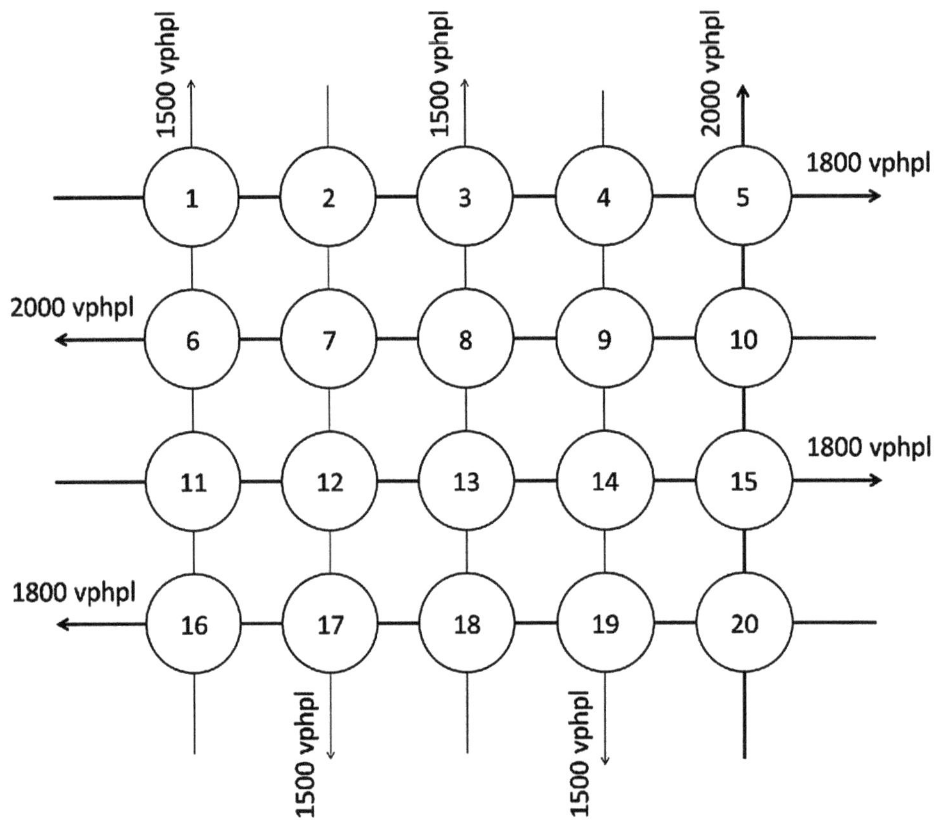

- Maximum green time: 80s
- Minimum green time: 20s
- Lost green time: 5s
- Speed limit: 40 ft/s
- Acceleration/ deceleration: 4 ft/ s2
- Saturation flow: 1800vphpl
- Estimated vehicle length: 25ft
- Starting shockwave speed: 16ft/s
- Stopping shockwave speed: 14ft/s

Both the ACO and GA approaches were executed for 30 times. The flows and queues are recorded at time interval of 5s.The total duration of oversaturation is 15 minutes. The number of vehicles in queue approaching the second signal from the first signal at the beginning of the downstream is known as the queue storage capacity constraint. The queue storage capacity constraints are satisfied for all the solutions obtained by the ACO and GA

A summary of the experiment results is shown in Table 1. It could be observed that the ACO is more effective in reducing signal congestion and offset using less computational power. The offset in this case is the time between the start of the green light of the traffic at one intersection and the start of the green light of another traffic light at another intersection. The ACO also shows no error during

Table 1. Attributes of ACO and GA

Ant Colony Optimization (ACO)	Genetic Algorithm (GA)
Requires large number of simulations	Requires large number of simulations
Lower objective functions obtained (better fitness)	Higher objective functions obtained
Less variability in objective function values	Unable to reduce offset to same degree
Ideal offset reduced roughly 3 to 2 seconds	Able to process larger number of vehicles
Process smaller number of vehicles	Higher Congestion
Lower Congestion	

the experimental process, with significantly lower variance among a set of random trials. Although it appears that the ACO is more superior in solving signal congestion problems, the result has also shown that the GA is more robust, since it is capable of handling a higher number of signals and obtaining higher objective functions. These findings are important as it indicates that the GA could be a better option for signal optimization if wireless power transmission is performed in an environment which is highly congested with signals.

In order to search for the most optimum solution, both the ACO and GAs are required to compute iteratively. Hence, these two approaches involve large amount of simulations, which require the collection and storage of a huge volume of data. Because of this reason, parallel computing has been commonly implemented to reduce the execution time, handling problems with a larger capacity, overcoming the signal timing problem and so on.

In ACO, a parallel ant colonies model has been developed. This parallel algorithm operates based on the synchronous master/slave paradigm. In this strategy, the master implements a central memory (or global memory which provides a place for storing the global knowledge obtained during the search process. The master coordinates the activities of all its slaves – which, in this case, are the ants. At the initial step, the master sends out the pheromone matrix to the ants. Each ant is then assigned a task created according to the pheromone levels and is responsible for constructing a complete solution. When all the solutions are constructed, they have to be evaluated. Thus, the ant carries out local search and the solution found is delivered back to the master for evaluation. The master collects all the solution and subsequently perform an analysis to identify the best solution. All these data are important for periodically updating the pheromone matrix as well as the best found solution. A simple illustration on how this system works is shown in Figure 11 (Talbi, Roux, Fonlupt, & Robillard, 1999).

For synchronous master/slave parallelization in GA, the master will first run the evolutionary algorithm. It will then broadcast the subsequent tasks to the slaves. Once the slaves receive the batches of individuals from the master, they will then perform crossover and mutation on the chromosomes. After performing evaluation process, the fitness is transmitted back to the master. The communication between master and worker in GA is shown in Figure 12. Clearly, parallel GA will only proceed to the next generation when it has completely received the fitness values for all the populations from the slaves. This strategy requires only the knowledge of the individual being evaluated and not the entire population, thus, no communication is needed between the master and the slave. Parallel GA differs from conventional GA in the aspect of computational time. Apart from achieving significant speed up in computational time, the rest of the properties are exactly identical with those in a conventional GA. In other words, parallel GA helps to promote genetic diversity while at the same time it is able to perform large-scale sequential

Figure 11. Master/Slave paradigm for ACO

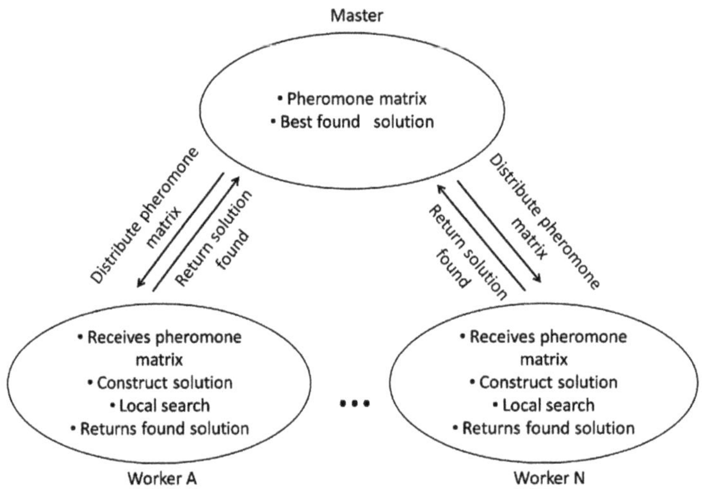

Figure 12. Master/Slave paradigm for GA

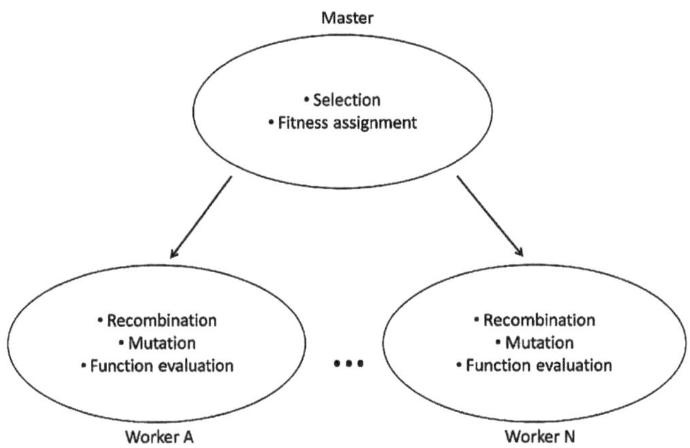

computation. It is worthwhile noting that although parallel GA can solve large optimization problems efficiently, the cycle can only be considered completed after all the fitness evaluations are finished. Therefore, the time required for completing an entire cycle will still be delayed by the slowest processor.

SUMMARY AND CONCLUSION

In this chapter, a review on the wireless power transmission (WPT) mechanisms is elucidated in detail. In general, the mechanisms can be classified into two categories, i.e. the (i) near-field and (ii) far-field mechanisms. When energy is harvested wirelessly within the distance of a wavelength (λ), the WPT mechanism falls into the near-field category. On the other hand, if energy is harvested at a distance greater than 2λ, then the mechanism belongs to the far-field category. In near-field, the energy at the load is usually induced via magnetic coupling. The wireless power transfer based on the far-field method is somewhat similar to the concept of wireless signal transmission. A pair of transmitter and receiver antennas is required in order to send and detect fields via free space propagation.

The technology of wireless power transmission (WPT) is still at its infancy. There are still challenging issues lying ahead, awaiting solution. One of the immediate issues which is to be looked into, is the efficiency of the system. Studies have shown that the efficiency of a WPT system deteriorates with the distance between the transmitter and the receiver antennas. It is to be noted that, when the signal travels through free space, it may take multiple paths to reach the receiving antenna. The efficiency of the WPT systems, especially those which operate in the far-field, is significantly affected since some of the electromagnetic energy may take a longer distance before coupling to the receiving antenna. The effect is particularly apparent if the signal is radiated from an omni-directional antenna. Some of the signals from the transmitting antenna may propagate in direct opposite direction from the receiving antenna. The efficiency of the WPT system may also be affected if the signals travel towards the direction where there is high interference, before reaching the receiving antenna. For example, the signals may be attenuated due to destructive interference from other signals or that the energy may be partially, if not fully, reflected or scattered by a highly conducting material, such as a metal plate. All these may result in considerable loss in the WPT system. To improve the efficiency of a WPT system, biologically-inspired algorithms could be applied to optimize the propagation path of the signals. In this chapter, the two most popular biologically inspired methods – the ant colony optimization (ACO) and the genetic algorithms (GAs) are reviewed and illustrated.

Ever since the advent of wireless power transmission WPT, the living standards of mankind have been remarkably enhanced. A mobile phone, for instance, can now be charged wirelessly and a circuit which runs on low power can now be turned on via far-field wireless charging. Due to its relatively low efficiency, the application of WPT systems is still rather restricted at the meantime. However, as the technology advances towards maturity, it is believed that lives would have been completely different then. Imagine how convenient lives would have become when electronic devices no longer have to be plugged to the power grid to be recharged or to be activated. Power could also be transferred to devices or areas where wire connection is infeasible. By then, wireless power transmission will most certainly become a key fuel for technological growth.

REFERENCES

Balasubramaniam, R. (2013). Wireless energy transfer using resonant inductive coupling. In *Proceedings of Sixth IRAJ International Conference* (pp. 44-47). Pune, India: Power Publishers.

Bertsimas, D., & Tsitsiklis, J. (1993). Simulated Annealing. *Statistical Science*, *8*(1), 10–15. doi:10.1214/ss/1177011077

Bidar, M., & Rashidy, K. H. (2013). Modified firefly algorithm using fuzzy tuned parameters. In *Fuzzy Systems (IFSC), 13th Iranian Conference* (pp. 1 - 4). Qazvin, Iran: IEEE.

Boaventura, A., Collado, A., Carvalho, N. B., & Georgiadis, A. (2013, March-April). Optimum Behavior. *IEEE Microwave Magazine*, *14*(2), 26–35. doi:10.1109/MMM.2012.2234631

Bou, E., Alarcon, E., & Gutierrez, J. (2012). A comparison of analytical models for resonant inductive coupling wireless power transfer. In *Progress In Electromagnetics Research Symposium Proceedings* (pp. 689-693). Moscow, Russia: The Electromagnetics Academy.

Buonanno, A., D'Urso, M., & Pavone, D. (2011). An ultra wide-band system for RF energy harvesting. in *antennas and propagation (EUCAP)*. In *Proceedings of the 5th European Conference* (pp. 388 - 389). Rome, Italy: IEEE.

Chen, Z., Kawasaki, S., & Carvalho, N. (2013, March-April). Wireless power transmission-The last cut of wires. *IEEE Microwave Magazine*, *14*(2), 22–24. doi:10.1109/MMM.2012.2234361

Cheng, D. K. (1989). *Field and Wave Electromagnetics*. Addison Wesley.

Dowhan, L., Wymyslowski, A., & Urbanski, K. (2009). Simulated annealing as global optimization algorithm used in numerical prototyping of electronic packaging. In *Thermal, Mechanical and Multi-Physics simulation and Experiments in Microelectronics and Microsystems, 10th International Conference* (pp. 1-5). Delft, Netherlands: IEEE.

Emile, A., Jan, K., & Wil, M. (2005). *Search methodologies*. Springer. doi:10.1007/0-387-28356-0

Farook, S., & Raju, P. S. (2013). Evolutionary hybrid genetic-firefly algorithm for global optimization. *International Journal of Computational Engineering & Management*, *16*(3).

Flynn, B. W., & Fotopolou, K. (2013, March-April). Rectifying loose coils: Wireless power transfer in loosely coupled inductive links with lateral and angular misalignment. *IEEE Microwave Magazine*, *14*(2), 48–54. doi:10.1109/MMM.2012.2234634

Girianna, M., & Benekohal, R. F. (2002). Dynamic signal coordination for networks with oversaturated intersections. *Transportation Research Record*, *1811*, 122–130. doi:10.3141/1811-15

Khripkov, A., Hong, W., & Pavlov, K. (2012). Integrated resonant structure for simultaneous wireless power transfer and data telemetry. *Antennas And Wireless Propagation Letters, IEEE*, *11*, 1659–1662. doi:10.1109/LAWP.2013.2238602

Kumar, R., & Jyotishree. (2012). Novel encoding scheme in genetic algorithms for better fitness. *International Journal of Engineering And Advanced Technology*, *1*(6), 214–218.

Kurs, A., Karalis, A., Moffatt, R., Joannopoulos, J. D., Fisher, P., & Soljacic, M. (2007). Wireless power transfer via strongly coupled magnetic resonances. *Science*, *317*(5834), 83–86. doi:10.1126/science.1143254 PMID:17556549

Luthra, J., & Pal, S. K. (2011). A hybrid firefly algorithm using genetic operators for the cryptanalysis of a monoalphabetic substitution cipher. In *Information and Communication Technologies (WICT), World Congress I* (pp. 202-206). Mumbai, India: IEEE.

Maulik, U., & Bandyopadhyay, S. (2000). Genetic algorithm-based clustering technique. *Pattern Recognition*, *33*(9), 1455–1465. doi:10.1016/S0031-3203(99)00137-5

Nadhirah, A., Mohd, A. O., Mohd, N. H., & Mohd, H. M. (2014). A review of firefly algorithm. *Journal of Engineering and Applied Sciences, 9*(10).

Nanda, B., & Das, G. (2014). Ant colony optimization. A computational intelligence technique. *International Journal of Computer & Communication Technology, 2*(6), 105–110.

Oyama, A., Obayashi, S., Nakahashi, K., & Nakamura, T. (1997). Transonic wing optimization using genetic algorithm. *AIAA Paper, 97*, 1854.

Prajapati, N. B., Agravat, R. R., & Hasan, M. I. (2009). *Comparative study of various cooling schedules for location area planning in cellular networks using simulated annealing. In Networks and Communications NETCOM'09* (pp. 146–150). Chennai, India: IEEE.

Putha, R., Quadrifoglio, L., & Zechman, E. (2012). Comparing ant colony optimization and genetic algorithm approaches for solving traffic signal coordination under oversaturation conditions. *Computer-Aided Civil and Infrastructure Engineering, 27*(1), 14–28. doi:10.1111/j.1467-8667.2010.00715.x

Sobha, G., Chittal, R. R., & Kumar, K. (2007). Medical applications of wireless networks. In *Systems and Networks Communications. ICSNC 2007. Second International Conference.* Cap Esterel, France: IEEE.

Strassner, B., & Chang, K. (2013). Microwave power transmission: Historical milestones and system components. *Proceedings of the IEEE, 101*(6), 1379–1396. doi:10.1109/JPROC.2013.2246132

Stützle, T., & Hoos, H. H. (2000). MAX--MIN ant system. *Future Generation Computer Systems, 16*(8), 889–914. doi:10.1016/S0167-739X(00)00043-1

Talbi, E. G., Roux, O., Fonlupt, C., & Robillard, D. (1999). Parallel ant colonies for combinatorial optimization problems. In J. Rolim (Eds.), Parallel and Distributed Processing, 11 IPPS/SPDP'99 Workshops, (LNCS), (Vol. 1586, pp. 239–247). Springer. doi:10.1007/BFb0097905

Tuovenin, T., Berg, M., Yazdandoost, K. Y., Salonen, E., & Iinatti, J. (2012). Reactive near-field region radiation of planar UWB antennas close to a dispersive tissue model. *Antennas and Propagation Conference (LAPC)* (pp. 1-4). Loughborough, UK: IEEE. doi:10.1109/LAPC.2012.6403037

Wei, X., & Zhi, L. (2010). The multi-objective routing optimization of WSNs based on an improved ant colony algorithm. In *Wireless Communications Networking and Mobile Computing (WiCOM), 2010 6th International Conference* (pp. 1-4). Cheng Du, China: IEEE. doi:10.1109/WICOM.2010.5601072

Yang, X. S. (2010). *Engineering optimization: An introduction with metaheuristic applications*. Wiley & Sons. doi:10.1002/9780470640425

Yang, X. S. (2013). Firefly algorithm: Recent advances and applications. *Int J. Swarm Intelligence.* DOI: 10.1504/IJSI.2013.055801

Yeap, K. H., Tham, C. Y., Nisar, H., & Loh, S. H. (2013). Analysis of Probes in a Rectangular Waveguide. In J. Rolf (Ed.), *Frequenz* (pp. 145–154). Berlin: Schiele und Schön. doi:10.1515/freq-2012-0117

KEY TERMS AND DEFINITIONS

Algorithm: Sequence of unambiguous computational steps or procedure for problem solving.

Antenna: A device that receives radio waves and converts it into electric energy or vice versa.

Coupling: (Electronics) Linking between two or more devices for the purpose of transmitting or receiving energy.

Crossover: (Genetic) Genetically exchanging between materials during meiosis.

Heuristic: A guide which is able to find solution based on experience yet not guaranteed to be optimal.

Induction: The alteration in magnetic flux produces an electromotive force.

Metaheuristic: A generalized algorithmic framework applicable to different optimization problems by slight modifications for specific problems adaptation.

Mutation: (Genetic) Alteration in DNA sequence that leads to the creation of genetic variation.

Offspring: An outcome or a product; descendants.

Pheromone: Chemical substances secreted or produced by insects as an attractant to the other members of the same species.

Rectification: (Electronics) Process of converting AC into DC.

Resonance: The tendency of a system oscillating or vibrating at greater amplitude subjected to a periodic force where its frequency is same or near to the natural undamped frequency of the system itself.

Wireless: Communication that does not require physical connection.

Chapter 3
Energy Harvesting Methods for Internet of Things

Vasaki Ponnusamy
Universiti Tunku Abdul Rahman, Malaysia

Yen Pei Tay
Quest International University Perak, Malaysia

Lam Hong Lee
Quest International University Perak, Malaysia

Tang Jung Low
Universiti Teknologi Petronas, Malaysia

Cheah Wai Zhao
Quest International University Perak, Malaysia

ABSTRACT

Internet of Things (IoT) has becoming a central theme in current technology trend whereby objects, people or even animals and plants can exchange information over the Internet. IoT can be referred as a network of interconnected devices such as wearables, sensors and implantables, that has the ability to sense, interact and make collective decisions autonomously. In short, IoT enables a full spectrum of machine-to-machine communications equipped with distributed data collection capabilities and connected through the cloud to facilitate centralized data analysis. Despite its great potential, the reliability of IoT devices is impeded with limited energy supply if these devices were to deploy particularly in energy-scarced locations or where no human intervention is possible. The best possible deployment of IoT technology is directed to cater for unattended situations like structural or environmental health monitoring. This opens up a new research area in IoT energy efficiency domain. A possible alternative to address such energy constraint is to look into re-generating power of IoT devices or more precisely known as energy harvesting or energy scavenging. This chapter presents the review of various energy harvesting mechanisms, current application of energy harvesting in IoT domain and its future design challenges.

DOI: 10.4018/978-1-4666-9792-8.ch003

INTRODUCTION

Internet of Things is gradually increasing the unity of people and things around us bringing a new trend towards "ambient intelligence". Ambient Intelligence takes into account of sensitive and adaptive electronic environment by responding to the needs of the things around (Aarts, & Wichert, 2009). According to Marcelo (2014), there are six fundamental building blocks that constitute the components of the Internet of Things:

- **Sensors:** For tracking and measuring activity.
- **Connectivity:** Internet or cloud infrastructure.
- **Processors:** Contain some computing power.
- **Energy Efficiency:** Difficult for access to charge and replace battery.
- **Cost-Effectiveness:** Easily available.
- **Quality and Reliability:** Need to operate in harsh environments for longer period.
- **Security:** Need to relay sensitive information.

Energy efficiency in IoT domain has becoming one of the key challenges for researchers as such devices are expensive to maintain especially being deployed in inaccessible locations where replenishment of energy is almost impossible. Since these devices should be designed to operate unattended for very long duration, constant energy supply should be taken as a major design consideration and thus, energy harvesting comes into place.

Energy harvesting is a process of scavenging or hunting energy from the environment such as solar, wind or other sources like body heat, foot strike or vibration and converting it into electricity. The harvesting rate determines the performance of wireless network and may typically increase or extend the lifetime indefinitely. This chapter aims at presenting several mechanisms and methods involved in energy harvesting in wireless and sensor network as they are becoming the prominent features for IoT. Various techniques of energy harvesting, including assessing the variety sources of energy, methods of wireless energy transfer and possible application scenarios will be discussed as shown in figure .1.

To the best of our knowledge, there is little survey conducted on the application areas of energy harvesting in IoT domain. Although there have been a number of excellent literature reviews surrounding energy harvesting for wireless sensor network (Hande et al., 2007; Alippi & Galperti, 2008; Eu et al., 2009; Tan et al., 2009; Simjee & Chou, 2008; Liu et al., 2006; Kumar et al., 2008; Mainwaring et al,. 2002) this chapter breaks new ground by examining potential energy harvesting application for IoT which is beyond wireless sensor network (WSN) alone as IoT can be any wireless communication enabled nodes. In doing so, it is hoped that the review provided in this chapter will lead future researchers to venture on energy efficiency for IoT rather than considering energy aware routing, better MAC protocols, battery replacement, higher battery capacity or low transmission range mechanisms.

The chapter is organized as follows: Section 2 surveys literature relevant to various energy harvesting and transmission techniques (see Figure 1). A complete discussion about energy harvesting application of IoT is given in Section 3. Section 4 outlines several design challenges and guidelines for future research in energy harvesting for IoT and section 5 concludes the chapter.

Figure 1. Energy harvesting domains

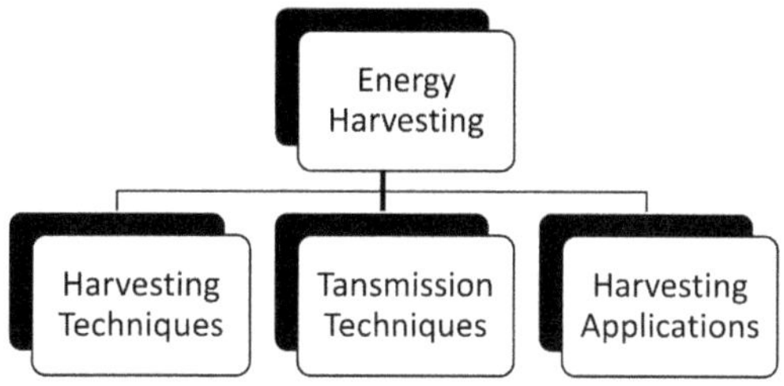

NETWORK OF THINGS FOR IoT

Providing Internet connectivity to everything is the key feature of IoT and this leads to three main categories as network of things as shown in figure. 2. Application domain identifies the areas of IoT applications that could range from industry, home, healthcare, commerce and more. Energy management for IoT becomes crucial as engineers are trying to figure out how these devices could be powered to provide uninterrupted service and that leads to energy harvesting, energy transmission and storage. Providing suitable infrastructure such as cloud computing for IoT is the next focus area. This chapter examines the energy management aspects and the application domain is restricted to energy harvesting applications only (see Figure 2).

ENERGY HARVESTING TECHNIQUES

There are various methods and sources of energy harvesting among which includes solar harvesting, piezo-electric harvesting, wind harvesting, motion energy harvesting, vibration harvesting and radio

Figure 2. IoT domains

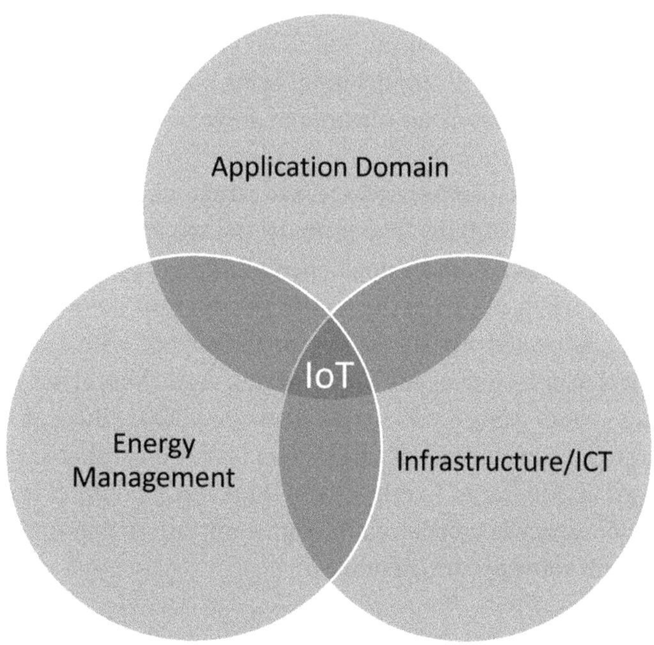

Figure 3. Energy harvesting techniques

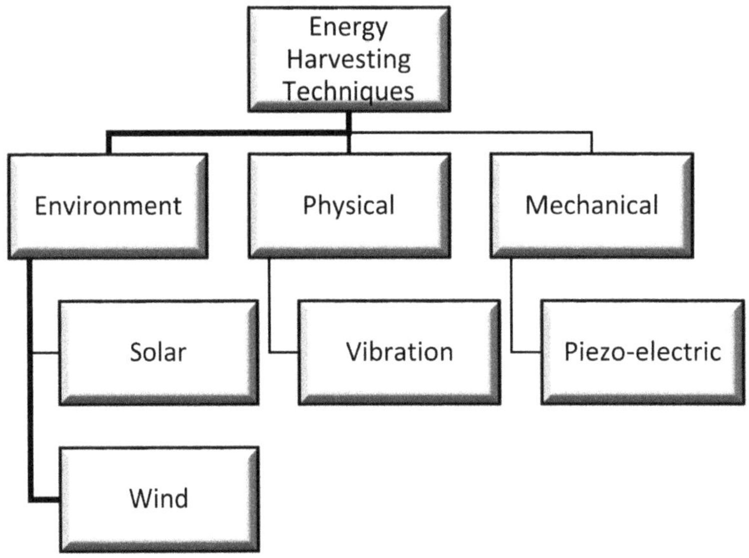

frequency harvesting (figure.3). These techniques differ in the sense that the source of each harvesting mode is different as some are derived from the environment and some of which are from mechanical and physical approach. The following section highlights some of the possible energy harvesting mechanisms with potential applications associated with it (see Figure 3).

Solar Energy Harvesting

Solar energy harvesting is an approach of converting solar energy into electricity which consists of monitoring, storing and transferring energy (figure.5). The harvesting capacity differs in terms of the type of solar panels, battery type and capacity, and complexity of recharge circuit. There are five different modes of solar energy harvesting namely, (i) Heliomote, (ii) Prometheus, (iii) Everlast, (iv) Sunflower, and (v) Ambimax. Of these, Heliomote (Kansal et al. 2007) is a single-storage solar energy harvesting scheme using a Mica2 platform. It has an energy monitoring component to track the amount and variance of energy extracted. Figure 4 shows an example of solar power wireless sensor network and the detailed design.

Prometheus (Jiang et al. 2005) is a double-storage harvesting scheme using the Telos B platform and it has both primary and secondary buffers. The primary storage charges the secondary storage when access energy is available and sensor nodes extract energy from the primary storage after which it falls back to the secondary buffer when energy at the primary buffer falls below a threshold value. Everlast (Sinje & Chou, 2006) is an integrated wireless sensor node equipped with a radio, micro-controller and energy harvesting sub-system, which does not use batteries. AmbiMax (Park et al. 2006) is a double-storage energy harvesting system using the Econode (Park et al. 2005) that can be used to harvest solar and wind energy. The design characteristic of this system is that it is capable of holding other sources like water flow and vibration. Similar to Everlast, AmbiMax has both the primary and secondary buffers to retain energy but the energy is transferred through hardware as opposed to Everlast whereby the energy is transferred through software (see Figure 5).

Figure 4. Adaptive system for optimal solar energy harvesting in wireless sensor network nodes (Alippi & Galperti, 2008)

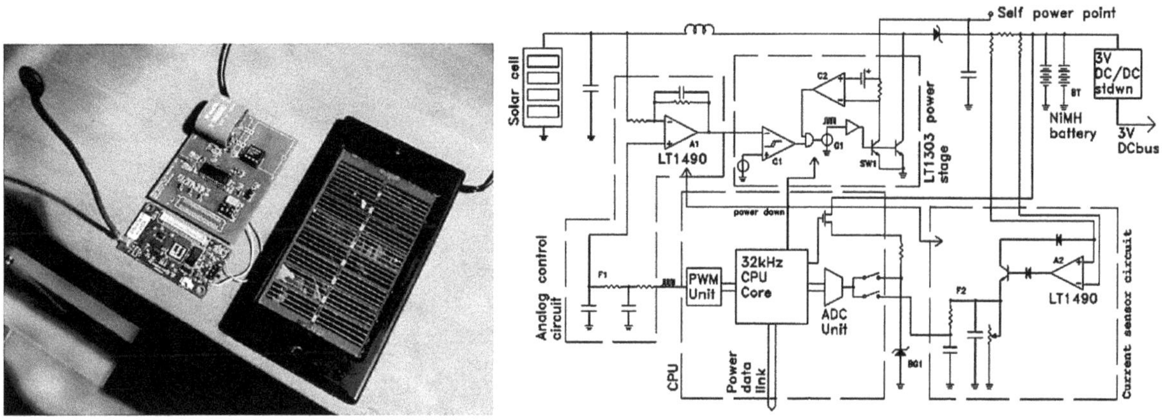

Figure 5. Solar Energy Harvesting Components

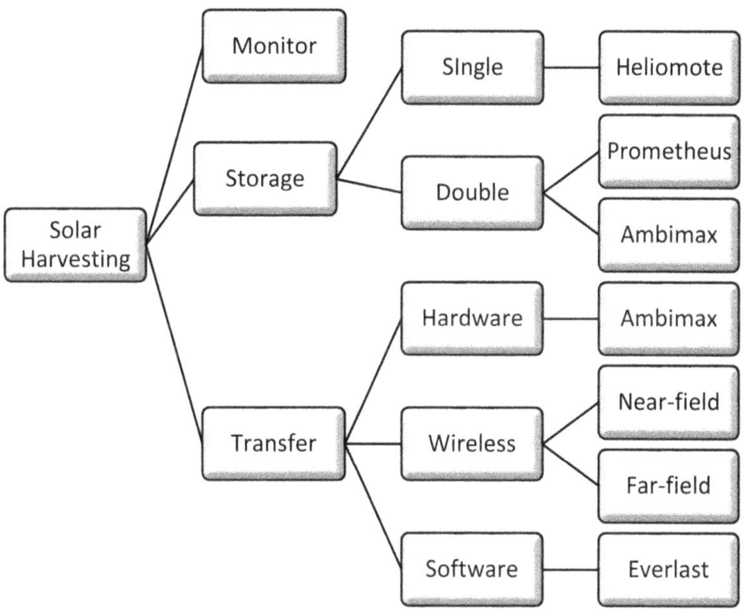

Vibration Energy Harvesting

Roundy (2003) stated that electricity generation through vibration is possible using several techniques such as electromagnetic, electrostatic, or piezoelectric vibrations. These vibration methods are used for inertial sensors as well as for actuators. The power output is usually determined by the magnitude and frequency of input vibrations and the size of the devices used. The type of vibrations in this research mainly focuses on vibrations in office buildings, manufacturing and plant environments and homes. Typical examples are door frames just after the door closes, washing machines, wooden decks with people

Table 1. Summary of energy harvesting by piezoelectric, electrostatic and electromagnetic

Piezoelectric	Electrostatic	Electromagnetic
Advantages No separate voltage source, no mechanical stops, highest energy density, voltages of 2 to 10 volts	Easier to integrate with electronics and Microsystems, Voltages of 2 to 10 volts.	No separate voltage source, no mechanical stops.
Disadvantages Micro fabrication processes are not compatible with standard CMOS processes and piezo thin films have poor coupling.	Separate voltage source needed, mechanical stops needed.	Max. voltage of 0.1 volts, difficult to integrate with electronics and microsystems.

walking etc. The researcher conducted a comparison of the three conversion techniques mentioned above to narrow down the range of design possibilities. Each technique has its own advantages and disadvantages, and each of these is suitable for different applications. Based on the research, electromagnetic converters are more suitable for larger systems or systems that require wider acceleration magnitudes. Piezoelectric converters demonstrate all of the advantages exhibited by electromagnetic converters with the additional advantage of providing useful voltages and higher practical energy densities. A comparison is summarized in table 1. More research on vibration based energy harvesting can be found at Ooi et al (2014), Mancelos et al (2014) and Ooi & Gilbert, (2014).

Besides harvesting energy from the environment, mechanical methods such as energy harvesting from kinetic forces such as human walking and the stress of piezoelectric materials incorporated into shoes are introduced. The bending ball of the foot has been designed as an approach using piezoelectric polyvinylidene fluoride (PVDF) which is strong and rather flexible. PVDF is manufactured into thin films by stretching them to align the polymer chains, and is also electrically poled. Silver electrodes are coated on it to form as capacitor plates to collect the stressed PVDF. Another approach is by using piezo-electric, e.g., PZT (Lead Zirconate Titanate) which is rigid and the piezo-electric effect is created through the crystal structure. Energy harvesting from walking, foot strikes and finger motions is performed by deformation of the PVDF and PZT. The piezoelectric effect is described as the relative link between the mechanical stress applied and the electrical energy generated as shown in figure. 6.

Figure 6. Piezoelectric Effect

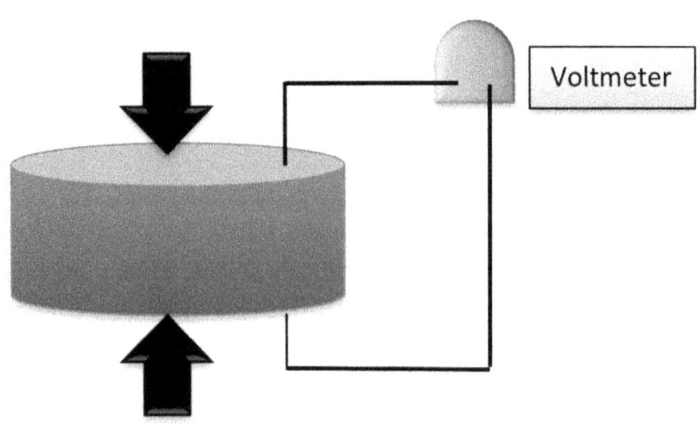

Wind Energy Harvesting

Wind energy harvesting is an approach of converting wind generated energy into electricity. The wind turbine is the basic element in generating energy using other components such as the vertical and horizontal axis of the turbine blades, the rotor and shafts. AmbiMax (Park et al. 2005) on Eco node is one of the implementations using wind energy harvesting mechanism. A biologically-inspired mechanism (Hobbs et al. 2012) inspired by tree swaying concept is applied to wind energy harvesting. A liner array of cylinder vibration attached to the piezoelectric energy transducers is proposed in this work.

ENERGY TRANSMISSION TECHNIQUES

Wireless energy transmission has emerged as an outstanding research area due to the need for infrastructure-less environment to support IoT. The ultimate goal of IoT would be achieved when there is no more need for wireless devices to be physically plugged to power source for energy. This limits the mobility of IoT devices mainly in remote locations such as forest and underwater. Wireless energy transmission works by two principles, near-field and far-field which is distinguished based on the transmission distance between the transmitting source and the receiving source (Balasubramaniam, 2013).

Mascareñas et al. (2006) explores a new alternative to the conventional battery constrained sensor nodes by exploiting mobile host. In this architecture (figure. 7), energy is transferred wirelessly to sensor nodes on an as-needed basis. The model used is similar to RF energy transfer to passive RFID tags. Besides, the mobile host is also responsible for collecting sensor data from nodes to perform centralized data aggregation and processing. The mobile host in this architecture is modified to be a remote controlled (RC) helicopter. The sensor nodes in this work are called THINNER in which its antenna is tuned to the RC helicopter antenna to receive the wireless energy as well as to send sensor data. The mechanism in this approach works by mobile host transmitting a 2.5 Ghz signal through a 14.5dB antenna and the 19dB patch antenna at the sensor nodes receive these signals. The RF-to-DC conversion and multiplication is performed before it is fed into a capacitor for storage purpose. THINNER is designed specifically to be powered by wireless energy transfer by a mobile host and employs a capacitance-to-digital converter for the need to store peak displacement values. The Xbee radio is engaged in order to communicate with unmanned aerial vehicle (UAV) and also to communicate with other sensors.

Figure 7. An Architecture of Wireless Energy Transmission

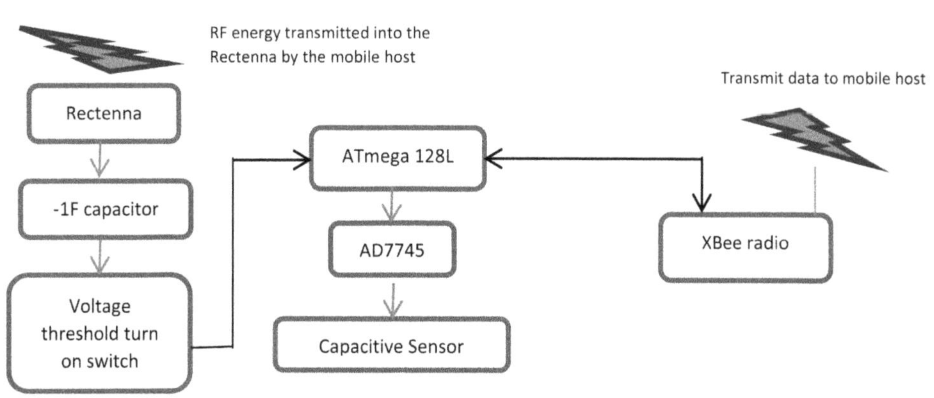

Figure 8. Wireless Energy Transmission Between The Antennas

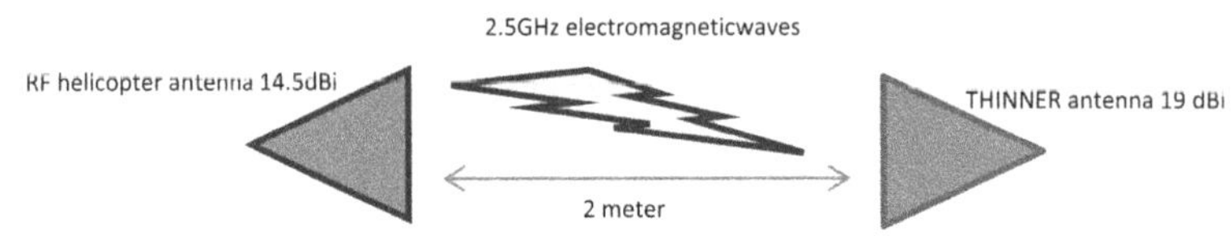

One of the main constraints of this approach is the high energy consumption of THINNER node as it is operated by power being transferred wirelessly due to free space loss and requirement for power at other components such as the capacitor and storage capacitor. Typically it takes approximately about 76-3-5s for the mobile node to be within the proximity of a sensor in order to transfer energy wirelessly for the nodes to operate properly. The RF energy is transferred at 2.5 GHz and this figure is chosen for the cost factor to reduce the hardware cost while increasing the cost to purchase high gain antennas. The distance between the THINNER antenna and the mobile host antenna is approximated at 2 meter so that the RF helicopter would be able to reach the antenna of the THINNER as shown in figure.8.

This work has contributed in the development of a novel capacitance-based wireless sensor node and peak displacement sensor. Through development and results obtained, it is feasible to deploy such mobile host for data collection and powering sensor nodes using wireless energy transmission. Moreover, this work can contribute to post-disaster applications such as earth quake and terrorist attack for rapid assessment results in life safety and economic reasons.

Separately, Aristeidis et al. (2008) analyses the ways in which wireless non-radiative mid-range energy is transferred. Rapid development of autonomous devices such as laptops, cellular phone, house-hold devices brings into attention the need to revisit the concept of wireless energy transfer. Some non-radiative mechanisms using magnetic induction is possible but restricted by distance and low power energy transfers. The authors investigate the feasibility of using long-lived oscillatory resonant electromagnetic modes with slowly-evanescent field patterns wireless energy transfer. This method works by the principle of resonant coupling in which two same-frequency resonant objects couple each other resulting in that energy can be coupled between objects in near field.

Lee, (2012) has conducted a research on energy harvesting for wireless sensor network. The author proposed a novel, plug through power monitoring system for commercial and home use. The device use current sense transformer to detect any appliance power consumption which wirelessly couples magnetic energy from appliance to output an electromotive force voltage. Then, the voltage signal will be relayed to the analog-to-digital converter of GINA radio mote, which transmit the data to laptop or mobile device wirelessly. It allows user to view real-time power usage by using handheld device. The

Figure 9. Power monitoring device (Lee, V. C., 2012)

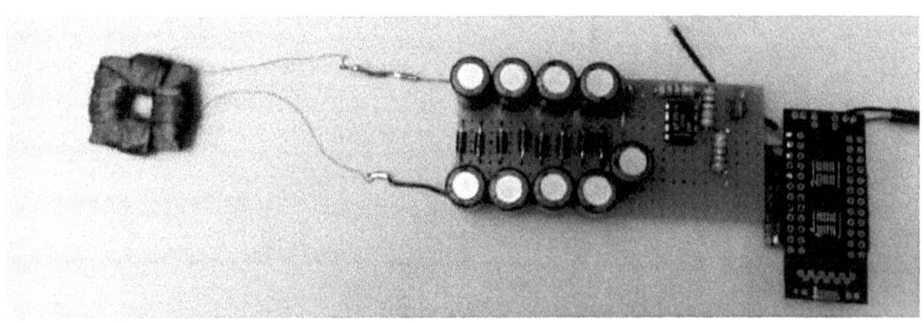

Figure 10. Energy harvesting for wireless sensor network (Lee, V. C., 2012)

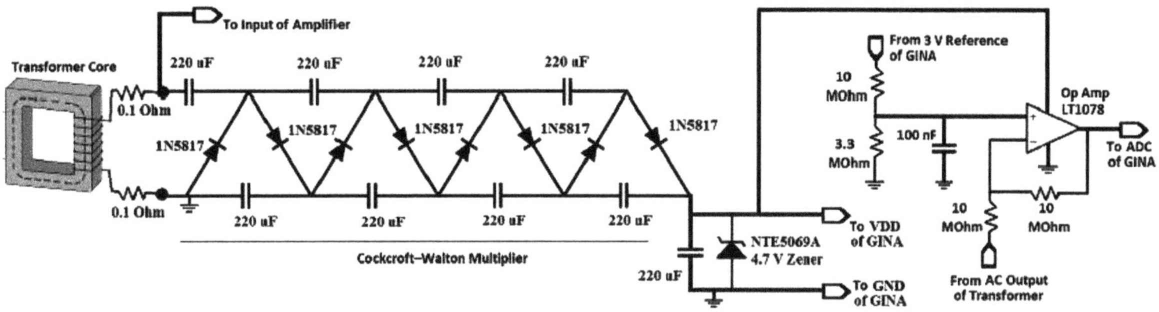

device is 10 times lower cost and 45 times smaller compared to other power monitoring devices in the present market. A power monitoring device based on energy harvesting is shown in figure.9 and the schematic circuit view is shown in Figure 10.

ENERGY HARVESTING APPLICATIONS

IoT applications are developed to create a smarter surrounding as smart environment/city, smart water, smart metering, eHealth, home automation and more. These applications are classified as wireless sensor network, body area network (wearables), and wireless electronics/networks (Figure 11) although sensor plays key role in IoT. One key aspect for these applications to be unique is to provide autonomic capabilities for energy management. The following section highlights some of the energy harvesting applications for IoT which provides autonomic capabilities in terms of energy management.

Energy Harvesting in Wireless Sensor Network

Alippi & Galperti (2008) has conducted a research on adaptive system for optimal solar energy harvesting in wireless sensor network nodes. They feel that it was finite resources for wireless sensor network that is limited by energy supply which can only be provided using battery. Therefore, the idea of energy harvesting mechanism has been taken for long time operational life with solar energy. They propose a low power maximum power point tracker (MPPT) circuit that is designed for wireless sensor nodes. The circuit will convey solar energy into rechargeable batteries in every situation even cloudy or under low light condition. The design is considered effective as it works in critical weather conditions where traditional solutions do not work. MPPT circuit was designed from power converter to digital control system and work in low power operation independently from nature of solar panel and battery type.

In another research carried out by Kansal & Srivastava (2003), a distributed framework for sensor network is presented to learn the energy environment and localized algorithm to use the information for sharing between nodes. This framework enables system to exploit energy resources efficiently and increase the lifetime. They use EEHF for solar energy based sensor network. It is a way towards development of optimal set of methods to completely task distribution in the network with available energy. In order to fully resolve the harvesting problem, it needs to develop distributed method to achieve maximum lifespan and requires to be tested on real-time system development.

Figure 11. IoT application domains

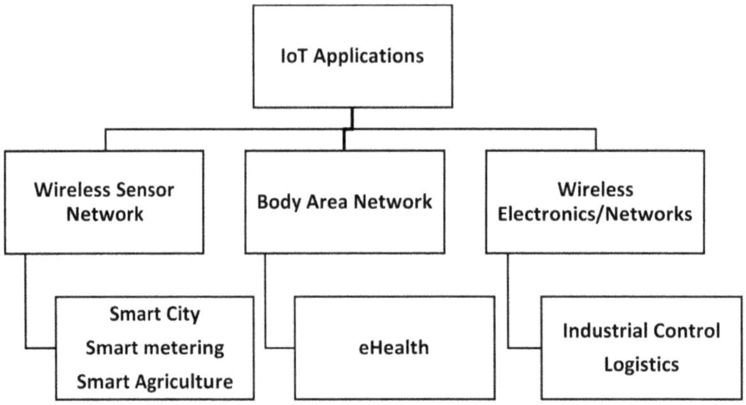

Besides, Seah et al. (2009) had published work on Wireless Sensor Networks Powered by Ambient Energy Harvesting (WSN-HEAP). Substantial research efforts have been used for designing energy-efficient networking protocol to maximize WSN lifespan. However, there are some WSN application required sensors to operate for longer duration after the implementation. For instance, in structural health monitoring for critical infrastructures and buildings where sensors are installed at locations almost inaccessible by humans, the alternative solution is to use ambient energy harvesting. Wireless sensor network uses ambient energy harvesting as a promising technology for many sensing application in reducing the need for battery replacement. However, there are challenges of designing wireless sensor node such as transmission power control, MAC protocol, routing and reliable of data delivery. Moreover, energy harvesting from environment is highly dependent on local environmental factors, therefore it requires further study to exploit this frontier so that energy harvesting can be optimum under any condition. Table 2 presents the achievable duty cycle by micaz with 10CM2 harvesting material.

A research on comparison of energy harvesting systems for wireless sensor network has been done by Gilbert & Balouchi (2008). In this chapter, it discuss about the characteristic and energy requirement of sensor network nodes, access range of ambient energy sources and figure out the characteristics of energy conversion devices. They proposed a way to normalize power density by comparing diverse source and conversion method. The conversion method mainly uses three energy harvesting device such as piezoelectric, electrostatic and electromagnetic. In term of considering published data, they found

Table 2. Wireless sensor networks powered by ambient energy harvesting (WSN-HEAP) Seah et al. (2009)

Technology	Power Density (μ W/cm^2)	Energy Harvesting Rate (mW)	Duty Cycle (%)
Vibration - electromagnetic	4.0	0.04	0.05
Vibration - piezoelectric	500	5	6
Vibration – electrostatic	3.8	0.038	0.05
Thermoelectric	60	0.6	0.72
Solar – direct sunlight	3700	37	45
Solar – indoor	3.2	0.032	0.04

Table 3. Comparison of energy harvesting systems for wireless sensor network (Gilbert & Balouchi (2008)

Device Number	Operating Mode / Materila	Output Power (μW)	Frequency (H_Z)	Amplitude (ms^{-2})	Normalised Power (μW)	Volume (mm^3)
1	Vibration - pz	2.1	80.1	2.3	0.5	125
2	Vibration – pz	210	120	2.5	28	1000
3	Vibration – pz	375	120	2.5	50	1000
4	Vibration – pz	0.6	900	9.81	0.0007	2
5	Vibration – es	3.7	30	50	0.005	750
6	Vibration – es	1052	50	8.8	27	1800
7	Vibration – es	70	50	9.2	1.7	32
8	Vibration – cm	0.3	4400	382	$4.7x10^{-8}$	5.4
9	Vibration – cm	180	322	2.7	7.7	840
10	Vibration - cm	4000	100	0.4	25000	30 000
11	PVDF shoe stave				1300	5000
12	PZT dimorph				8400	1700
13	Em shoe insert				60000	56000
14	Em shoe insert				90000	97500
15	EAP shoe insert				1 000000	50 000
16	Thermoelectrc				50	41
17	PV cell outdoor				20000	500 000
18	PV cell indoor				1500	500 000

that there is large range of power levels and device volume but similar power densities are archived for best performing device in every size range. Table 3 presents a comparison of energy harvesting devices.

Park et al. (2008) has conducted a work on energy harvesting for structural health monitoring sensor networks. The work reviews the development of energy harvesting for structural health monitoring (SHM) sensing system. The system is beneficial as it can improve the reliability, safety and ownership cost engineering systems by autonomously monitoring the condition and sensing the damage before entering into critical state. However, the development of energy harvesting for large-scale source such as wind turbines exist due to high commercial products but energy harvesting for embedded sensing system is still in fancy stage. The authors argue that major drawback in the field of energy harvesting is energy generated by harvesting device is too small to direct power for most of the devices.

Kansal et al. (2007) has done research on power management in energy harvesting sensor networks. The work reviews the importance of power management in sensor network and how energy harvesting implemented on powering sensor networks. Energy harvesting has an energy neutral mode to supplement energy supply in battery powered systems. The authors discuss some challenges that is faced in this mode and modelling the characteristics of energy generation sources and loads. It presents theoretical foundations, various practical methods to optimize performance in energy neutral mode.

Rabaey et al. (2000) has worked on PicoRadio Supports Ad Hoc Ultra-Low Power Wireless Networking. PicoRadio application uses sensor monitoring network for environmental control in office buildings, warehouse inventory, patient monitoring and diagnostic, interactive toys etc. PicoNodes were used as

Figure 12. PicoRadio supports ad hoc ultra-low power wireless networking (Rabaey et al. (2000)

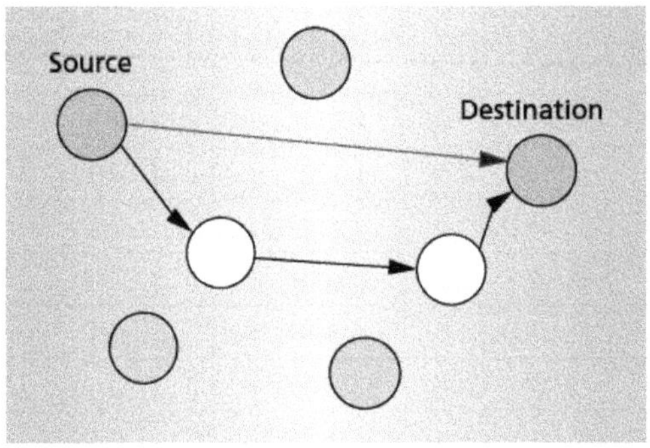

self-contained and self-powered by using energy scavenging. Harvesting ambient energy will work only when applicability within environment envision for PicoNodes and size constraint of one cm chip. In the implementation of wireless ad hoc, sensor, monitor and actuator network, they found a problem which is minimizing energy consumption. The authors only proposed an ultra-low power node by optimizing all layers of protocol. The author argues that sending data through multiple intermediate hops is more energy efficient than sending through one longer hop (Figure 12).

Energy Harvesting in Body Area Networks

A study on Energy Harvesting and Remote Powering for Implantable Biosensors has been done by Olivio et al. (2011). Various popular energy harvesting techniques for implantable biosensors have been conducted. Among the areas explored were kinetic, thermoelectric, fuel cells, infrared radiation, low frequency magnetic field and inductive links. Inductive links are considered viable for implanted biosensors because communication between the internal sensors and external control devices are performed easily. High frequency inductive link produces higher data rates and have higher misalignment tolerance as compared to low frequency inductive links. The connection of multiple thermocouples placed in series (left) in electrical and placed parallel in therma forms a thermopile as shown in figure 13. Although this can be an advantage but this technique has lower efficiency and requires higher transmission power.

Jia & Liu (2009) conducted a research on human power-based energy harvesting strategies for mobile electronic devices. The authors proposed energy harvesting through daily activities such as joint rotation and enforcement of body weight for mobile or medical device as shown in figure 14 . This technique seems to provide longer energy capacity since most of the mobile devices consume lesser energy. Therefore energy can be continuously harvested from human body and converted to mobile device and probably these devices no longer need recharging in the future. However, a good wearable energy harvesting system requires light weight, high power density, good regulation ability, convenient power delivery and minimum user effort. Currently, the existing technology is still unavailable to achieve the above objective. Seyedi & Sikdar (2208) have explored efficient energy transmission techniques for body sensor network using energy harvesting capabilities. An effective decision making strategy via scheduling on the transmission mode is proposed to improve the coverage quality. Three different approaches of energy transmission scheduling with varied energy levels were proposed for body sensor network with energy harvesting capability.

Figure 13. Energy harvesting and remote powering for implantable biosensors (Olivio et al., 2011)

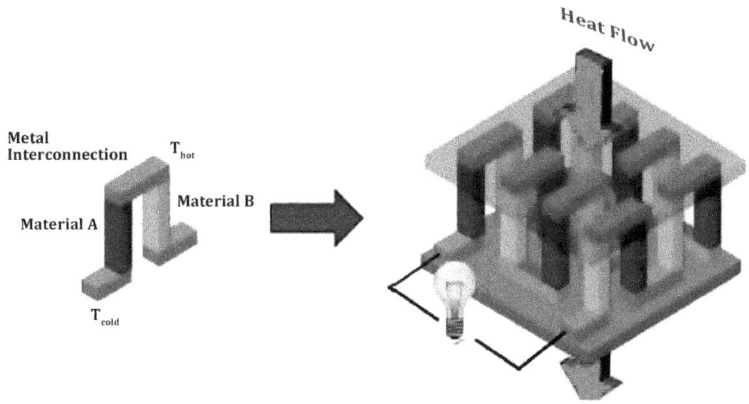

Figure 14. human power-based energy harvesting strategies for mobile electronic devices (Jia & Liu, 2009)

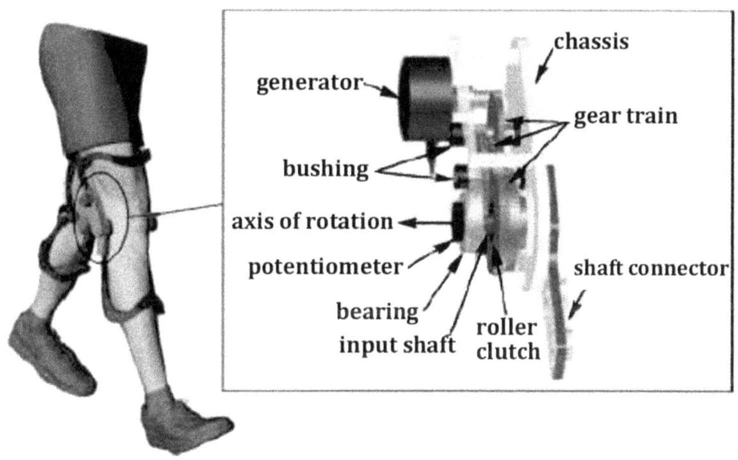

Energy Harvesting in Wireless Electronics

Energy harvesting for mobile and wireless electronic has been conducted by Paradiso & Starner (2005). The authors mentioned that battery energy is slowest trend in mobile computing in 1990. However, the revolution of battery form factor, its energy density unable to scale exponentially. To overcome the trend, on-going power management developments allow battery-powered electronics to live longer. It is not enough, as energy harvesting is the best approaches to scavenge energy from waste heat or vibration. A lot of work has been carried out on low-power electronics, wireless standards and miniaturization of devices and technology to boost the trend in wireless sensor network and mobile devices. There are several techniques to harvest energy from different environment to power electronics, and devices limit the power yield in pervasive computing. As such, Figure 15a shows the magnetic generators embedded in shoes and a strap-on overshoe managed to produce about an average of 250 mW during normal walk. Whereas Figure 15b reveals how an assembly hosting twin motor-generators embedded into a shoe that managed to produce about 60 mW.

Figure 15. Energy harvesting for mobile and wireless electronic (Paradiso & Starner, 2005)

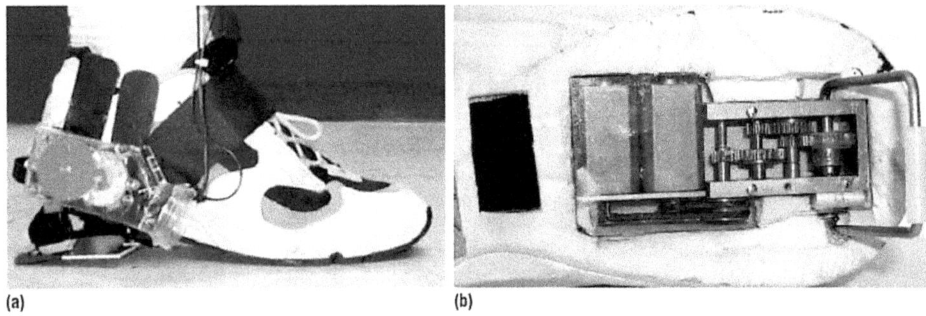

Figure 16. Energy harvesting vibration sources for microsystem applications (Beeby et al., (2006)

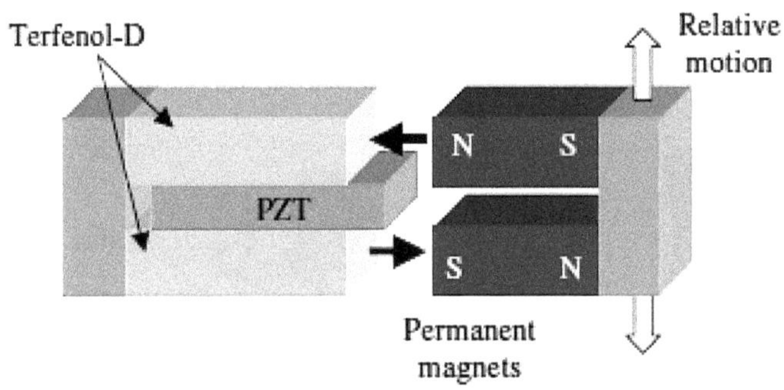

Beeby et. al. (2006) has worked on energy harvesting vibration sources for microsystem applications. It is state-of-the art in vibration energy harvesting for wireless and self-powered microsystems. The vibration-powered generator is similar like spring and mass system. The generator consists of three main transduction concepts which are piezoelectric, electromagnetic and electrostatic (Figure 16). Vibration-powered wireless sensor is very beneficial as it can be used in ambient intelligence, medical implants and smart clothing.

A research on powering MEMS portable devices with non-regenerative and regenerative power supply systems with special emphasis on piezoelectric energy harvesting systems has been conducted by Cook et al. (2008). Their work focuses on approaches for powering MEMS using non-regenerative and regenerative power supplies. The work also reviews on the fundamental of piezoelectric energy harvesting and future trend of piezoelectric energy harvesting technology. The authors found that most of the MEMS-scaled piezoelectric energy harvesting device does not produce enough current to power the systems directly. To solve the problem, a hybrid system consisting of piezoelectric energy harvesting by improving the charge density using suitable material and a proper storage devices is proposed. Moreover, the authors propose to reduce the device size and increase the number of coupled devices in order to allow for higher frequency oscillation of piezoelectric devices to increase output power. A schematic view of the thermoelectric system is shown in Figure 17.

Rocha et al. (2010) has conducted a research on energy harvesting from piezoelectric material that is fully integrated into footwear. Their research is about using piezoelectric polymers that is capable of harvesting energy via human walking and shoe fabrication process capable of producing and collecting

Figure 17. Powering MEMS portable devices with non-regenerative and regenerative power supply systems with special emphasis on piezoelectric energy harvesting systems (Cook et al. (2008))

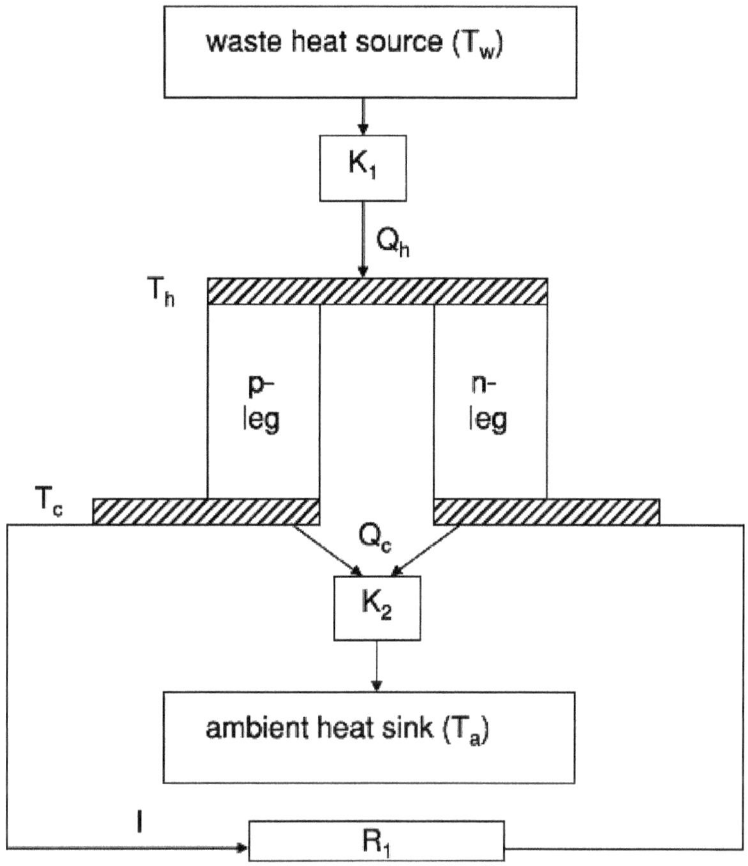

energy. Figure 18a shows the first protoptype developed using two piezoelectric polymer films above the shoe sole. The researchers used electronic β-PVDF as energy harvesting element that is injected into bicolored sole together with electronic needed to improve the energy transfer and storage efficiency. The authors also included electrostatic generator to increase energy harvesting (Figure.18b). It was of value by using electro-active polymers because such technology is low cost, flexible and can be easily integrated into elements such as shoes and clothes.

Mitcheson et al. (2008) has worked on energy harvesting from human and machine motion for wireless electronic devices. Their work reviews about the principle and state-of-art in motion driven tiny energy harvester and trends, existing application and future work. The motion driven energy harvester uses micro generator to apply direct use of forces and inertial forces working on proof mass. The authors discovered that there are two existing companies offering electromagnetic harvester which is Perpetuum and Ferro Solution, it was similar to the device but their proposal has greater potential. Practical utilization of harvesting devices is depending on the efficiency of power processing and control circuitry. Lastly, an effective motion energy harvesting requires combination of complete wireless system design and power aware operation.

Figure 18. Energy harvesting from piezoelectric material that is fully integrated into footwear (Rocha et al., 2010)

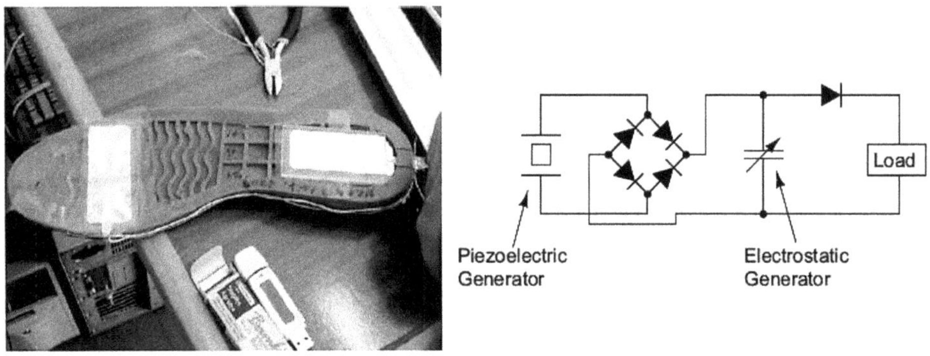

DESIGN CHALLENGES AND GUIDELINES FOR FUTURE RESEARCH

Various energy harvesting mechanisms have been examined for IoT domain in previous sections. Among the methods, solar energy harvesting is widely popular due to the availability of solar source. Solar energy could be harvested during daylight and stored in secondary storage for night or indoor use. Solar energy harvesting is limited in countries that have limited daylight such as Alaska and Sweden during certain seasons. Besides that it is limited for indoor use such as tunnel where sunlight may not be able to penetrate. But this can be overcomed by using secondary storage whereby harvested solar energy is stored in a secondary buffer for future use. However, the reliability of such approach depends on the storage capacity and energy transmission mechanism to such indoor environments. Vibration and wind energy harvesting works on the same mechanism of mechanical and physical elements of movements. Piezoelectric is widely used in these two approaches but some researchers argue that the energy generated by this material is too small to directly powering the devices. Although using secondary energy storage could solve the issue, capacitor-based storage could only provide short power surge. Sodano et al. (2004a) has proposed that rechargeable battery could be used together with piezoelectric materials to substitute capacitor.

Wireless power transmission technology has been there for some time, but it is still at its premature stage due to the challenges and limitations of such technology. Wireless power transmission is highly restricted by the distance between the antennas. Based on Choi et al. (2011), the transmission efficiency of wireless energy transmission resonators decreases proportionally over distance irrespective of the receiving coil size. Researchers are working on ways to reduce the distance whilst increasing the transmission efficiency. This is possible by improving the coupling efficiency of receiving circuit and employing magnetic resonance. Although distance can be the pitfall in energy transmission efficiency, advancements in microprocessor technology has increased power efficiency thus reducing power consumption. This can substantiate that small volume of energy generated could be able to power wireless devices. The challenges and design principles of such technologies can be found at Gungor & Hancke (2009).

CONCLUSION

In the nutshell, energy harvesting is the process of deriving energy from external sources, capturing and storing the energy into wireless autonomous devices or sensor networks. Energy harvester provides less power for low-energy electronics. Its energy can come from different sources such as heat, vibration, radio wave or kinetics motion. There are three main types of energy harvesting material which is piezo-electric, thermoelectric and pyro electric. Researchers have proven that energy harvesting can be used to replace batteries in medical field. The heartbeat, body glucose or internal organ motion can serve as alternative sources to produce small amount of energy.

"Internet of Things via Energy Harvesting" is a famous topic and current trend that is being explored by many researchers. The main challenge for design engineers is how to fully manage the power to IoT nodes by eliminating the last wire. One such potential application is solar-powered street lighting system (Ramachandran. T et al, 2014). Energy harvesting is an ideal solution for IoT devices as it is renewable and self-sustained once it is deployed. However, such energy is not available all the time so the IoT devices should be designed in such a way to be able to be adaptable and autonomous. Therefore, mimicking biologically-inspired (Saleem et al. 2010, Dorigo et al 1996; Boonma and Suzuki, 2007) autonomous system capabilities could be a potential research in energy harvesting for IoT.

REFERENCES

Aarts, E., & Wichert, R. (2009). *Ambient intelligence*. Springer Berlin Heidelberg.

Alippi, C., & Galperti, C. (2008). An Adaptive System for Optimal Solar Energy Harvesting in Wireless Sensor Network Nodes. *IEEE Transactions on Circuits and Systems*, *55*(6), 1742–1750. doi:10.1109/TCSI.2008.922023

Aristeidis, K., Joannopoulos, J. D., & Marin, S. (2008). Efficient wireless non-radiative mid-range energy transfer. Elsevier.

Balasubramaniam, R. (2013). *Wireless energy transfer using resonant inductive coupling*. In *Proceedings of Sixth IRAJ International Conference* (pp. 44-47). Pune, India: Power Publishers.

Beeby, S. P., Tudor, M. J., & White, N. M. (2006). Energy harvesting vibration sources for microsystems applications. *Measurement Science & Technology*, *17*(12), R175–R195. doi:10.1088/0957-0233/17/12/R01

Boonma, P., & Suzuki, J. (2007). BiSNET: A biologically-inspired middleware architecture for self-managing wireless sensor networks. *Computer Networks*, *51*(16), 4599–4616. doi:10.1016/j.comnet.2007.06.006

Choi, J., Cho, J. K., & Seo, C. (2011). Analysis on Transmission efficiency of wireless energy transmission resonator based on magnetic resonance. In Microwave Workshop Series on Innovative Wireless Power Transmission: Technologies, Systems, and Applications (IMWS), 2011 IEEE MTT-S International (pp. 199-202). IEEE. doi:10.2528/PIERM11050903

Cook-Chennault, K. A., Thambi, N., & Sastry, A. M. (2008). Powering MEMS portable devices—a review of non-regenerative and regenerative power supply systems with special emphasis on piezoelectric energy harvesting systems. *Smart Materials and Structures*, *17*(4), 043001. doi:10.1088/0964-1726/17/4/043001

Dorigo, M., Maniezzo, V., & Colomi, A. (1996). Ant system: Optimization by a colony of cooperating agents. *IEEE Transactions on Systems, Man, and Cybernetics. Part B, Cybernetics*, *26*(1), 29–41. doi:10.1109/3477.484436 PMID:18263004

Eu, Z. A., Tan, H. P., & Seah, W. K. G. (2009). *Routing and relay nodeplacement in wireless sensor networks powered by ambient energy har-vesting. In Proc. of the IEEE Wireless Communications & Networking Conference*. WCNC.

Gilbert, J. M., & Balouchi, F. (2008). Comparison of energy harvesting systems for wireless sensor networks. *International Journal of Automation and Computing*, *5*(4), 334–347. doi:10.1007/s11633-008-0334-2

Gungor, V. C., & Hancke, G. P. (2009). Industrial wireless sensor networks: Challenges, design principles, and technical approaches. *Industrial Electronics. IEEE Transactions on*, *56*(10), 4258–4265.

Hande, A., Polk, T., Walker, W., & Bhatia, D. (2007). Indoor solar energyharvesting for sensor network router nodes. *Microprocessors and Microsystems*, *31*(6), 420–432. doi:10.1016/j.micpro.2007.02.006

Hobbs, W. B., & Hu, D. L. (2012). Tree-inspired piezoelectric energy harvesting. [tree inspired]. *Journal of Fluids and Structures*, *28*, 103–114. doi:10.1016/j.jfluidstructs.2011.08.005

Jia, D., & Liu, J. (2009). Human power-based energy harvesting strategies for mobile electronic devices. *Frontiers of Energy and Power Engineering in China*, *3*(1), 27–46. doi:10.1007/s11708-009-0002-4

Jiang, X., Polastre, J., & Culler, D. (2005). *Perpetual Environmentally Powered Sensor Networks*. In *Fourth International Symposium on Information Processing in Sensor Networks*.

Kansal, A., & Srivastava, M. B. (2003). *An environmental energy harvesting framework for sensor networks*. In *Low Power Electronics and Design, 2003. ISLPED'03, Proceedings of the 2003 International Symposium*. doi:10.1145/871506.871624

Kansal, A., Hsu, J., Zahedi, S., & Srivastava, M. B. (2007). Power management in energy harvesting sensor networks. *ACM Transactions on Embedded Computing Systems*, *6*(4), 32, es. doi:10.1145/1274858.1274870

Lee, R.-G., Chen, K.-C., Chiang, S.-S., Lai, C.-C., Liu, H.-S., & Wei, M.-S. (2006). A Backup Routing with Wireless Sensor Network for Bridge Monitoring System. In *Annual Conference on Communication Networksand Services Research*. IEEE Computer Society.

Lee, V. C. (2012). *Energy Harvesting for Wireless Sensor Network*. (Doctoral dissertation). University of California, Berkeley, CA.

Mainwaring, A., Culler, D., Polastre, J., Szewczyk, R., & Anderson, J. (2002). Wireless Sensor Networks for Habitat Monitoring. In *Proc. 1st ACM International Workshop on Wireless Sensor Networks and Applications*. ACM.

Mascarenas, D. L. (2006). *Development of an impedance-based wireless sensor node for monitoring of bolted joint preload.* (M.S. Thesis). Dept. of Structural Engineering, University of California, San Diego, CA

Mancelos, N., Correia, J., Pires, L. M., Oliveira, L. B., & Oliveira, J. P. (2014). Piezoelectric Energy Harvester for a CMOS Wireless Sensor. In *Technological Innovation for Collective Awareness Systems* (pp. 470–477). Springer Berlin Heidelberg. doi:10.1007/978-3-642-54734-8_52

Mitcheson, P. D., Yeatman, E. M., Rao, G. K., Holmes, A. S., & Green, T. C. (2008). Energy harvesting from human and machine motion for wireless electronic devices. *Proceedings of the IEEE*, *96*(9), 1457–1486. doi:10.1109/JPROC.2008.927494

Ooi, B., & Gilbert, J. (2014). Design of wideband vibration-based electromagnetic generator by means of dual-resonator. *Sensors and Actuators. A, Physical*, *213*(0), 9–18. doi:10.1016/j.sna.2014.03.037

Ooi, B. A., Aziz, R., & Gilbert, J. (2014). Analytical and numerical modelling for dual-resonator vibration-based electromagnetic generator. *5th International Conference on Intelligent and Advanced Systems (ICIAS)* (pp. 1-5). Kuala Lumpur: IEEE. doi:10.1109/ICIAS.2014.6869443

Olivo, J., Carrara, S., & De Micheli, G. (2011). Energy harvesting and remote powering for implantable biosensors. *IEEE Sensors Journal*, *11*, 1573–1586.

Paradiso, J. A., & Starner, T. (2005). Energy scavenging for mobile and wireless electronics. *Pervasive Computing, IEEE*, *4*(1), 18–27.

Park, C., & Chou, P.H. (2007). AmbiMax: Autonomous Energy Harvesting Platform for Multi-Supply Wireless Sensor Nodes. *3rd Annual IEEE Communications Society on Sensor and Ad Hoc Communications and Networks.*

Park, G., Rosing, T., Todd, M. D., Farrar, C. R., & Hodgkiss, W. (2008). Energy harvesting for structural health monitoring sensor networks. *Journal of Infrastructure Systems*, *14*(1), 64–79. doi:10.1061/(ASCE)1076-0342(2008)14:1(64)

Rabaey, J. M., Ammer, M. J., da Silva, J. L. Jr, Patel, D., & Roundy, S. (2000). PicoRadio supports ad hoc ultra-low power wireless networking. *Computer*, *33*(7), 42–48. doi:10.1109/2.869369

Ramachandran, T., Jung, L. T., Ponnusamy, V., & Hudaya, A. (2014, June). ZigBee wireless communication for monitoring renewable street light system. In *Computer and Information Sciences (ICCOINS), 2014 International Conference on* (pp. 1-6). IEEE.

Rocha, J. G., Goncalves, L. M., Rocha, P. F., Silva, M. P., & Lanceros-Mendez, S. (2008). Energy harvesting from piezoelectric materials fully integrated in footwear. *Industrial Electronics. IEEE Transactions on*, *57*(3), 813–819.

Roundy, S. J. (2003). *Energy scavenging for wireless sensor nodes with a focus on vibration to electricity conversion.* (Doctoral dissertation). University of California, Berkeley, CA.

Saleem, M., Di Caro, G. A., & Farooq, M. (2011). Swarm intelligence based routing protocol for wireless sensor networks: Survey and future directions. *Information Sciences*, *181*(20), 4597–4624. doi:10.1016/j.ins.2010.07.005

Seah, W. K., Eu, Z. A., & Tan, H. P. (2009). Wireless sensor networks powered by ambient energy harvesting (WSN-HEAP)-Survey and challenges. In *Wireless Communication, Vehicular Technology, Information Theory and Aerospace & Electronic Systems Technology, 2009. Wireless VITAE 2009. 1st International Conference on.*

Seyedi, A., & Sikdar, B. (2008). Energy Efficient Transmission Strategies for Body Sensor Networks with Energy Harvesting. In *Proc. of the42nd Annual Conference on Information Sciences and Systems (CISS).* doi:10.1109/CISS.2008.4558613

Simjee, F. I., & Chou, P. H. (2008). Efficient Charging of Supercapacitors forExtended Lifetime of Wireless Sensor Nodes. *IEEE Transactions on Power Electronics*, *23*(3), 1526–1536. doi:10.1109/TPEL.2008.921078

Simjee, F., Pai, H., & Chou, P. H. (2006). Everlast: Long-life, Supercapacitor operated Wireless Sensor Node. In *Proc 2006 International Symposium on Low Power Electronics and Design*. ACM. doi:10.1145/1165573.1165619

Sodano, H. A., Inman, D. J., & Park, G. (2005). Generation and storage of electricity from power harvesting devices. *Journal of Intelligent Material Systems and Structures*, *16*(1), 67–75. doi:10.1177/1045389X05047210

Tan, H. P., Eu, Z. A., & Seah, W. K. (2009). Impact of Power Control in Wireless Sensor Networks Powered by Ambient Energy Harvesting for Railroad Health Monitoring. In *Proc. of the 2nd Intl Workshop on Applications of Ad hoc and Sensor Networks* (AASNET). doi:10.1109/WAINA.2009.113

Werner-Allen, G., Lorincz, K., Ruiz, M., Marcillo, O., Johnson, J., Lees, J., & Welsh, M. (2006, March-April). Deploying a Wireless Sensor Network on an Active Volcano. *IEEE Internet Computing*, *10*(2), 18–25. doi:10.1109/MIC.2006.26

Internet of Thing. (1994-2014). Retrieved from Tech Target web site: http://whatis.techtarget.com/definition/Internet-of-Things

Internet of Thing (IoT). (n.d.). Retrieved from Cisco web site: http://www.cisco.com/web/solutions/trends/iot/overview.html

Manyika, J. (2013). *Disruptive technologies: Advance that will transform life, business and the global economy*. Retrieved from McKinsey & Company web site: http://www.mckinsey.com/insights/business_technology/disruptive_technologies

Internet of Things. (2014). Retrieved from Microsoft web site: http://www.microsoft.com/windowsembedded/en-us/internet-of-things.aspx

Burrus. (2014). *The Internet of Things Is Far Bigger Than Anyone Realizes*. Retrieved from Wired web site: http://www.wired.com/2014/11/the-internet-of-things-bigger/

Sander & Saulny. (2007). *Bridge collapse in Minneapolis Kills at least 7*. Retrieved from New York Times web site: http://www.nytimes.com/2007/08/02/us/02bridge.html?_r=0

Chapter 4
Analysis of Energy Loss in Superconducting Waveguides

Kim Ho Yeap
Universiti Tunku Abdul Rahman, Malaysia

Kee Choon Yeong
Universiti Tunku Abdul Rahman, Malaysia

Choy Yoong Tham
Wawasan Open University, Malaysia

Humaira Nisar
Universiti Tunku Abdul Rahman, Malaysia

ABSTRACT

In this chapter, the characteristics of the propagation of waves in superconducting waveguides are investigated. To compute the propagation constant, the complex conductivity of the superconductor is incorporated into the set of characteristic equations which describes the propagation constant of waves in the waveguide. An important outcome from this analysis is that superconducting waveguides are shown to behave like a lossless waveguide, exhibiting literally lossless behaviour at frequencies above the cutoff and below the gap frequency. Above the gap frequency, however, the waveguide loses its superconductivity, giving attenuation which increases in correspond to frequencies. The result suggests strongly that superconducting waveguides can be applied in receiver systems to minimize the loss of propagating signals.

INTRODUCTION

Wave guiding structures such as circular and rectangular waveguides and microstrip transmission lines are widely used in telecommunication receiver systems to channel and couple signals to the mixer circuits. In order to ensure that the received signal is converted in the mixer circuit with minimum loss, accurate and versatile mathematical formulations are used as a guide to compute and predict the losses in the guiding structures.

DOI: 10.4018/978-1-4666-9792-8.ch004

Figure 1. Block diagram of a heterodyne receiver

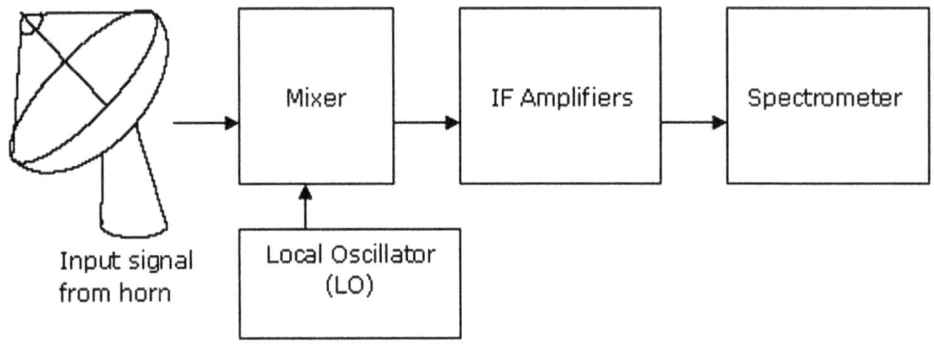

Figure 1 shows the functional block diagram of a typical heterodyne receiver which comprises a receiving antenna, a local oscillator LO, a mixer circuit, an intermediate frequency IF amplifier, and a spectrometer (Chattopadhyay *et al.*, 2002; Kraus, 1986; Yeap, Tham, Nisar, & Loh, 2013). The RF signal from the antenna is directed down to the receiver system via mirrors and beam waveguides (Paine, Papa, Leombruno, Zhang, Blundell, 1994). At the front-end of the receiver system, the RF signal is channeled and coupled to a mixer circuit via hollow waveguides and microstrips. A heterodyne mixer is commonly implemented to down convert the RF signal to an intermediate frequency IF signal. After going through multiple stages of amplification, the IF signal is fed to a data analysis system such as an acousto-optic spectrometer. The data analysis system will be able to perform Fourier transformation and record spectral information about the input signal.

The front-end receiver noise temperature T_R is determined by a number of factors. These include the mixer noise temperature T_M, the conversion loss C_{Loss}, the noise temperature of the first IF amplifier T_{IF}, and the coupling efficiency between the IF port of the junction and the input port of the first IF amplifier η_{IF}. Walker *et al.* (1992) has performed a comparison among different waveguide receivers. In their analysis, it is found that the value of T_R for the 230 GHz system is a factor of 3 to 4 less than that achieved with the 492 GHz system. The decrease in system performance at 492 GHz is due to the increase of C_{Loss} and T_M by a factor of approximately 3.

Since the input power level of the telecommunication signal could be quite small, i.e. of the order of 10^{-18} to 10^{-20} W (Shankar, 1986), it is therefore of primary importance to minimize the conversion loss C_{Loss} of the mixer circuit. One way is to ensure that the energy of the LO and, in particular, the RF signals is channeled and coupled from the waveguides to the mixer circuit in a highly efficient manner. To minimize the loss of the propagating signals, the availability of a highly efficient wave guiding structure is, of course, central to the development of the receiver circuits.

Although most waveguides implemented in radio receivers are made of copper, the attenuation level exhibited in standard metallic waveguides such as copper may actually degrade the detection of signals at such weak intensity (Winters and Rose, 1991). Superconducting waveguides, on the other hand, feature low transmission losses and dispersion level below the gap frequency f_g. A theoretical study is, therefore, performed in this chapter to analyze the propagation of electromagnetic waves in superconducting waveguides.

BACKGROUND

A perfect conductor exhibits lossless condition, where the tangential electric field E_t and the normal derivative of the tangential magnetic field $\frac{\partial H_t}{\partial a_n}$ at the boundary of the wall are zero. In contrast with a perfect conductor, field penetration occurs at the superconducting walls. In order to account for the field penetration, an alternative boundary condition based on the penetration depth of the Meissner effect has been suggested to study the wave properties of superconducting rectangular (Wang, Qiu, & Yalamanchili, 1994; Yalamanchili, Qiu, & Wang, 1995; Ma, 1998), circular (Ma, 1995), and parallel-plate (Ma, 1999) waveguides. In the work of these authors, the boundary condition for the longitudinal magnetic field H_z of a TE mode is given by,

$$\frac{\partial H_z}{\partial a_n} - \frac{1}{\lambda_L} H_z = 0, \tag{1}$$

where λ_L known as the London penetration depth, is a measure of the distance of magnetic field penetration into the superconductor. An important implication of this theoretical study is that the dominant mode for a rectangular waveguide is found to have switched from TE_{10} to TE_{11}; while that for a circular waveguide has switched from TE_{11} to TE_{01}. The formulation developed based on (1) predicts that the cutoff frequencies f_c for both TE_{10} and TE_{11} of rectangular and circular waveguides, respectively, increase when the temperature T drops below the critical temperature T_c, therefore resulting in the change of dominant modes in superconducting waveguides. Yassin *et al.* (2001) has performed an experimental validation on the above theory using a superconducting circular waveguide. The experimental result, however, shows that the work reported in Wang *et al.* (1994), Yalamanchili *et al.* (1995), Ma (1998), Ma (1995), and Ma (1999) turned out to be invalid. The mode order in a superconducting waveguide remains the same as those found in a perfectly conducting waveguide.

Yassin, Tham, and Withington (2003) have performed a theoretical analysis on superconducting circular waveguides based on incorporating the complex conductivity of a superconductor into Stratton's equation (Stratton, 1941). The complex conductivity was found by solving Mattis-Bardeen's equation (Mattis & Bardeen, 1958), derived from the Bardeen-Cooper-Schrieffer microscopic theory of superconductivity – the BCS theory. Winters and Rose (1991) have performed a study on the attenuation in superconducting rectangular waveguides. Since Stratton's equation fails to be applied in the case of rectangular waveguides, the method proposed by Winters and Rose (1991) was, therefore, based on the approximate power-loss method and the phenomenological two-fluid model. Since the power-loss method assumes lossless field's expression in the waveguide, it fails to account for the penetration depth in a superconductor. Moreover, the two-fluid model does not account for the existence of the gap energy in its formulation and therefore is not able to indicate clearly the transition from superconducting to normal states. As concluded by Kautz (1978), due to its ability in describing the relationship of quasiparticles and Cooper-pairs with the energy gap, Mattis-Bardeen's equations were found to be more realistic to be applied in the case of superconductors.

In this chapter, a new approach to investigate the propagation of waves in superconducting rectangular waveguides is presented. In the analysis, the complex conductivity of a superconductor based on Mattis-Bardeen's equation is incorporated into the transcendental equations formulated for rectangular waveguides by Yeap, Tham, Yassin, and Yeong (2011a). The characteristic equations which describe the propagation constant of waves in rectangular waveguides, allows the penetration of fields into the wall material; whereas, Mattis-Bardeen's equation takes into account the existence of the energy gap. Hence, the incorporation of the two in this chapter provides a more realistic study of superconductivity effect in waveguides – in particular, rectangular waveguides. In the results and discussion section, a comparison between the loss in superconducting and normal waveguides is made. It is worthwhile noting that the results obtained using the this method actually show that the cutoff frequencies f_c of the dominant modes remain unchanged when T drops below T_c. This agrees with the experimental result of Yassin *et al.* (2001), further confirming the validity of the new method and therefore disproving the theoretical findings in Wang *et al.* (1994), Yalamanchili *et al.* (1995), Ma (1998), Ma (1995), and Ma (1999).

SUPERCONDUCTORS

Superconductivity denotes a remarkable state transition of a material which is characterized by the disappearance of electrical resistance and the complete expulsion of magnetic flux. The temperature at which the superconducting state transition occurs is known as the critical temperature T_c. A superconductor exhibits zero DC resistivity and diamagnetism at temperatures below T_c. At temperatures above the critical temperature a superconductor behaves as a normal metal.

The Physics of Superconductors

The understanding of superconductivity was advanced by Bardeen, Cooper, and Schrieffer (1957), through their microscopic theory of superconductivity, known as the Bardeen–Cooper–Schrieffer (BCS) theory. To explain the phenomenon of superconductivity in a simple manner, one can imagine a negatively charged electron passing by positively charged ions in the lattice of the superconductor. Due to the attraction of the charges, the positively charged atoms distort toward the electron. This in turn causes phonons or packets of sound waves to be emitted which forms a trough of positive charges around the electrons. As depicted in Figure 2, before the electron passes by and before the lattice springs back to its normal position, a second electron is drawn into the trough. The forces exerted by the phonons overcome the electrons' natural repulsion, allowing the electrons to pair up. The coupled electrons are known as Cooper pairs (Cooper, 1956).

The Semiconductor Picture of the Superconductor

The condensation of electrons out of a continuum of allowed energy values into Cooper pairs at a single energy level also gives rise to an energy gap (Δ) at the Fermi surface. This energy gap is orders of magnitude less than the Fermi energy, typically about one millielectron volt, compared to Fermi energies of several electrons volts. The energy gap Δ is the average energy per electron of a Cooper pair, relative to the continuum. The binding energy of a Cooper pair is thus 2Δ, this being the minimum energy required to break the pair. The situation in a superconductor can be thought of as analogous to a semiconductor,

Figure 2. A negatively charged electron passes between positively charged atoms in the lattice causes the atoms to be attracted inward

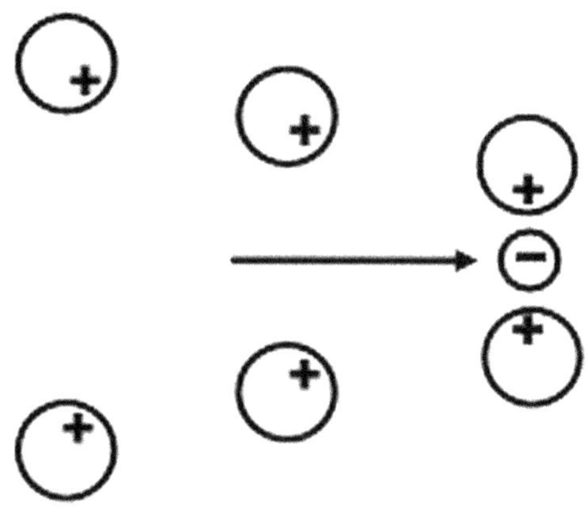

with both having an energy gap at the Fermi surface. When a superconductor is at finite temperatures below the critical temperature, T_c, thermal energy and incident radiation can break Cooper pairs. The electrons from the broken pairs are known as excited quasiparticles, which behave as normal electrons with well specified momenta. Since the binding energy between paired electrons is 2Δ, absorption of incident radiation is possible for field frequencies of $f > \frac{\Delta}{(\pi\hbar)}$, where $\hbar$ is the reduced Planck's constant. Frequency $f_g = \frac{\Delta}{(\pi\hbar)}$ is thus referred to as the gap frequency of a superconductor.

In the superconductor, the electronic states in the immediate neighbourhood of the Fermi energy E_F have their energy pushed away from E_F. The result is a range of energies from $E_F - \Delta$ to $E_F + \Delta$ in which there are no quasiparticle states. Immediately above and below are more states than in the normal metal. These are the states that would have been in the gap if the superconductor were a normal metal. In the ground-state superconductor shown, there is an empty continuum of states available above the gap which is analogous to the conduction band in an intrinsic semiconductor. Below the gap is a filled valence like band. This density of states description of a superconductor which shows the presence of an energy gap reminiscent to that of an intrinsic semiconductor, and is thus called the semiconductor picture of the superconductor.

The Complex Conductivity

The equations for the complex conductivity, i.e. $\sigma_1 - j\sigma_2$, of a superconductor have been developed by Mattis and Bardeen from the microscopic analysis of BCS superconductor weak coupling theory, as shown below (Mattis & Bardeen, 1958)

$$\frac{\sigma_1}{\sigma_n} = \frac{2}{\hbar\omega}\int_{\Delta}^{\infty}\left[f(E) - f(E+\hbar\omega)\right]\frac{E^2+\Delta^2+\hbar\omega E}{(E^2-\Delta^2)^{1/2}\left[(E+\hbar\omega)^2-\Delta^2\right]^{1/2}}dE$$
$$+\frac{1}{\hbar\omega}\int_{\Delta-\hbar\omega}^{-\Delta}\left[1-2f(E+\hbar\omega)\right]\frac{E^2+\Delta^2+\hbar\omega E}{(E^2-\Delta^2)^{1/2}\left[(E+\hbar\omega)^2-\Delta^2\right]^{1/2}}dE \quad , \quad (2a)$$

$$\frac{\sigma_2}{\sigma_n} = \frac{1}{\hbar\omega}\int_{\Delta-\hbar\omega,-\Delta}^{\Delta}\left[1-2f(E+\hbar\omega)\right]\frac{E^2+\Delta^2+\hbar\omega E}{(\Delta^2-E^2)^{1/2}\left[(E+\hbar\omega)^2-\Delta^2\right]^{1/2}}dE\,, \tag{2b}$$

where σ_n is the normal conductivity and $\Delta = \Delta(T)$ the energy-gap parameter. The function,

$$f(E) = \frac{1}{1+\exp\left(E/kT\right)}, \tag{3}$$

gives the Fermi-Dirac statistics and k is the Boltzmann's constant. The first integral in Equation (2a) describes the effect of the thermally excited quasiparticles. The second integral denotes the generation of quasiparticles by fields with frequencies f corresponding to energies above the gap energy. Thus, the second integral is zero for $\hbar\omega < 2\Delta$. Since σ_2 indicates the contribution due to the Cooper pairs, the lower integration limit in Equation (2b) becomes $-\Delta$ when $\hbar\omega > 2\Delta$. Δ depends on temperature and is obtained from the relation (Kautz, 1978)

$$\ln\left(\tilde{\Delta}\right) = -2\int_0^{\infty}\left(E^2+\tilde{\Delta}^2\right)^{-1/2}\left\{1+\exp\left[\left(\pi/\gamma_E\tilde{T}\right)\left(E^2+\tilde{\Delta}^2\right)^{1/2}\right]\right\}^{-1}dE\,. \tag{4}$$

where $\tilde{\Delta} = \dfrac{\Delta(T)}{\Delta(0)}$, $\tilde{T} = \dfrac{T}{T_c}$, and $\gamma_E = 1.781$ is the Euler's constant.

SUPERCONDUCTING RECTANGULAR WAVEGUIDES

In this section, the characteristics of electromagnetic waves propagating in a rectangular waveguide with finite conducting walls are illustrated. Since the characteristic equations developed for lossy rectangular waveguides in Yeap *et al.* (2011a) have been experimentally validated, these equations shall be extended further to the case of a superconductor here. The method used to determine the propagation constant k_z is developed based on matching the tangential fields at the boundaries of the waveguide with the electrical properties of the wall material. Here, in order to present a complete scheme, the derivation of the transcendental equations in Yeap *et al.* (2011a) is briefly outlined for convenience.

Fields in a Rectangular Waveguide

For waves propagating in a practical rectangular waveguide with imperfectly conducting walls, as shown in Figure 3, a superposition of TM and TE waves is necessary to satisfy the boundary condition at the wall (Stratton, 1941; Yeap, Tham, Yassin, & Yeong, 2011b). The longitudinal electric and magnetic field components E_z and H_z, respectively, can be derived by solving Helmholtz's homogeneous equation in Cartesian coordinate. Applying the method of separation of variables (Cheng, 1989), the following set of longitudinal field equations is obtained (Yeap *et al.*, 2011b)

Figure 3. The cross section of a rectangular waveguide

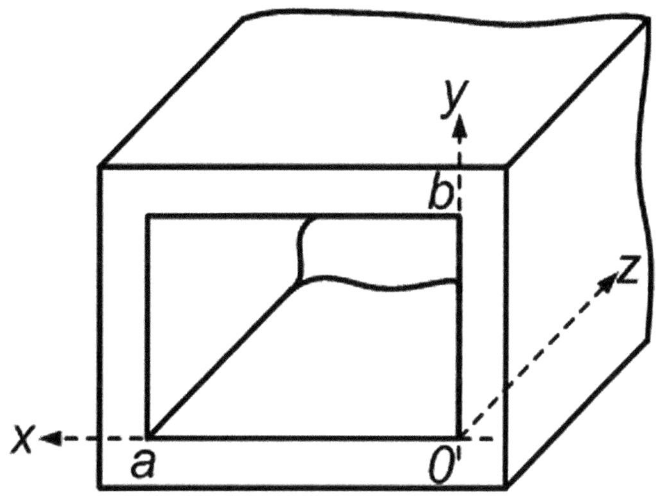

$$E_z = E_0 \sin\left(k_x x + \varphi_x\right)\sin\left(k_y y + \varphi_y\right), \tag{5}$$

$$H_z = H_0 \cos\left(k_x x + \varphi_x\right)\cos\left(k_y y + \varphi_y\right), \tag{6}$$

where E_0 and H_0 are constant amplitudes of the fields, $\varphi_x = \frac{\left(m\pi - k_x a\right)}{2}$ and $\varphi_y = \frac{\left(n\pi - k_y b\right)}{2}$ the field's penetration factors, and k_x and k_y are the transverse wavenumbers, in the x and y directions, respectively. The integers m and n in the penetrations factors define a possible mode for TE_{mn} and TM_{mn} waves. The relationship between the propagation constant k_z and the transverse wavenumbers, i.e. k_x and k_y, is given as

$$k_z = \sqrt{k_0^{\,2} - k_x^{\,2} - k_y^{\,2}} \tag{7}$$

where k_0 is the wavenumber in free space. Here, $k_z = \beta_z - j\alpha_z$ is a complex number, where the real component β_z represents the phase constant and the imaginary component α_z represents the attenuation constant of the waves in the waveguide.

The transverse field equations can be solved by substituting (5) and (6) into Maxwell's source free curl equations (Yeap *at al.*, 2011b)

$$H_x = \frac{j}{h^2}\left[k_z k_x H_0 + \acute{E}\varepsilon_0 k_y E_0\right]\sin(k_x x + \varphi_x)\cos(k_y y + \varphi_y) \ , \tag{8}$$

$$H_y = \frac{j}{h^2}\left[k_z k_y H_0 - \acute{E}\varepsilon_0 k_x E_0\right]\cos(k_x x + \varphi_x)\sin(k_y y + \varphi_y) \ , \tag{9}$$

$$E_x = -\frac{j}{h^2}\left[k_z k_x E_0 - \acute{E}\mu_0 k_y H_0\right]\cos(k_x x + \varphi_x)\sin(k_y y + \varphi_y) \ , \tag{10}$$

$$E_y = -\frac{j}{h^2}\left[k_z k_y E_0 + \acute{E}\mu_0 k_x H_0\right]\sin(k_x x + \varphi_x)\cos(k_y y + \varphi_y) \ , \tag{11}$$

Propagation in a Superconducting Rectangular Waveguide

The ratio of the tangential component of the electric field E_t to the tangential component of the magnetic field H_t, at the conductor surface is given by (Yeap *et al.*, 2009; Tham, McCowen, & Towers, 2003)

$$\frac{E_t}{a_n \times H_t} = \sqrt{\frac{\mu_c}{\varepsilon_0 - j\frac{\sigma_c}{\omega}}} , \tag{12}$$

where μ_c and σ_c are the permeability and conductivity of the wall material, repectively, ε_0 the permittivity of free space, ω the angular frequency, and a_n is the outward normal from the wall. It is to be noted that for non-magnetic materials, μ_c is equivalent to the permeability of free space μ_0.

Substituting the tangential fields (8) to (11) at the $x = a/2$ and $y = b/2$ surfaces of Figure 4 and solving the determinants for the coefficients of the fields, the following pair of transcendental equations is obtained

$$\left[\frac{j\omega\mu_0 k_y \tan\left(k_y b + \varphi_y\right)}{k_x^{\,2} + k_y^{\,2}} + \sqrt{\frac{\mu_c}{\left(\varepsilon_0 - j\frac{\sigma_c}{\omega}\right)}}\right]\left[\frac{j\omega\varepsilon_0 k_y \cot\left(k_y b + \varphi_y\right)}{k_x^{\,2} + k_y^{\,2}} - \sqrt{\frac{\left(\varepsilon_0 - j\frac{\sigma_c}{\omega}\right)}{\mu_c}}\right], \tag{13a}$$

$$= \left[\frac{k_z k_x}{k_x^{\,2} + k_y^{\,2}}\right]^2$$

$$\left[\frac{j\omega\mu_0 k_x \tan\left(k_x a + \varphi_x\right)}{k_x^{\,2} + k_y^{\,2}} + \sqrt{\frac{\mu_c}{\left(\varepsilon_0 - j\frac{\sigma_c}{\omega}\right)}}\right]\left[\frac{j\omega\varepsilon_0 k_x \cot\left(k_x a + \varphi_x\right)}{k_x^{\,2} + k_y^{\,2}} - \sqrt{\frac{\left(\varepsilon_0 - j\frac{\sigma_c}{\omega}\right)}{\mu_c}}\right]. \tag{13b}$$

$$= \left[\frac{k_z k_y}{k_x^{\,2} + k_y^{\,2}}\right]^2$$

Figure 4. Attenuation from cutoff to submillimeter wave frequencies, for TE_{10} mode in a Nb rectangular waveguide at room temperature (solid line) and T = 4.2 K (dotted line)

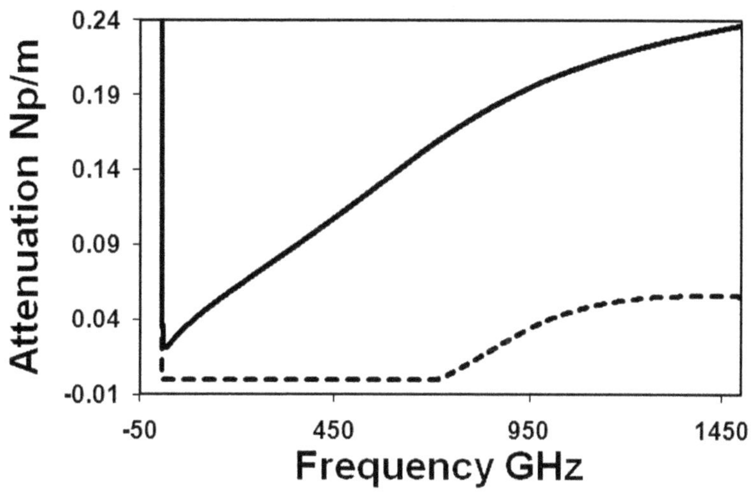

Substituting the complex conductivity of the superconductor in (2) into (13), the following characteristic equations can be obtained

$$\left[\frac{j\omega\mu_0 k_y \tan\left(k_y b+\varphi_y\right)}{k_x^{\ 2}+k_y^{\ 2}}+\sqrt{\frac{\mu_c\omega(\omega\varepsilon_0-\sigma_2+j\sigma_1)}{(\omega\varepsilon_0-\sigma_2)^2+\sigma_1}}\right]\left[\frac{j\omega\varepsilon_0 k_y \cot\left(k_y b+\varphi_y\right)}{k_x^{\ 2}+k_y^{\ 2}}-\sqrt{\frac{\omega\varepsilon_0-(\sigma_2+j\sigma_1)}{\omega\mu_c}}\right]=\left[\frac{k_z k_x}{k_x^{\ 2}+k_y^{\ 2}}\right]^2, \tag{14a}$$

$$\left[\frac{j\omega\mu_0 k_x \tan\left(k_x a+\varphi_x\right)}{k_x^{\ 2}+k_y^{\ 2}}+\sqrt{\frac{\mu_c\omega(\omega\varepsilon_0-\sigma_2+j\sigma_1)}{(\omega\varepsilon_0-\sigma_2)^2+\sigma_1}}\right]\left[\frac{j\omega\varepsilon_0 k_x \cot\left(k_x a+\varphi_x\right)}{k_x^{\ 2}+k_y^{\ 2}}-\sqrt{\frac{\omega\varepsilon_0-(\sigma_2+j\sigma_1)}{\omega\mu_c}}\right]=\left[\frac{k_z k_y}{k_x^{\ 2}+k_y^{\ 2}}\right]^2, \tag{14b}$$

The solutions for the transverse wavenumber kx and ky in (14) can be numerically solved for. By substituting both values into (7), the phase β_z and attenuation constants α_z for a particular mode in the waveguide can then be found.

RESULTS AND DISCUSSION

To investigate the attenuation of the dominant modes for waves propagating in superconducting waveguides, the transcendental equations in (1.14) are numerically solved using a root-searching algorithm from NAG. The attenuation constant of the dominant TE_{10} mode in a Nb rectangular waveguide with dimensions 2.29 × 1.02 cm^2 below and above the critical temperature T_c of 9.2 K is computed and plotted in Figure 4. As can be clearly seen, at frequencies f below the gap frequency f_g, the superconducting waveguide operating at $T = 4.2$ K behaves exactly like a perfectly conducting waveguide. The attenuation diverges significantly at cutoff frequency f_c. Above cutoff, the superconducting waveguide exhibits practically lossless attenuation. To explain this phenomenon, the complex conductivity of the superconducting Nb at 4.2 K has been computed using Mattis-Bardeen equation in (1.2). As can be observed in Figure 5, σ_1 which indicates the effect of the quasiparticles, is negligible at frequencies below f_g, explaining the lossless result below f_g in Figure 5. Above f_g of approximately 716.45 GHz, σ_2 decreases gradually toward zero, while σ_1 approaches the value of σ_n, implying that Cooper-pair breaking takes place above f_g. With the increase of quasiparticles, the random collision of quasiparticles with the lattice structure can thus be expected to become more frequent, resulting in higher conduction loss at frequencies above f_g. Considering that THz signals from astronomical sources are extremely weak, the result suggests that superconducting waveguides that operate at frequencies below the gap frequency can be applied in Superconductor Insulator Superconductor or SIS receivers to minimize the loss of such signals.

It is worthwhile noting that, although BCS theory has been widely accepted in describing the behaviour of superconductors, it has assumed that the electronic states at the immediate vicinity of the Ferni level E_F have their energy pushed away from E_F (Wengler, 1992). Hence, no quasiparticle state exists within the gap energy. Recent findings show, however, that this may not be exactly the case. Experimental measurements suggested that intragap states may actually exist inside the energy gap (Mitrovic and Rozema, 2008; Noguchi, Suzuki, Endo, & Tamura, 2009; Noguchi, Suzuki, & Tamura, 2011; Noguchi, Naruse, & Sekimoto, 2012; Noguchi, Naruse, & Sekimoto; 2013). Since Mattis-Bardeen's formulation

Figure 5. The real part σ_1/σ_n (dashed line) and the imaginary part σ_2/σ_n (solid line) of the normalized complex conductivity of Nb at 4.2 K

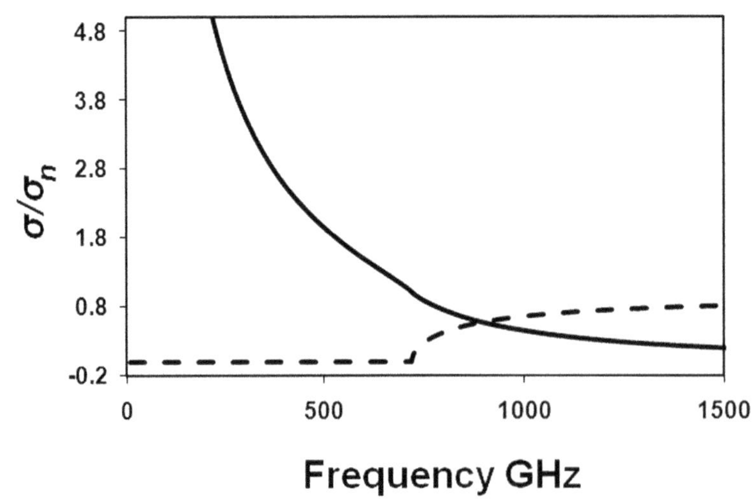

is developed from the BCS theory, quasiparticles in the intragap states fail to be incorporated in the derivation of the complex conductivity. To give a more accurate prediction of loss in the superconducting waveguide, taking into consideration loss due to the random collision of the quasiparticles in the intragap states is, therefore, necessary.

SUMMARY

A new analysis on wave propagation in superconducting rectangular waveguide is presented. The complex conductivity is calculated using Mattis-Bardeen's equation, developed from BCS theory. The attenuation constants are computed by substituting the values of the complex conductivity into the transcendental equations formulated based on the boundary matching method by Yeap *et al.* (2011a).

It is found from the results that at frequencies below the gap frequency, the waveguide exhibited almost lossless transmission behaviour; while above the gap frequency, Cooper pairs breaking dominates and the loss increases considerably. An important implication of this study is that, it suggests that superconducting waveguides – particularly, those that operate at frequencies below the gap frequency, can be applied in receiver systems to minimize energy loss of propagating signals.

REFERENCES

Bardeen, J., Cooper, L. N., & Schrieffer, J. R. (1957). Theory of superconductivity. *Physical Review*, *108*(5), 1175–1204. doi:10.1103/PhysRev.108.1175

Chattopadhyay, G., Schlecht, E., Maiwald, F., Dengler, R. J., Pearson, J. C., & Mehdi, I. (2002). Frequency multiplier response to spurious signals and its effects on local oscillator systems in millimeter and submillimeter wavelengths. *Proceedings - Society of Photo-Optical Instrumentation Engineers*, *4855*, 480–488.

Cheng, D. K. (1989). *Field and Wave Electromagnetics*. Addison Wesley.

Cooper, L. N. (1956). Bound electrons pairs in a degenerate Fermi gas. *IEEE Transactions on Microwave Theory and Techniques*, *104*, 1189–1190.

Kautz, R. L. (1978). Picosecond pulses on superconducting striplines. *Journal of Applied Physics*, *49*(1), 308–314. doi:10.1063/1.324387

Kraus, J. D. (1986). *Radio Astronomy*. Cynus-Quasar.

Ma, J. (1995). Properties of high-T_c superconducting circular waveguides with Meissner boundary. *International Journal of Infrared and Millimeter Waves*, *16*(1), 147–157. doi:10.1007/BF02085853

Ma, J. (1998). TM-properties of HTS's rectangular waveguides with Meissner boundary condition. *International Journal of Infrared and Millimeter Waves*, *19*(3), 399–408. doi:10.1023/A:1022599324690

Ma, J. (1999). Wave propagation properties in high-temperature superconducting parallel-plate waveguides. *IEEE Microwave and Guided Wave Letters*, *9*(5), 183–185. doi:10.1109/75.766758

Mattis, D. C., & Bardeen, J. (1958). Theory of the anomalous skin effect in normal and superconducting metals. *Physical Review*, *111*(2), 412–417. doi:10.1103/PhysRev.111.412

Mitrovic, B., & Rozema, L. A. (2008). On the correct formula for the lifetime broadened superconducting density of states. *Journal of Physics Condensed Matter*, *20*(1), 015215. doi:10.1088/0953-8984/20/01/015215

Noguchi, T., Naruse, M., & Sekimoto, Y. (2012). RF conductivity and surface impedance of a superconductor taking into account the complex superconducting gap energy. *Physics Procedia*, *36*, 318–323. doi:10.1016/j.phpro.2012.06.166

Noguchi, T., Naruse, M., & Sekimoto, Y. (2013). Contribution of quasiparticles in the subgap states to the surface impedance of superconductors. *IEEE Transactions on Applied Superconductivity*, *23*(3), 1501404. doi:10.1109/TASC.2013.2240761

Noguchi, T., Suzuki, T., Endo, A., & Tamura, T. (2009). Contribution of the imaginary part of the superconducting gap energy on the SIS tunneling current. *Physica. C, Superconductivity*, *469*(15-20), 1585–1588. doi:10.1016/j.physc.2009.05.218

Noguchi, T., Suzuki, T., & Tamura, T. (2011). Subgap tunneling current at low temperature in Nb/Al-AlN/Nb SIS junctions. *IEEE Transactions on Applied Superconductivity*, *21*(3), 756–759. doi:10.1109/TASC.2010.2089033

Paine, S., Papa, D. C., Leombruno, R. L., Zhang, X., & Blundell, R. (1994). Beam waveguide and receiver optics for the SMA. *Proceedings of the 5th International Symposium on Space Terahertz Technology*. University of Michigan.

Shankar, N. U. (1986). *Application of digital techniques to radio astronomy measurements*. (Doctoral dissertation). Available from Raman Research Institute Digital Repository. (2289/3391)

Stratton, J. A. (1941). *Electromagnetic Theory*. New York: McGraw-Hill.

Tham, C. Y., McCowen, A., & Towers, M. S. (2003). Modelling of PCB transients with boundary elements/method of moments in the frequency domain. *Engineering Analysis with Boundary Elements*, *27*(4), 315–323. doi:10.1016/S0955-7997(02)00119-4

Walker, C. K., Kooi, J. W., Chan, M., Leduc, H. G., Schaffer, P. L., Carlstrom, J. E., & Phillips, T. G. (1992). A low-noise 492 GHz SIS waveguide receiver. *International Journal of Infrared and Millimeter Waves*, *13*(6), 785–798. doi:10.1007/BF01011595

Wang, Y., Qiu, Z. A., & Yalamanchili, R. (1994). Meissner model of superconducting rectangular waveguides. *International Journal of Electronics*, *76*(6), 1151–1171. doi:10.1080/00207219408926024

Wengler, M. J. (1992). Submillimeter waves detection with superconducting tunnel junctions. *Proceedings of the IEEE*, *80*(11), 1810–1826. doi:10.1109/5.175257

Winters, J. H., & Rose, C. (1991). High-T_c superconductors waveguides: Theory and applications. *IEEE Transactions on Microwave Theory and Techniques*, *39*(4), 617–623. doi:10.1109/22.76423

Yalamanchili, R., Qiu, Z. A., & Wang, Y. (1995). Rectangular waveguides with two conventional and two superconducting walls. *International Journal of Electronics*, *78*(4), 715–727. doi:10.1080/00207219508926205

Yassin, G., Jung, G., Dikovsky, V., Barboy, I., Kambara, M., Cardwell, D. A., & Withington, S. (2001). Investigation of microwave propagation in high-temperature superconducting waveguides. *IEEE Transactions on Microwave Theory and Techniques*, *11*, 413–415.

Yassin, G., Tham, C. Y., & Withington, S. (2003). Propagation in lossy and superconducting cylindrical waveguides. *Proceedings of* the *14th International Symposium on Space Terahertz Technology*.

Yeap, K. H., Tham, C. Y., Nisar, H., & Loh, S. H. (2013). Analysis of probes in a rectangular waveguide. *Frequenz J. of RF-Engineering and Telecommunications*, *67*, 145–154.

Yeap, K. H., Tham, C. Y., Yassin, G., & Yeong, K. C. (2011a). Attenuation in rectangular waveguides with finite conductivity walls. *Radioengineering J.*, *20*, 472–478.

Yeap, K. H., Tham, C. Y., Yassin, G., & Yeong, K. C. (2011b). Propagation in lossy rectangular waveguides. In A. A. Kishk (Ed.), *Electromagnetic waves propagation in complex matter* (pp. 255–272). Intech.

Yeap, K. H., Tham, C. Y., Yeong, K. C., & Yeap, K. H. (2009). A simple method for calculating attenuation in waveguides. *Frequenz J. of RF-Engineering and Telecommunications*, *63*, 236–240.

Section 2

Presenting Energy Harvesting in WSN and Its Perspectives

Chapter 5
Different Resources Consumption of Renewable Energy

Soobia Saeed
Institute of Business and Technology, Pakistan

ABSTRACT

Electricity consumption will encompass a large converse about connected with international electricity demand while in the next 2 decades. Newly, this improving rate connected with fossil fuels and also issues about the environmentally friendly consequences connected with gas emissions get renewed the attention in the progress connected with alternative electricity resources. Renewable Energy Sources and Climate Change Modify Minimization offers a good estimation on the chapter for the technological, scientific, environmentally friendly, financial and also societal aspects of this factor connected with six renewable energy (RE) options for the minimization connected with weather adjust. This functioning chapter on environmentally friendly Energy Solutions and Local climate Change Minimization presents an assessment on the literature for the scientific chemical, technological, environment, economic in addition to social areas of the contribution connected with six environmentally friendly energy (RE) sources on the mitigation connected with climate alter. This chapter is definitely an overview of presentation of the Local climate Change Minimization expansion on the essential results. Considering this significant component of Renewable Energy Sources can be reduce carbon dioxide, there is an international relating to reducing carbon emissions. Due to the fact most of the United Nations wanted to greenhouse gas (GHG) emissions is carbon dioxide, there is a can be a global concern on minimizing carbon emissions. Emissions of greenhouse gases (GHGs) resulting from the provision of the services of one have contributed significantly to improve the historical concentrations of greenhouse gases to the atmosphere of MIT. The IPCC (AR4) concluded that "most of the observed global climate improving as it is very likely that as a result of the improvement observed in the concentrations of anthropogenic gases mit techniques this mid of 20th century confirms Recent Files the use of fossil power accounts for most of the international anthropogenic GHG emissions". Emissions always grow, in addition to CO2 concentrations of it had increased to more than 390 ppm, or perhaps 39% above pre-industrial levels, by holding from 2014-5.

DOI: 10.4018/978-1-4666-9792-8.ch005

There are many options for reducing GHG emissions from energy system while satisfying the desire for global energy services. Some of these possible alternatives, such as energy conservation and competition, switching fossil fuel, RE, nuclear, plus carbon capture and hard drive (CCS) was evaluated from the AR4. A full assessment related to any profile minimization options will likely involve an evaluation of respective potential alongside minimization with his bargain with sustainable development as well as all associated risks, and costs. This phase will focus on the role that this display technology related to RE can participate in within the portfolio related to mitigation alternatives. In this sense, the only policies can be given to reduce emissions of carbon dioxide, to improve the implementation of green energy, and such encouraging technological innovation. At inclusion, supporting components, such as feed-in tariffs, rules Renewable side view in addition to tax insurance policies are used by governments to help develop green energy generation in addition to the implementation of the efficiency of energy use save energy. In this chapter, the various insurance policies could possibly be placed on reducing carbon emissions, for instance improving green energy deployment and also significant technologies. A pair of main clarifications may be realizing to scale back carbon emissions and also overcome the issue connected with weather adjust: exchange fossil fuel having green electricity options wherever possible and also enhancing energy proficiency. In this chapter, many of us discuss most up-to-date performance connected with technology intended for improving green electricity deployment and also electricity work with proficiency.

1. INTRODUCTION

Electrical energy usage can comprise an increasing reveal involving international energy demand above the subsequent two full decades. Recently, soaring energy price ranges and concerns regarding the environment effects involving emissions involving greenhouse gas fossils include renewed involvement in the particular development involving alternate energy sources. Renewable energy is already regarded a more appealing energy resource with regard to nuclear energy due to the absence of hazards and problems. The primary part of greenhouse gas is actually carbon dioxide; there may be international issue in relation to cutting down carbon dioxide emissions. With this impression, the different plans may be put on the particular decrease involving carbon dioxide emissions, including increasing the particular rendering involving environmentally friendly energy and the advertising involving technologies. A couple of key solutions may be carried out to cut back laser emissions and conquer the situation involving climate adjust: exchanging fossil heats up having environmentally friendly energy sources wherever possible and strengthen energy productivity. This chapter discuss with regards to innovative distinct alternatives for enhance the effective use of renewable energy and also productivity in energy use. Moderately a couple of nations have begun to introduce offices by which utilize renewable good quality hotspots for energy era. The significance associated with option energy sources all meets upward with environmental change difficulties of the exorbitant make use of fossil powers. You'll locate a few essential sparks by which invigorate the off shoot of renewable good quality advancements: energy solidity, monetary effects and also co2 partial lazer outflow decrease (Rask, 2009). The expression "elective vitality" portrays any kind of vitality other than the customary wellsprings associated with quality, including hydropower. Of late the center will keep on being in renewable quality sources. As indicated by simply Global Energy Firm portrays two critical worldwide patterns that may describe the design of renewable innovations in the medium term. Firstly, as renewable drive advancements scale upwards, from an aggregate worldwide cause of 1, 454 giga-watts (GW) inside 2011 to simply just two, 167 GW inside 2017, they additionally should spread out topographically. Second, the more the past number of years of high fossil quality asset vitality utilization has driven renewable advancements in the event that you are progressively aggressive more than an expense premise for their options in a few countries around the globe and circumstance. Agreeing with a specific end goal to Worldwide Energy Agency reckonings, wind is a standout amongst the most aggressive sort related together with renewable vitality innovation one of numerous different choices, if nearby conditions like financing, CO outflow amounts and fossil quality asset costs demonstrate viable (Organization for Financial Co-operation notwithstanding Development, 2010). Indeed, even through since you that have two essential ideas inside of your advances: vitality gives building, which consider option sources connected with renewable vitality like wind and sun oriented force and vitality capability designing. It ought to be noticed that changing the capacity division and redesigning customary vitality having renewable vitality is transformative connected to mechanical change along together with shaping markets. The changing procedure for a few types of earth inviting vitality, for example, the wind and sun oriented may happen after 2020, paying little heed to in the event that they development cost of utilization is typically firmly expanding on your one decade from now. Similarly, renewable vitality markets are not effectively shaped in view of expense inconveniences furthermore the financing of fossil gets more sultry (Muneer, 2007).

2. RENEWABLE ENERGY AND CLIMATE CHANGE

Aspiration for energy, as well as associated products and services, to meet besides the monetary societal development and also improved human being welfare, besides health, is actually increasing. All organizations need power services to satisfy basic human being needs like lighting, cooking, space convenience, mobility and also communication) plus serve productive treatments. Considering somewhere around 1850 world-wide fossil fuel use and also characteristics on the fossil powers, oil and also gas increased to enhance the power, resulting in a rapid expansion of beam of light fractional beam of light (CO2). [Figure 1] Types of greenhouse gas (GHG) due to the provision of a person get contributed significantly on the historical increase in every greenhouse gasoline concentrations in the atmosphere MIT. According Assessment Report on the IPCC Final (AR4) accomplished that "predominantly inside observed raise in international conditions because the mid-20th century is probably because of the observed increase above the concentrations of greenhouse fumes anthropogenic." New data state that the usage of fossil fuels accounts for the majority of international anthropogenic GHG emissions. Emissions generally grow and CO2 levels of mit received risen above 390 ppm, or possibly 39% above pre-industrial levels, for the final outcome of this participation of 2, 010. You will observe multiple options the reduction of GHG emissions inside energy program although still meet international demand particularly in regards to energy providers. Most of most of these possible options, including power conservation with regards to efficiency, fossil power resources move, RE, nuclear and H2O as well as the capture and storage (CCS) is it being evaluated within just IE4. A comprehensive assessment of almost any single shareholding of mitigation solutions involve an assessment of individual mitigation potential together with its info to sustainable development together with all this risks and associated prices (Bain, 2010) . Relating Assessment Report within the IPCC Remaining (AR4) done that "predominantly on the inside observed raise in overseas conditions since the mid-20th century is most likely because of the observed enhance above your concentrations connected with greenhouse toxins anthropogenic." New data state that the

Figure 1. Renewable energy resource technologies

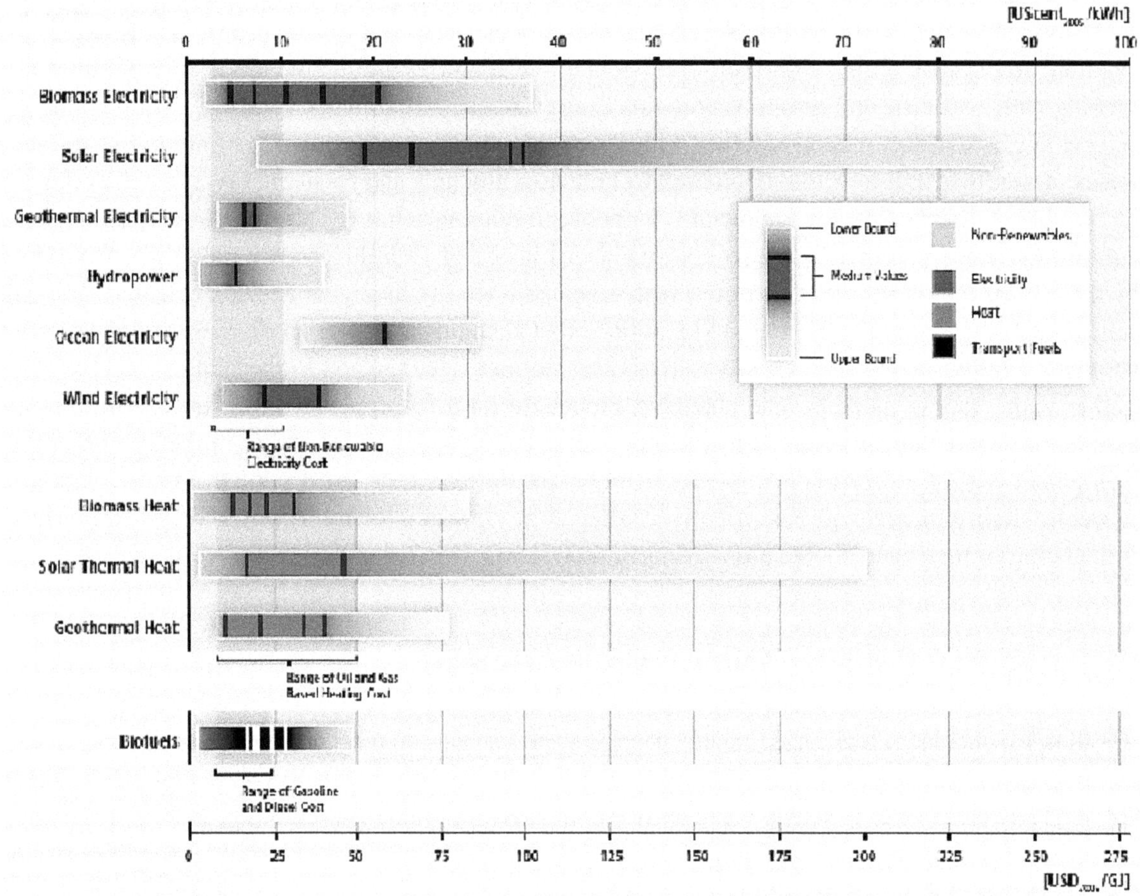

effective use of fossil energy sources accounts for most of overseas anthropogenic GHG emissions. Emissions normally grow and CO2 amounts of mit obtained risen earlier mentioned 390 ppm, or possibly 39% earlier mentioned pre-industrial ranges, for the last outcome in this participation connected with 2, 010. You will observe multiple selections the reduction of GHG emissions on the inside energy system although still meet overseas demand particularly in terms of energy suppliers. Most of most of these possible selections, including power conservation with regards to efficiency, fossil power resources precede, RE, nuclear and H2O as well as the capture and storage (CCS) is it being evaluated in a matter of IE4. An extensive assessment of almost any single shareholding connected with mitigation remedies involve a assessment connected with individual mitigation potential as well as its details to environmentally friendly development as well as all this specific risks and associated price ranges (Bain, 2010).

3. RENEWABLE ENERGY RESOURCE TECHNOLOGIES

Ones supply of renewable energy that is continuously intensifying. A high level of investment has recently done in recent years, plus the advancement of engineering offers to eligible countries to provide additional cost effective renewable energy. It foretasted that the quality of the producing countries mentioned above 100 megawatts (MW) with respect to renewable energy increases significantly by 2017. Caused by other negative externalities irreversible besides returning to conventional production of vitality, is crucial to promote the development of renewable energy in addition to providing technologies. These technologies are in all probability, are not comparable using conventional fuels in relation to production costs, but often are comparable if many of us consider their associates, such as its environmental and social externalities. In addition, it should be noted that regular scale enterprises could play an essential role in reducing the production cost of the system. Warning expenses and distribution, as well as technologies do not differ much in the list of conventional and organic energies. Here the key points about the development around key technologies of renewable energy supply that often occur are discussed below (Van Kooten, 2008):

1. Bioenergy,
2. Solar Energy,
3. Geothermal Energy, and
4. Hydropower.

3.1Bioenergy

Bioenergy can be created from an assortment of biomass feedstocks, as well as materials of high rural and animals waste; backwoods estates short cycle; force edits; the natural division of metropolitan strong waste; and other natural waste streams. During a mixed bag of procedures, these materials can be utilized specifically to create vitality or warmth, or used to make vaporous, fluid, or strong energizes might. The decision of bioenergy advances is wide and specialized development fluctuates significantly. Examples of technology you can find include small boilers and also large scale heating systems dependant on domestic pellet, and output of ethanol from sugars and starch. The combined cycle power plants advanced biomass gas if incorporated cation and transport fuels dependant on lignocellulose are examples of technologies which have been in a pre-commercial stage, while the production associated with liquid biofuels from algae and a few other methods of natural conversions they are investigation and development (R & D) phase. Bioenergy engineering have programs in centralized in addition to decentralized environments with the traditional use of biomass within developing countries will be the current almost all widespread iphone app. Bioenergy usually features a constant or controllable efficiency. Bioenergy assignments usually depend on the option of method of getting local in addition to regional gas, but recent developments obtained show reliable biomass in addition to liquid biofuels are usually significantly traded internationally (Bain, Third. L., 2010).Bioenergy can be embedded in complex approaches biomass systems global foods, feed and fiber creation and forest products and also waste management and spend. Perhaps most important, bioenergy comes with an intimate and critical role from the media everyday activities of immeasureable people in creating nations throughout the world. Figure 3 shows the kinds of biomass useful for bioenergy inside developing places and produced nations. Expanding bioenergy manufacturing may not need sophisticated supervision cant land and in addition water work with; increased productivity regarding international foods commodity, supply, linens, forest products and energy; change of substantial improvements, plus a far more refined understanding of the societal, energy in addition to environmental interactions in the production and using bioenergy complex. In '08, biomass offers about 10% (3 50 EJ and year) when using the global supply primary vitality. Main uses of biomass directly enter two extensive categories: (1) small efficiency traditional biomass including wood, straw, manure fertilizers are used together with belonging to the cooking, lighting and heating in the room, usually by poorest countries in construction. This biomass is burned to produce most serious negative impacts throughout the lives and health. Increasingly, coal is every second assistance energy in rural areas to be able to create productive chains. For indicator in the extent of traditional utilization associated with biomass, Figure 2 illustrate which cause global primary energy output from traditional industrial wood biomass will involve parallel world. (2) Greater productivity in using modern bioenergy quick solids, liquids and gases to realize that secondary energy carriers that is included in the heat, electricity, combined temperature and strength (CHP) in addition to transport heats up from unique substantial. Liquid biofuels contain ethanol and moreover biodiesel likewise for transportation by road with this particular planet and a few industrial operates by using (J. P oker. Galdo, 1997). Biomass extracted gases, largely methane, anaerobic digestion of yard waste and also treatment affecting municipal sound waste (MSW) are usually widespread to build electricity, temperature in addition to both. An important contribution of one's services depends on solids, for example motherboards, pellets, previously applied reclaimed wood and many others. It consists of heating programs district warming and trouble heating, and also. The modern-day bioenergy predicted total principal way of obtaining biomass will be 11. 3 EJ and 12 weeks and supply to extra energy end-use. Consumers can be estimated at 6. EJ and year. In addition, an industry, such as pulp industry foodstuff and paper, forestry and eats about 7. 7 EJ of biomass each year, mainly as a method to produce steam for industrial processes (EIA, 2009).

Figure 2. Biomass sources of energy

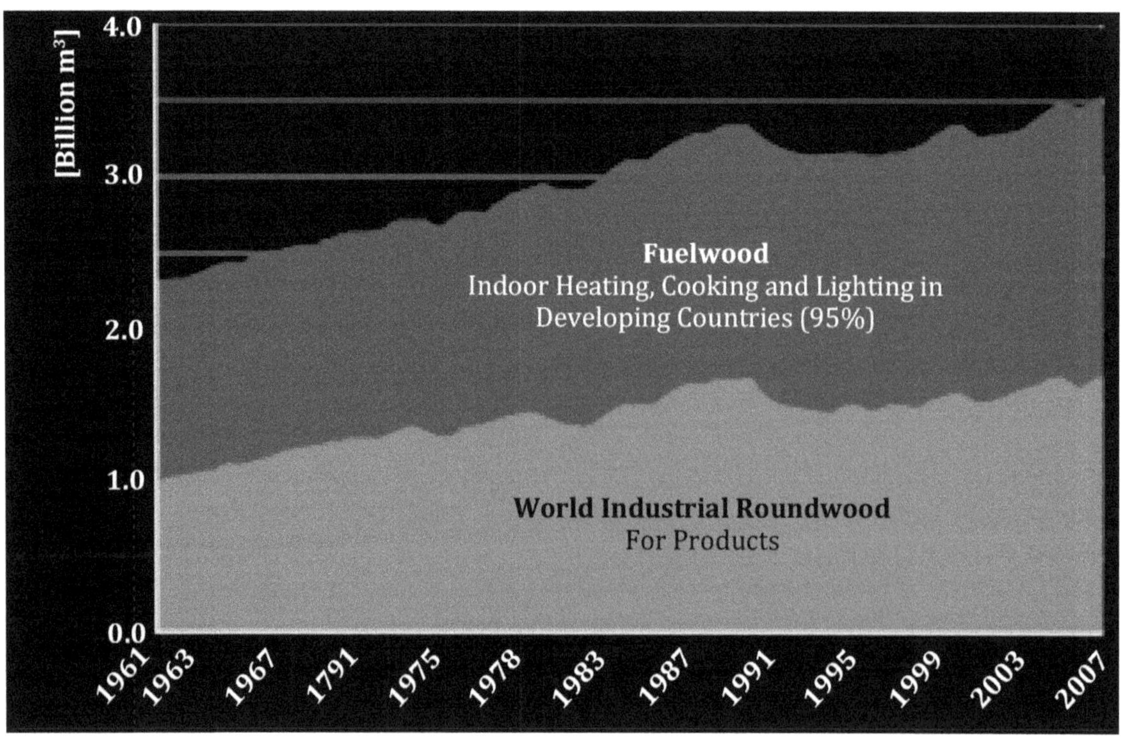

3.2 Solar Energy

Technologies to harness solar irradiation facility for you to present electricity via photovoltaics along with concentrating solar panel (csp) to be able to develop energy winter (uniform heating and cooling), to be able to meet your right illumination wishes and, potentially, to help provide fuels the idea may be taken intended for transportation along with purposes. the maturity of your technology connected with photovoltaic applications vary via R & D (eh ;. gary, fuels designed throughout solar energy), an relatively mature, with regard to example, csp, to help mature equally solar home heating and in line with silicon wafer (pv). Many, but not the many technologies are usually modular throughout nature, allowing effortless use both national inside decentralized energy tactics. solar energy will be inconsistent and, for you to varying degrees, unknown, while ones temporal profile of the production connected with solar energy inside numerous ailments sensibly correlate nicely needs tenacity.Energy storage cold weather gives the button in order to improve ones output control intended for a lot of regarding you csp as well as pv technologies similar to straight home heating. Right solar energy technologies usually are regarding various other natures. filling up as well as adjusting the various methods of which a person employ energy as heating, electricity and also fuel electric power, that happen to be children related technologies. the actual team aims at a few most significant types: (a) thermal dom, containing both additionally encouraged non-aggressive heating regarding buildings, everyday in addition to business driven mineral water solar heating, method heating hottest action intended for market; (b) photovoltaic (pv) power generation coming from direct change from daylight in order to electric power through photovoltaic fabric; (C) concentrated solar power (CSP) bringing in electric power through the optical focus of solar energy pertaining to high temperature water or perhaps delivers heat in order to drive a good car applications, and in addition electromechanical generators; as well as (d) pv power output methods, your current intention involving employing solar energy in order to supply helpful fuels. Solar energy at the term refers on the base involving power to its an individual RE technologies This draw towards the power of your sun directly.Specific renewable technologies, like wind and also ocean thermal, solar panel considered soon after being absorbed regarding the earth along with became otherwise.(In ones rest regarding the actual part, ones "direct" adjective applied in order to solar energy is often suppressed Just like proving for you to become understood). Solar energy what you mean around the precise emitted thermal radiation for the outdoor stage. Merely outside your current atmosphere connected with Earth, this radiation, known Equally solar radiation, carries a good magnitude This average 1 367 W / m2 intended for single individual perpendicular to the rays of any sun surface. in ground level (usually erection Conditions Demands As sea level with the sun directly overhead), your current irradiance will be attenuated in the atmosphere related to 1, 000 interior clear sky Conditions W / m2 throughout a great short night out associated with noon from an song called 'full sun'. Internal electromagnetic waves away by the atmosphere, your durability of a sun can be done within wavelengths covering anything from about 0.20 for you to two microns. Probably your own solar irradiance is sent from the sunlight (K.-M. Chung, 2011). This reaches about the sun directly with out remaining scattered of approximately inside. Irradiance, this particular "beam" and also is capable of becoming concentrated by, in addition to decorative mirrors as well as lenses, this is the lowest you may take cloud cover inside pieces. Irradiation is called diffuse irradiance. The sum of a diffuse irradiance beams along with refers to the global solar radiation. The theoretical potential of solar energy, that will indicates The amount of radiation on the surface of the Earth (land as well as sea) is usually theoretically shown intended for energy production; It is usually estimated at 3.9.106 EJ / year. This type involving number, clearly intended for illustrative purposes easily demand all use almost all available land in addition to sea location That has 100% conversion efficiency. a great further convenient metric would be the technical potential; this requires evaluating your own small group of land That is regarding hassle-free EMPLOY regarding conversion machines The item have more realistic financial transaction effi efficiency. Estimates of a technical potential involving energy solar farm incorporate 1 575 for you to 49, 837 EJ / year, ie exactly about two to 100 times your own consumption connected with initial electricity in the world with (Y. Lu, 2010).

3.3 Solar Fuel Manufacture

Fuel technologies convert solar PV technology throughout chemical fuels like hydrogen, synthetic gas and liquid just like methanol and diesel engine. Your three easy fuels your current sun, that could possibly help work alone or maybe in any kind of combination, routes are: (1) electrochemistry; (2) photochemical / photo-biological; and also (3) thermochemical. For the first route, hydrogen is generated from electrolysis driven practice derived solar power This has been generated by just about any technique PV or perhaps CSP. Electrolysis associated with water is actually an old and also properly understood technological know-how, usually reaching 70% efficiency electricity alteration coming from hydrogen. With the second route, solar photons are designed to drive photobiological photochemical reactions as well as such time, these kinds of items in which are fuels: It is generally mimic, what plants in addition to creatures do. Alternatively, ones semiconductor material can be employed being an anode for absorbing sunlight decrease your own photoelectrochemical cells, that will usually additionally quickly create hydrogen through decomposing water (Harvey, 2006). At the third route, your current solar heat via high temperature (as considered with any kind of ground station receiver CSP-receiver) for driving a endothermic chemical good effect of your gasoline can be used. Next, your own reagents will probably include combinations connected with water, CO2, coal, biomass and also gas. Products, that will generally constitute your own photovoltaic fuel, is actually almost any (or combinations) of an following: hydrogen, synthesis gas, along with methanol and also dimethyl ether synthesis oil. When the fossil fuel is consumed to be a reagent, your calorifi c values exceed anyone solutions within reagents, so The item less fossil fuel must be burned pertaining to energy release similar. Solar fuel additionally synthesized via hydrogen and also photovoltaic CO2 to write hydrocarbons compatible in existing energy infrastructureCurrently, solar fuel production is actually for the pilot plant stage. Pilot plants within the power variety over 300 in order to 500 kW end up being produced regarding reducing thermal carbo particularly in relation to zinc oxide, sauna and steam methane reforming involving petcoke respect gasifi cations. A good 250-kW reactor steam reforming can be operating with Budgets (Harvey, 2006).

3.3.1 Photovoltaic Electricity Generation

In the Simplest terms, published thin silicon semiconductor material, with regard to example is placed towards the Sol this sheet, known to be a cell comprises two shaped layers via introducing impurities on the resultant silicon in to of any n-type layer that has a layer involving p-type to form a great junction interface. Cell impressive solar photons develop electron-hole pairs that are spaced apart through a internal electric field of your union. This creates negative charges using an Individual interface side in addition to optimistic charges are generally towards the reverse side. This separation of a resulting requires creates stress. Whenever both sides of an personal lighted cell, are generally linked with a fill current flows about the same side of a unit via The load to the various other side by the battery electric power generation. [3. 3. 3] numerous photovoltaic technologies happen to be able to be designed within parallel. for the counter available PV technological know-how will include wafer-based crystalline silicon photo voltaic and thin-fi movie technological know-how copper indium/ gallium disulfi of / (di) selenide (CIGS), cadmium telluride (CdTe) meager fi lm silicon (undefined, microcrystalline silicon in addition to), alongside the color sharpened sunlight based boards. with expansion, perceptible for the market concentration photovoltaic concepts, by which very high efficiency tissue (such as Equally gallium arsenide (GaAs) based materials) fit in to your current windows concentrate emphasis as well as additional manifolds with regard to example Fresnel lenses. Mono furthermore multi-crystalline (now and then called "polycrystalline") of photovoltaic silicon wafers, including tape advancements would be the prevailing innovation to the photovoltaic business, that has a 2009 piece of the pie of around 80% (IEA, 2007).Top EFI deficiencies conducted coming from different mobile or portable kinds include a great large quantity connected with in excess of 40% intended for cells Based on GaAs concentration, information on 25% in respect for you to monocrystalline, 20% compared to multicrystalline and also cigs, 17% respect to cdte, in addition to pertaining to 10% involving amorphous silicon. generally, groups regarding cells are mounted side through side in a good sheet see-thorugh (usually glass) and also connected throughout sequence to help application form an "module" inside dimensions all tips on how to 1 m coming from 1 m (Martinot, 2010). With considering efi deficiencies, this is ticks to name between mobile or perhaps portable efi sciences (quoted above) plus the efi module sciences; your own latter usually are 50 to be able to 80% of a first. Manufacturers carry on to be able to improve performance along with reduce service fees during automation, cell processing faster along with inexpensive along with high-performance producers. The performance of any modules is typically guaranteed through companies for you to 20 in order of 30 years. the make use of photovoltaic energy for you to useful energy consists of in excess of simply just your own cells as well as modules; The specific PV system, with regard to example, often authored of your inverter to convert your own right current of a cells for the ability associated with CA to help become compatible with recognized networks in addition to devices. Intended for software outside your current network, your current program will certainly include storage models such as batteries. It is recording to post these kind of much extra reliable devices, reduce cost, along with extend its life to be able to become comparable within it of a modules such as Agricultural solar energy systems are generally classified as a few most significant types: non-network along with gridconnected. Networking methods are themselves grouped within two ways: distributed and also centralized. The distributed system consists of these types of small local resistance plants, a number of primarily supplying electricity in order to a good consumer at site, as well as the remaining electricity for the grid. Centralized particular, moreover, seed technique works just as one expansive force. Off-framework frameworks are normally devoted to help an awesome individual or little gathering with respect to clients and also for the most part call for electric storage element or even in the same way backup power. These kinds of systems have signifi cant potential inside non-electrified parts. Today your current Focus power generation solar power:CSP technologies designed electricity through paying attention sunlight to be able to heat the medium (Gevorkian, 2012). The item will then be taken either directly or indirectly inside a great hassle-free heat engine very to be able to drive a good electric turbine. CSP effortlessly uses the part of your beam connected with sunlight, consequently their maximum benefit tends to help become restricted to help a good small geographical range. your current hub provides sunlight to be able to a good point (focal point) Any time consumed throughout systems within central receiver or maybe dish to the line (line approach) Whenever considered within programs or maybe linear Fresnel minimum. The same technique incase additionally be used in order to drive thermo substances processes for energy production, as defined below. Concentrator's valley extended lines relative for you to parabolic reflectors which follow your current movement of an sun concentrate solar irradiation at the range around 75 to help 100 situations a heat-collecting component (HCE) mounted lower your focal sequence reflector. HCE can be a blackened inner tube like any selective surface. IN addition to a glass outer pipe that has a space between only two evacuated. In present commercial designs, a change oil heat is circulated through the steel pipe where truly heated (to almost 400 ° C), but systems using other materials heat transfer as current cast or direct sodium Steam is being demonstrated(Gevorkian, P., 2012). You're second type involving approach line system, ones linear method Fresnel mirror as well as parallel prolonged strips obtained with the hub, again, with a fixed linear device. sole of a 2 systems focus point, ones real central receiver (also known In the same way your 'Power Tower ") uses a great multitude connected with mirrors (heliostats) in the grass, each sun two-axis tracking to help focus your current sun rays on an point, and also a great high tower. ones central level to a great receiver, comprising either fixed inverted cavity as well as / as well as pipes the fluid circulates heat transfer tooth heads. Can achieve temperatures high (up at the least sole 000 ° C) ones associated with current line approach, that makes it possible for ones engine for you to convert heat "at least within theory" extra heat to power collected. In the second type associated with point process focus, your current hub plate, single parabolic reflector (as opposed for you to a great multitude regarding reflectors) sun tracking within 2 axes may be used for its concentration. The plate is usually created solar ray in an receiver is actually not fixed, but moves from the plate, being no greater than a length plate away. Temperatures within the actual receiver engine will probably reach just as high immediately as 900 ° C. in individual embodiment popular concept, a Stirling engine driving the generator is actually mounted for the utility approach. Stirling plate devices are generally comparatively small, usually build 10-20 kW, but could be excess at the field configuration in order to recognize superior productivity power-station Just like principal (Gipe, 1995). Many four involving the various CSP plants possesses advantages and disadvantages. All four continues to be developed in addition to demonstrated. an mouse clicks advantage regarding csp technologies (except belonging to be able to dishes) can be being capable to shop thermal energy after it\'s collected with this receiver and also previously going to the high temperature of a engine. Taken storage media include molten salts, compressed air along with steam accumulators (for sole short term storage), persists solid ceramic high temperature, your phase change resources, graphite, and high genuine temperature. Commercial solar thermal power plants are usually made in thermal storage features to achieve 15 hours, allowing manageable CSP for you to give the energy (Gordon, 1987).

3.4 Geothermal Energy

Geothermal heat is usually an application form connected with energy created in addition to stored at the Earth. That is still taken throughout history for the bathroom, heating and kitchen. Geothermal energy is made from radioactive corrosion, within achievement 4 temperatures, 000C with the core of any planet Earth. While geothermal energy is usually exhibited worldwide, This really is called carefully assessed the geothermal gradient The item indicates whether or not a good region is really a privileged place with regard to enactment. Rateat steps your own temperature increases to its reason how the depth of your planet earth increases. regarding example, the average geothermal gradient throughout France can be 4 ° C / 100 m within number 10 / 100m in the Alsace region inside Personal on a couple of ° C / 100 m inside the mountains of any Pyrenees. Within Iceland plus the volcanic regions, your own gradient will probably reach Just as high As 30 ° C / 100 m (Lund, 1995). Your own geothermal gradient is usually not lone your sole tool used to measure your availability involving geothermal energy. concerning permeability rocks, of which determines your rate regarding heat flowing on the surface; This can be acknowledged As another mouse clicks step about the availability connected with geothermal energy. Geothermal energy features an big advantage than wind in addition to solar energy This really is available from Whenever through the season. approximated CO2 emissions which are generated because of the geothermal resources tend to be 55 g / kWh make use of information from just about any statement regarding 73% associated with geothermal power plants. There is plausibility that the worth could be really decreased to zero if geothermal fluid is reinjected into the ground. A sum of 24 nations is at present being utilized as geothermal force plants. The majority of the world's geothermal limit (GW) can be introduced in eight global territories: United Claims (V. Lukosevicius, 2000). The Philippines (right around 2 0) furthermore Indonesia (February 1), South America (0 9), Italy (0 9), totally New Zealand (0 8), Iceland (0 7) and in addition Japan (0. 5) (BP, 2012). It genuinely is foreseen that geothermal ability to increment with advances in the co-generation of geothermal vitality, utilizing water through coal and oil wells all through 2015. Fridleifsson and Freeston (1994) researched the advancement with the geothermal vitality as per encounters from around the world, and are required to full speculation costs from 15 for you to 20 billion dollars in geothermal vitality worldwide over the next decade. Notwithstanding, this assessment enhanced five times in the IEA report, which appraises that the aggregate speculation expenses to get 104 billion dollars worldwide between 2010-2020 so you can meet the destinations perfect (IEA, 2007). Fridleifsson practice and Freeston, who view self as adequacy geothermal condotions diverse time furthermore capacity limit, geothermal vitality, could well be utilized by every base furthermore crest burden force plants. They guaranteed to which natural issues made by the freedom of steam furthermore gas, and additionally challenges from the stream, it could be affected by cutting edge building. To make utilizing innovation of geothermal vitality accessible in the nations that procure, troubles of deficiency of subsidizing and comprehension of the framework should be determined. Murphy and Niitsuma talk about techniques to balance the higher expenses of geothermal power as a result of attractive natural advantages. That they proposed actualizing financial approach measures to fortify the feasibility, including the usage of a carbon as opposed to weight and adaptation of some incredible advantages of geothermal vitality (Murphy & Niitsuma, 1999). Prompting Okazaki, japan, Indonesia and the Philippines as nations around the globe with high rates of geothermal increment, halfway because of bolster arrangements founded by its overseeing figures. Reif, T. Utilizes a factual and information used in Iceland to help assess the speculation expenses taught to manufacture an innovation geothermal quality plant. The aggregate expense of procurement in geothermal vitality establishments inside of the scope of 20-60 MW, evaluated for 1 267 USD/ kW in an endlessly more known field and 1 440 USD/ kW in an obscure field(Reif, T., 2008).b Lund and is the reason. Assessed the general programming of geothermal power with information from 72 nations. Their outcomes demonstrated that the utilization of cement 273, 372 TJ (terajoule)/ year regarding geothermal vitality, and expecting uncover 6. 06 × 109 J of one's for each clasp or barrel of oil could spare 128. 9 trillion barrels of oil or 19. 2 million tons for every year. They additionally contended that extra staff may be a sparing of 2 million barrels (41. 6. 2 million tons) contrasted and fuel oil and 7, 000, 000 heaps of carbon outflows if the utilization of geothermal warmth pumps is alluded to as amid the cooling mode. Frick connected a specialized examination in the life cycle for enhanced geothermal quality era (EGS) having low temperature stores that will evaluate the impact of the land conditions inside of nature framework (Frick, & Schröder, G., 2010). The outcomes demonstrate the way that life cycle of a double geothermal force plant is subordinate upon the materials and vitality inputs. Consequently, the effective access in your base store boring is overweight to have minimal ecological effect. There is confirmation that maybe less positive geothermal warmth and power can bring about a procedure of vitality alongside cnhance solidness. In a comparable study and Saner examined the essentialness utilization and outflows of nursery gasses after which you can connected an existence cycle appraisal to take a gander at the ecological effects as a result of area source savings introduced contrasted all together with traditional warming strategies. Purkus furthermore Barth (2011) explored the German dialect geothermal segment, with expand exposure from the significance associated with political assist, in addition to the general conditions from the electricity market (Purkus and also Barth, 2011) .They argued that despite the disadvantages associated with high investment costs almost no risk of heat from geothermal engineering, the advantage associated with flash could not allow him to become reliable base-load power. Kaya et is the reason. (2011) investigated the reinjection in geothermal fields with data via 91 plants (worldwide) found which geothermal and reinjection plan should reduce the possibility of contamination of groundwater. Results showed the response of different strategies for geothermal reservoirs reinjection determined through geothermal technology. In the most recent study, now the status from the (global) geothermal energy and found the high temperature technologies (Flash, plus dry steam) generation technologies incorporate most developed geothermal energy. These folks grow define four models, as well as 1FMP (single flash) 2FMP (double flash), 3FMP (triple flash) plus DSMP (dry steam), to manage various geothermal techniques. The results showed the DSMP model receives the best amount NPV (1 013 united states dollar 6, 000) and the difficulty TIR (22. 8%). In addition, it is estimated that the expenses associated with electricity to turn into a model (Nawaz, & Tiwari, 2006).

3.5 Hydropower

Hydropower makes use of energy which involves far more water to reduce elevations, especially to reach electricity. Pursuits include hydroelectric dams together with reservoirs river management, in addition to current jobs, as well as protecting a procession in the dimensions of the project. This assortment gives hydropower to meet up with urban desires large centralized in addition to decentralized non-urban desires. Hydro technologies are generally mature. Hydroelectric things to do ranging manipulate a company temporarily. By comparison, the controllable output given by hydroelectric facilities that may have reservoirs allows you to meet maximum power demands and turn into a help to balance the particular electrical systems that may have large portions relative age group RE variable. The function on hydroelectric reservoirs commonly refl ects its many works by using, for illustration, water, irrigation in addition to Drought flaood control and navigation, as well as thoroughly clean energy present(IEA (2010).

3.6 Ocean Energy

Water power leads to the potential, kinetic, chemical energy along with power involving sea water, along with this really is transformed to provide the electricity, heat, or perhaps drinking bottled water (Charlier, 2003). A great wide variety of technologies will be achieved, like dams respect tidal range, sub turbines for you to ocean currents as well as tides in addition heat exchangers for its conversion of thermal energy associated with water, and also a great menu of models to help harness ones vitality involving waves as well as salinity gradients. Water technology, from the exception of the genuine tidal dams is actually near your own phases of an pilot in addition to demonstration project, and also requires quite a few additional research in addition to development. A number of any technologies have changed profi energy production l'ensemble des within some other degrees of predictability (eg, waves, tides and also the current selection), whilst others can give almost constant operation or maybe possibly controllable for such as ocean thermal as well as salinity gradient (ETSAP, 2010).

3.7 Wind Energy

Wind power uses the specific kinetic energy in the stirring air. The main request of significance to help climate change mitigation ought to be to create electricity from large wind generators located on the property (land) or perhaps fresh seasonal or perhaps (offshore) water. Onshore wind technology was being made and deployed a huge scale. Offshore wind power technologies will probably continued technical progress. Wind power will be variable along both equally, to some quantity, unpredictable, but experience with comprehensive studies as of many provinces has demonstrated that the incorporation connected with wind energy normally involves not pose insurmountable barriers approach (Blanco, 2009).

3.8 Mutual Heat and Power

Consolidated warmth or maybe joined with force (CHP) is the system of warmth and power together. It is required to get an extensive pick up in effectiveness for every source independently. All force dispersion organizations novel power easily to manage or sauna. For right around 30-40% of the aggregate vitality heap of any nation pays for warming, cogeneration is a proficient utilization of fuel every time a piece of the imperativeness is removed, while the residual heat. Capture some or all residual energy as a byproduct involving heating. In Reykjavik end users can easily invest in both branches, together with the thermal energy into energy (Tester, 2005). Undoubtedly one of cogeneration can be particularly cogeneration unit Avedore, Denmark multifuel grow (NGOs and Natowitz, 2009). Shipley ET age. (2008) estimated the increase in cogeneration capacity America and 20% in 2030 simply cause a reduction 5. 3Quads (quadrillion British temperate units) and the use of only one of 848 MMT fractional laser laserlight. Depending on the results, America more than 1. 9 Quads of energy resource consumption and emissions of 248 million metric numerous CO2 by using Generation Company saves. Based around the WEO (2012) Registration, the average efficiency of the vegetation of energy will be a 41% worldwide, with practically 60% of the primary energy is converted to heat loss (IEA, 2012E). CHP could extensively alter the residual heat standard in a positive financial significance for professional technical or business waste and heating constructions. It is anticipated that the new elements of cogeneration could help the energy efficiency of its highest and not 85%. (Madiment and Tozer, 2002) investigated the effective use of the combined temperature and cooling power (CCHP) for supermarkets in England and directly compared with the energy content saving price label / capital while using conventional engineering. The results suggest that CCHP provides a significant level of primary energy despite the reduction of CO2 released compared to conservative schemes, but also intended to compete with the most useful technologies regarding long help. They argued that innovative technologies such as fuel cells can deliver updates over the efficiency of the CCHP force belonging to refrigeration. Hawkes and Leach looked profitable trading strategies associated with three alternative micro-cogeneration engineering such as Sterling engine, gas engine with the fuel system of solar panels dependent solid oxide (SOFC) for residential use in the United together. They assessed the financial and ecological attributes of the technologies mentioned above led by the heat, electricity brought with techniques minimum load operation (Leach, 2007). The effects showed the greatest reduction in CO2 emissions following approach cheap operating costs were good high SOFC cost computer, the same. An individual considered this export volume in relation to electricity aggregates reveal the micro-cogeneration unit as a virtual power plant (VPP) through participation in the energy of low-cost sector. Thcy found that the export capacity connected with micro-cogeneration systems depends largely around the technical parameters, energy prices along with affiliate demand profile. Based on the real model applied, the marginal price was conjectured with just a small cogeneration system is above the placement price of high season. Kiviluoma and Meibom (2010) studied the impact associated with variable power generation through wind turbines and electricity stored in electric vehicles the opportunity to improve the elasticity of its influence grating. Using the results, cogeneration units could actually be viable to produce more elastic in terms of invention techniques and using heat energy options. Is investigated this contribution Christidiset high temperature storage to boost cogeneration units in liberalized areas of energy, the implementation of particular design which measures the ability of custody of probable heat and optimal tax. They concluded that the separation of electricity production and heat stipulate could make available a stretch of cost recovery time for storage devices within the proposed program of energy (Tsatsaronis, 2012).

CONCLUSION

The actual improvement regarding climate change will be simply just regarding the main challenges of the last century. Transition to help the single end user exactly who supports the higher power green global energy program incase be a great integral section of a reply involving humanity to live. Your chapter offers a great tick's basis for these types of chapters generally transition. The particular a complete in addition to reliable contribution to the debate at whether or not renewable energy can easily solve your climate problem with a financialally attractive method. That is a model intended for further development of the renewable energy sector, as well as absolutely establishes the role within minimizing climate change everything. Ongoing queries about climate change have intended renewable energy a switch towards the overall collection of world energy consumption component. Technology regarding renewable energy might reduce CO2 emissions from replacing fossil fuels fractional power generation industry and the transport sector. Because of its negative in addition tO irreparable externalities at the production associated with conventional energy, which is to be vital regarding development, along with promote renewable energy technologies provide. power generation while in friendly energy sources on the environment must end up being updated to reduce the unit cost associated with energy and also make sure these are generally compatible using a competitive substitute regarding conventional energy sources. A couple main solutions might be implemented to lessen CO2 emissions and to overcome the particular dilemma involving climate alter: the substitution of fossil heating system by alternative energy sources regularly and strengthen energy performance, no matter what kind. In this specific chapter, we should instead be, wind flow, geothermal along with hydroelectric sun places, due to important contribution in the energy generated through the production involving renewable energy and gives sources. Renewable is increasing on the global level. Following extreme increase in oil prices as well as effects about the prices involving coal along with propane, the lots of investment it has recently been made in the recent past on alternative energy. These advancements in technologies have permitted countries to generate renewable power in larger plus more profitable levels.

Due to the negative along with irretrievable externalities linked to conventional energy taking out along with consumption, you will need to promote along with develop option vitality produce and usage. The IEA expects developing a positive growth involving alternative power means. They become substitutes related to fossil stimulates and decrease emissions. At any time, some option technologies might not be comparable that you can conventional stimulates, the area of manufacturing costs along with transmission, but may very well be comparable in the case we look at the many beneficial externalities, for example the environment along with social effects. Also, understand that should turn out to be financial regions of degree can play an essential role in reducing the cost of manufacturing program. Costs along with transmission technology and presentation will not differ considerably from typical and alternative energy options. We have at the moment presented detailed home elevators key technology of option development energy supply, such as hydro, the wind, solar, geothermal as well as length. The emphasis may be on latest production potential, in addition on the estimated, along with development expenses sink potential. Hydropower may be the largest source of renewable energy for energy generation throughout the world. Despite its important factor of energy generation, its growth is difficult on account of high expense costs as well as relocation in the initial set cost in the environment and the population. Hydro power is attractive as a result of combined way to obtain water with regard to agriculture, people, and recreation along with industrial employ. Moreover, you will keep water along with energy that can be used for inserts electrical power generation basic and high two. Having access to financial, politics and market risks, the focus with the allocation connected with resources along with ecological neighborhood concerns are viewed as obstacles to the development connected with hydropower potential. The capacity with the installed blowing wind power in addition has been growing, especially in countries just like China, the USA all, Germany and Denmark. Advantages connected with wind electrical power plants including contracts along with turnkey installing within a short period of period, lower when compared with nuclear along with hydropower investment decision, economies connected with production in lean muscle, lack connected with fuel prices and low operation along with maintenance prices. The issues connected with the aid of intermittent blowing wind energy contain wind electricity and added cost for that transmission of energy to buyers. Generation cost is determined by the location, viability about the minimum speed required for wind generators. China is promoting a volume of only solar panel technology, decreased era load as a result of availability connected with cheap labour and the general public subsidies. Geothermal energy has been used to be able to record more than bathroom, home heating and cooking food. The geothermal gradient, as well as the rock permeability app determines their particular financial viability. Unlike wind along with solar, geothermal solar panel is consistently available over the year, despite the technological know-how has some adverse environment effects. Improving electricity efficiency is surely an important method to reduce electricity use and as a consequence CO_2 emissions, so that you can overcome neighborhood problem rescheduling. We outlined the state of the art methods connected with technical along with financial viability within the usage of green electricity, plus to be able to use along with combined replacement inside the first part of this evaluation. In the particular latter part of energy efficiency technologies are discussed. The efficiency connected with electricity networks is sccn at distinct levels, including power creation, transmission, demonstration and consumption. To do this, different engineering are Energy Efficiency, as well as electric vehicles, combined temperature and electrical power, virtual electrical influence plants. Each one of the technologies discussed in depth and when compared their activities. This technological innovation was critical of green energy and efficiency provided complete and useful information which they can use in doing the determination from the different actors in the present rapid current market development. Each technological innovation has its advantages and disadvantages vary through location, availability, ease connected with technology suppliers, financial restrictions and environmental items to consider. Each metropolis, region or even nation has different first conditions in which conclude the mixture of action that may produce less costly costs, while minimizing the damage a result of the atmosphere. Therefore, there is absolutely no single solution for all energy must have and the situation but a remedy of ideal location between some achievable renewable options.

REFERENCES

Andersen, P. H., Mathews, J. A., & Rask, M. (2009). Integrating private transport into renewable energy policy: The strategy of creating intelligent recharging grids for electric vehicles. *Energy Policy*, *37*(7), 2481–2486. doi:10.1016/j.enpol.2009.03.032

Asif, M., & Muneer, T. (2007). Energy supply, its demand and security issues for developed and emerging economies. *Renewable & Sustainable Energy Reviews*, *11*(7), 1388–1413. doi:10.1016/j.rser.2005.12.004

Benitez, L. E., Benitez, P. C., & Van Kooten, G. C. (2008). The financials of wind power with energy storage. *Energy Financials*, *30*(4), 1973–1989.

Bain, R. L. (2007). World Biofuels Assessment, Worldwide Biomass Potential: Technology Characterizations" NREL/MP-510-42467. *National Renewable EnergyLaboratory, Golden, CO, USA*, *140*, 144–150.

DeMeo, E. A., & Galdo, J. F. (1997). *Renewable Energy Technology Characterizations. TR-109496.* Washington, DC: U.S. Department of Energy and Electric Power Research Institute.

EIA. (2009). *Energy Consumption by Manufacturers—Data Tables*. EIA.

McGowin, C. (2008). *Renewable Energy Technical Assessment Guide. TAG-RE: 2007, Electric Power Research Institute*. Palo Alto, CA: EPRI.

Lund, J. W. (1995). Onion dehydration. *Transactions - Geothermal Resources Council*, *19*, 69–74.

Radeckas, B., & Lukosevicius, V. (2000). Klaipeda Geothermal demonstration project. In *Proceedings World Geothermal Congress 2000.*

Reif, T. (2008). Profitability analysis and risk management of geothermal projects. *Geo-Heat Center Quarterly Bulletin*, *28*(4), 1–4.

Purkus, A., & Barth, V. (2011). Geothermal power production in future electricity markets— A scenario analysis for Germany. *Energy Policy*, *39*(1), 349–357. doi:10.1016/j.enpol.2010.10.003

Frick, S., Kaltschmitt, M., & Schröder, G. (2010). Life cycle assessment of geothermal binary power plants using enhanced low-temperature reservoirs. *Energy*, *35*(5), 2281–2294.

Fridleifsson, I. B., & Freeston, D. H. (1994). Geothermal energy research and development. *Geothermics*, *23*(2), 175–214.

Murphy, H., & Niitsuma, H. (1999). Strategies for compensating for higher costs of geothermal electricity with environmental benefits. *Geothermics*, *28*(6), 693–711. doi:10.1016/S0375-6505(99)00018-8

Nawaz, I., & Tiwari, G. (2006). Embodied energy analysis of photovoltaic (PV) system based on macro- and micro-level. *Energy Policy*, *34*(17), 3144–3152. doi:10.1016/j.enpol.2005.06.018

Chang, K.-C., Lin, W.-M., Lee, T.-S., & Chung, K.-M. (2011). Subsidy programs on diffusion of solar water heaters: Taiwan's experience. *Energy Policy*, *39*(2), 563–567. doi:10.1016/j.enpol.2010.10.021

Han, J., Mol, A. P. J., & Lu, Y. (2010). Solar water heaters in China: A new day dawning. *Energy Policy*, *38*(1), 383–391. doi:10.1016/j.enpol.2009.09.029

Harvey, L. D. D. (2006). *A Handbook on Low-Energy Buildings and District-Energy Systems: Fundamentals, Techniques and Examples*. Sterling, VA: Earthscan.

IEA. (2007). *Renewables for Heating and Cooling – Untapped Potential*. Paris, France: International Energy Agency.

Zhang, X., Ruoshui, W., Molin, H., & Martinot, E. (2010). A study of the role played by renewable energies in China's sustainable energy supply. *Energy*, *35*(11), 4392–4399. doi:10.1016/j.energy.2009.05.030

Gevorkian, P. (2012). *Large Scale Solar Power Systems: Construction and Financials*. Cambridge University Press.

Gipe, P. (1995). *Wind energy comes of age* (Vol. 4). Wiley.

Gordon, J. (1987). Optimal sizing of stand-alone photovoltaic solar power systems. *Solar Cells*, *20*(4), 295–313.

Blanco, M. I. (2009). the financials of wind energy. *Renewable & Sustainable Energy Reviews*, *13*(6-7), 1372–1382. doi:10.1016/j.rser.2008.09.004

ETSAP. (2010). *Marine Energy Technology Brief E13 - November, 2010, Energy Technology Systems Analysis Programme*. Paris, France: International Energy Agency.

Charlier, R. H. (2003). Sustainable co-generation from the tides: A review. *Renewable & Sustainable Energy Reviews*, *7*(3), 187–213. doi:10.1016/S1364-0321(02)00011-4

Kerr, D. (2007), Marine energy. *Philosophical Transactions of the Royal Society London, Series A (Mathematical, Physical and Engineering Sciences)*, *365*(1853), 971-92.

Ngô, C., & Natowitz, J. (2009). *Our energy future: resources, alternatives and the environment*. Wiley. doi:10.1002/9780470473795

Maidment, G., & Tozer, R. (2002). Combined cooling heat and power in supermarkets. *Applied Thermal Engineering*, *22*(6), 653–665. doi:10.1016/S1359-4311(01)00117-X

Hawkes, A., & Leach, M. (2007). Cost-effective operating strategy for residential microcombined heat and power. *Energy*, *32*(5), 711–723. doi:10.1016/j.energy.2006.06.001

Christidis, A., Koch, C., Pottel, L., & Tsatsaronis, G. (2012). The contribution of heat storage to the profitable operation of combined heat and power plants in liberalized electricity Markets. *Energy*, *41*(1), 75–82. doi:10.1016/j.energy.2011.06.048

Chapter 6
Solar Powered Smart Street Light with Maintenance Service System

Thinaharan Ramachandran
Segi University, Malaysia

Vasaki Ponnusamy
Universiti Tunku Abdul Rahman, Malaysia

Noor Zaman
King Faisal University, Saudi Arabia

ABSTRACT

Advancement of technology brings advantages to human race to certain extent. Unfortunately, some of the technological inventions not only brings advantages, but also disadvantages. One of the disadvantages is the pollution of carbon emission in the environment from the non renewal energy resources. Therefore, to prevent pollution from getting worse, the environmental-friendly aspect is suggested to be included in new inventions nowadays. Thanks to the invention of solar panel, solar energy is able to become a renewable energy source that can be applied to various applications. Considering the increase of electricity consumption every year in Malaysia, solar powered street lighting system is proposed in order to reduce the burning of fossil fuels to generate electricity, hence reducing air pollution. The proposed street lighting is companied with automatic switching mechanism to further reduce the electricity consumption of street lights. The system will also be able to provide energy measurement to determine the efficiency of street light. Security features like motion detection and camera module interfaced to the street light for surveillance monitoring in a residential area to overcome intruders and crime such as theft. Theft is now major issue in Malaysia and in many developing and developed countries. The news reported that many people lost their belongings due snatch theft and averagely 16 people go missing every day. Overall design of the solar powered street lighting systems with the security features will help people live in green and protected environment. The implementations of client server connection and Remote Method Invocation (RMI) in wireless network system ease the monitoring process of the entire street lights in a residential area for giving continuous service to the community and environment.

DOI: 10.4018/978-1-4666-9792-8.ch006

INTRODUCTION

Street light is necessary lighting system in many urban areas and in connecting highways. People have difficult vision on the road while driving during night time without street light. In certain areas, street light supports the businesses like night stalls and for people enjoying the night lives and view in cities and parks. Street light symbolizes of safety for many of us because walking at night in brighter place is safer than in a darker place.

There has been decades, where the existence of street light in urban areas. The first most modern street light implemented in Austrian Empire in 1853 (Scott, 1904). The street light is powered with the Kerosene but after some time the carbon arch lights substituted due to the risk of kerosene. The kerosene type of street light caused fire incidences in many places. For example in the year 1880 in New York, two of every five fire incidences are caused by kerosene powered street light (Caponetto, 2008). The carbon arc light has both the carbon-made electrodes and alternative current (AC) power in equal rates. The combination of carbon and AC powered in street light is the most widely used electric type street light and it is named as 'First City of Light' in Tamworth, New South Wales in Australia due to the first location to implement the arc lights as the city's first street lights in 1888 (Liu, 2009). Although the carbon arc street light substituted with Kerosene Street light but it still has its disadvantages. Commonly it is not user friendly because the energy consumption of this type street light is poor therefore high usage of human resource to replace carbon rods, which works for very short period of time. The invention of incandescent light bulb an evolution takes place in street light design. The incandescent light bulb does not required a carbon electrode, just by heating filament in a high temperature with electric current, incandescent light bulb produces light without carbon electrodes. However the incandescent light bulb solves the insufficient of energy consumption in the carbon arc street light but 5% of energy used in visible light, which reduce the efficiency. Therefore, High Pressure Sodium (HPS) lights introduced in a street light design and it is widely used in many street light systems (Jherniak, 2012). HPS has high photopic in a least power consumption, which solves the inefficiency issue and hence making it a more reliable than the previous type of street light.

Today, most everything has build in wireless module to connect between things and the users conveniently. Mostly wireless communication service applied in SCADA telemetry control and process control system. The remote control device with infrared enable controls and establishes synchronization between devices in a short range. Global Positioning System (GPS) uses satellite to locate position in a globe. It is now clear that wireless communication changed the impression of telecommunication which is used to be difficult to install, expensive and sophisticated for establishing connection between any devices, human or equipments. Not only that, the wireless communication system changed people's lifestyle and assured that the possibility to complete a task without extensive travelling. Therefore the wireless communication provides another opportunity for making street lights connected in a network structure for the purpose of managing and improving maintenance service which currently unavailable. Researchers have worked out few frameworks to have street lights connected to wireless communication medium like GPRS or GSM but the implementation costing will be burdening the city municipal council. Nodes are devices with the control mechanism and connectivity medium. Control nodes are implemented with the variety of wireless links such as IEEE 802.15.4, low-power IEEE 802.11, or IEEE 802.15.1 (Bluetooth), which give limited radio range but accepts large number of nodes in a cluster via suitable

Figure 1. General view of WiFi for smart street light

routing protocols' (Galperti, 2008). Zigbee (IEEE 802.15.4), WiFi (IEEE 802.11), and Bluetooth (IEEE 802.15.1) are suitable wireless communication technology for connecting things in an urban area than using GPRS or GSM transmission.

In this project proposal, as shown in Figure 1 WiFi technology will be implemented in the solar based smart street light design with remote service management system. Besides that, the project also involves the design of street lighting system. The design of the smart street light system comprises of three major aspects which is the implementation of smart street light control system, the use of an environment friendly energy resource and to provide a surveillance security system. The microcontroller connects and controls sensors, voltage and current measurement. The controller will control the camera and the motion detection for monitoring environment in a residential area against theft, crime or malicious attacks. The overall idea of this project proposal is to implement street lighting system with sustainability, consistency and reliability.

PROBLEM STATEMENT

Smart street light system in Malaysia is designed based on previous reliability standards and does not get benefit of latest technological developments. The unsophisticated control mechanism of a street light system can incur cost to the electric board for neglecting the street light lamp turned on during the day time. HPS light is not appropriate for night lighting because white light source creates double driver peripheral vision and increase driver brake reaction time at least 25% (al). Therefore, use of LED light for the source of light can be considered. In between year 2000 and 2009 there is 20% increase of electricity production in Malaysia and it is projected that there will be extra 3% of increase every year (Leonowicz). The demand will increase more fossil fuels to produce electricity. Burning fossil fuels like oil, coal and gas will produce by-products such as $CO2$ and $SO2$. These by products cause air pollution when released into the air. Substituting renewable energy will prevent air pollution and also help to sustain the increase of electricity demand in Malaysia. Malaysian Institute of Road Safety Research reported that road accidents are caused by malfunctioned street lights (Bayar, 2011). The present street lighting systems in Malaysia do not have an automatic report mechanism to notify malfunctions in the street light. Therefore appropriate maintenance and report service must be implemented to provide consistent rectification service immediately.

OBJECTIVES

- To implement solar panel as the primary source energy to recharge street light DC battery.
- To conduct efficiency monitoring of street light based on energy measurement from the solar and battery.
- To implement a security system using the motion detector and camera to report for unforeseen situation in the surrounding area like theft and intrusion.
- To initiate a remote method invocation registry service for each street light to ease the process of notifying the status of each street light that connected in network structure.

FUNCTIONAL REQUIREMENTS

- The system must able to check the LED light status either in on operating or non-operating mode from the current sensor that connected to the.
- The system must switch on or off LED street lights automatically based on the environment intensity measurement.
- The system must be able to check voltage from the solar panel to ensure sufficient voltage produced to charge the DC battery.
- The system must able to check the battery voltage to ensure sufficient storage available to use during dark time.
- The system must trigger the motion sensor when there is suspicion intrusion and simultaneously trigger camera to capture image of the suspicion.
- Remote Method Invocation (RMI) is established to register the individual street light in a network in its registry. Each time a health check required for each street light the RMI is invoked to obtain the data from the server.
- The system must create a health log file with the contents of report on the status of LDR, motion sensor status, current sensor, storage battery voltage, solar voltage and lamp post code. The log file is send over WiFi wireless connection medium to the central operation unit for monitoring the individual status of the street light.

NON-FUNCTIONAL REQUIREMENTS

- The system must maintain consistency and sustainability of the street light via remote maintenance service.
- The system to operate in low power energy consumption using renewable energy resources.
- The system must connect in expandable wireless distributed network.

LITERATURE REVIEW

Immanuel Schweizer, Nils Fleischhacker, Dirk Bradler, Max Muhlhauser and Thorsten Strufe (al I. S.) Proposed an approach called Solar-aware Distributed Flow (SDF) to maximizing the sampling rate in solar wireless sensor networks. SDF allows each node to predict the harvested energy and compute a sustainable flow and control its local neighborhood. Hence, feasibility can be achieved in solar sensor networks as SDF approach is able to overcome energy constraints, which is a hard bound on sensor networks. The notion of the energy consumption rate is also introduced to allow perceptual operation. SDF achieved over 80% of the theoretical optimum based on extensive simulation.

(Galperti, 2008) proposed Maximum Power Point Tracker (MPPT) circuit that is specifically designed for wireless sensor network. MPPT is a power transferring circuit that conveys solar energy into rechargeable batteries for the network. Therefore, wireless sensor network that deploys MPPT no longer rely on battery (which is a finite resource) as energy supply. An ad hoc adaptive algorithm is introduced to maximize energy transfer from the solar cell to the batteries, thus keeping the MPPT electronics in optimal working point and achieve high efficiency. The implementation is particularly effective even in critical weather conditions. As MPPT is designed for low power operations, experimental results showed that high energy transfer rate is generated when the cell is in shadow or the weather is cloudy, where existing solutions for wireless sensor network nodes do not produce energy in such situations.

(al T. V.) Proposed to utilize solar power in wireless sensor networks and extend LEACH (a well-known cluster-based protocol) for networks to become solar-aware. The aim of this protocol is to extend the life-time of the sensor network by preferably choosing solar-powered nodes to perform the energy intensive task if being a cluster head. Simple heuristics that has shown good results in test cases is used to choose cluster head. The simulations showed that making LEACH solar-aware increases the lifetime of a sensor network in typical scenarios. A handover scheme that allows changes of cluster heads during steady-state phase of the LEACH protocol was also presented and evaluated for solar-aware version of the centralized LEACH protocol.

(R, 2013) Proposed GSM based autonomous street illumination system for efficient power management. GSM is used as the communication medium between the concentrator controller and control center. Microcontroller C8051F350 is installed on each street light to control various sensors and GSM SIM300 module is installed for wireless communication. The main challenge of hardware is to produce low cost and tiny sensor nodes. Power theft control was also proposed to be added into the system.

(D, 2013) Proposed that the wireless streetlight control system with Jennic Wireless Microcontroller JN-5139 which is an IEEE802.15.4 based wireless microcontroller as communication between street lights and monitoring software. The duration of the operation of street lights is taken into concern. The project also aimed to increase frequency band for network nodes to get maximum possible data rate. Result showed that the duration of street lights operation can be monitored through implementing light dependent sensors. The life of lamps is extended due to efficient control of duration of operation. Higher data rate of transmission is also achieved using JN-5139 microcontroller which has higher frequency band up to 2.4 GHz compared to ISM frequency band which has only some tens of MHz.

(De Dominicis C.M, 2011) Proposed an intelligent street lighting and monitoring system can maximizes energy saving and minimizes light pollution as well as taking account of the environment and safety standards needed in all traffic condition. Energy saving is achieved making light source into minimum illumination. Settings can be adjusted to suit various environmental conditions. Light level can be predefined according to requirement provided.

(Mohamaddoust R Haghighat AT, 2011) Proposed a dynamically lighted roadway in which lighting can be adjusted to any three levels (0.2cd/m2, 1cd/m2, and 2.0cd/m2) depending on the amount of traffic, time of day, and weather conditions. 900MHz is used as frequency band of the system wireless communication. This results in lower data communication rate. The evaluation of DYNO (Dynamic Public

Lighting) concluded that low level (0.2cd/m2) can be applied under low traffic volume and favorable weather conditions.

(Riener, 2011) Mentioned that the conventional street light system is installed in places where no passerby available during night time therefore it is a waste of electricity power. Introducing flexible lighting system like light emitting diode with wireless internet connection availability solves the issue of wasting energy unnecessarily.

Wireless communication system should be fast reacting and the overall system design must be reliable to achieve the objective of the concept of smart street lighting. The street light system is design with a dynamic switching based on the pedestrian availability. The system makes synchronize connection with the user mobile smartphone to localize the location of the user at the present. This feature performs like a hidden surveillance system that reports to the respective enforcement security to ensure safety. The system periodically sends the location and configuration information to the smart street light server. The smart street light lamps are controlled by the ZigBee radio device which receives control message from the smart street light server via multi hop routing. This smart street light proposal confirmed that this revolutionized smart street lighting is suitable to be installed in suburban areas with less pedestrian availability. Also this system contributes to the reduction of CO_2 emission because the LED lamp is only turned on whenever required. The outcome of the experiments shows objects like trees obstruct wireless communication between lampposts. In this research, GPS is used for turning on and off the street lamp therefore the interference causing inaccurate global positioning and leads to unexpected lighting effects. The research is said novel because the smart street light system consists of the pedestrian localization information for safety purpose. The researcher proposed LED light because the problem of mercury and sodium steam bulbs has long switching interval time. This bulb take some time to reach the full light intensity and vice versa takes some time to turn off complete even the main switch is off. This issue is now tackled with the LED lamp. LED lamp is benefiting because of low power, short transition time, long life time, durable and comes in multiple colors. LED is energy saving, less expensive and Siemens has given energy contracting scheme to the municipal council to install LED street lamp for free and takes money saved from lower energy utilization. In Budapest, Siemens replaced 33 000 traffic lights with LED's with low installation cost than the saving obtained from the reduced energy consumption. Even in less developed countries, OSRAM has installed houses with free LED lamps and collects payment from the United Nations subsidy, in which the world preparing for the reduction of CO_2 emission. In the process of reducing CO_2 the street lamps should turned off when it is not necessary, because the energy used by the street lights will definitely emits CO_2. The feature of GPS location based services is a novel in this research because the safety of pedestrian is assured while in place of less pedestrian. Low cost radio technology like Zigbee allows individual lamp post to form network structure on a reliable connection for exchanging data. The wireless sensor network in a multi-hop routing enables the wireless network to cover entire city with smart street lamps. The street lamp system composed of ZigBee motes, central server with a master ZigBee configuration, which set as radio base station, server application design with administrator rights to configure, a web application designed installed in the server to administrate lamppost lighting zones and the end user device with the localization capability. Mobile application software is designed to install on user's smartphone who lives in the vicinity permanently to interact with the security system. The smart street light system is designed in a way for user to interact with the street light for requesting more light intensity if the intensity is dimmer. In terms of privacy and security, the interacting user is requested for administration key, which is temporary key for all configuration and the user trigged operation is protected by strong authentication mechanism. The architecture of Smart

Street light is design as expandable and scalable system. The system is possible to extend for mega city wide connection of Smart Street lights. The researcher uses the radial zones around thc user to determine which street lamps should be switched on or off. The radial zone function is easy to implement but the disadvantage is can triggers fault alarm if the target stays around the street lights near a building. The wireless street lamp control demand for a fully wireless communication between users, street lights and the background intelligence. The researcher use Zigbee for wireless connection because the distance between each lamppost is only 30 to 50 m and Zigbee with multi-hop routing capability allows fully connected network structure to access multiple nodes at same time. The system is implemented in a desktop PC with a Linux or a Window plus Cygwin operating system. The requirement is determined by the size of the user location data, facility, number of street lamps. On the application side, the smart street light application is written in Java code supported by Apache Web Server providing internet access from an administrator or end-user devices. The ATMega 128L based Zigbee node is installed on the lamppost. The TinyOS is used for linking the Ethernet interface between the smart street light and server. HTTP request is received from the respective smartphone but the size of the message is not a constraint because it is small in size. The link between the hardware and software is established from the user ID. The researcher has given the user ID as UUID for privacy reason. The database consists of user's safety location in timestamp. The connection of user end device with the server is to only control two main tasks. Provides interface connection between the user and end device and the server for transmitting user position of safety zone. Another task is to assist the administrator interface its functions like lamppost definition from the end device and prepare them to store in database.

The smart street light server connects to the Zigbee nodes to poll spatial database periodically to get the light posts that need to turn on, assemble the corresponding command packets and sends over to the Zigbee nodes. A specific Zigbee node act as base node to establish link between the lamp post. This specific Zigbee encode incoming and outgoing information into Ethernet. The Zigbee on the lamppost have two major functions. The wireless nodes functions as multi-hop therefore each node acts as router or relay station for packets that not designated to them. The other function of the Zigbee is to turn on and off LED lamp with respective brightness. The brightness is increased when the user enters the zone and the light fade once user leaves the zone. The administration application deals with the assigning of position information to lamppost. The administration key is entered and verified for user authentication rights. A successful registration the administrator can add, delete or update street lamps in the database. Administration interface is secured with temporary administration key. All the street lamps are registered using the administrator end device. The administration process maps the physical address of the Zigbee node on the lamppost with its GPS location. Zigbee cannot capture the GPS location therefore need the mapping of physical address with the GPS. The lamppost position is entered manually because the lamppost location is fixed. The location data transmitted to the smart street light server via HTTP connection that stored into a spatial database. Web application is used for maintaining infrastructure zones. The Google Earth KLM file provides sufficient accurate GIS functionality to successfully accomplish this task.

DESIGN METHODOLOGY

The design methodology comprises of hardware and software mechanism required in developing a solar energy smart street light system.

OVERALL HARDWARE CONTROLLER DESIGN

Figure 2 illustrates the architecture of a control system in the smart street light system. The control system architecture has an 8 bits microcontroller unit, rechargeable battery, circuit for recharging battery, lamp post and solar panel. The motion sensor works simultaneously with the camera module. The camera will start capturing the event when there is a trigger from the motion sensor. The purpose of the motion sensor is to send out trigger when there is suspicion subject noticed around the area of the street light. Obviously the placement of the motion sensor is very crucial because placing it in a lower, middle and higher position may not function to trigger as required, therefore experiments will be conducted to evaluate the position of motion sensor in a street light to justify an exact accuracy of trigger. The microcontroller required at least 5VDC to operate and Arduino microcontroller has built in regulator to regulate 12VDC input voltage from an adaptor. The light sensor, (LDR) captures the light intensity of the environment and turns on or off the LED lamp post. The LDR will only sends a high state to the microcontroller to turns the relay circuit to turn on the LED lamp after 7.05pm. LDR sensor functions against resistance and light intensity. An experiment to find the average nominal resistance will trigger relay circuit to turn on the LED lamp at 7.05pm, during cloudy climate or when other causes of the climatic changes. Since SPST type of relay used in this architecture therefore the same relay will switch off the LED lamp in the day time after 7.00am or when the sufficient brightness reduces the prescribed nominal resistance. A hall current sensor measures the current supplied to the LED lamp. If there is high

Figure 2. Overall controller architecture of smart street light system

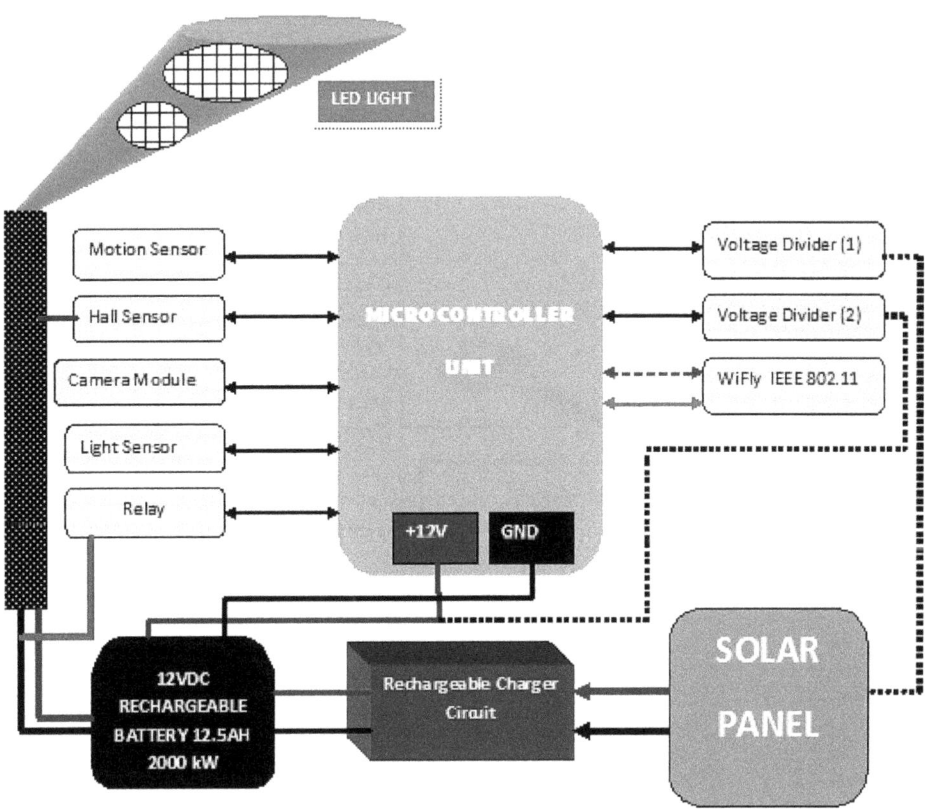

surge current measured in the LED lamp post the controller will determines that there is malfunction of LED lamp and send the record to the remote coordinator unit. A voltage divider circuit that connected to the microcontroller from the storage battery and solar panel measures the differences of voltage from the reference voltage in order to determine the capacity of the storage battery and the performance of the solar panel. The WiFly is an IEEE802.11 WiFi unit that has the similar outlook of a XBEE wireless RF module. The WiFly is the unit for sending in and out the data from the microcontroller RAM to the coordinator for measuring the overall performance of the individual street light. Logs file which created from the firmware code consists of health check profile of a street light. The rechargeable circuit is for charging the rechargeable storage battery. The microcontroller has the feature of 10 bit ADC sampling rate. The hall sensor, light sensor and voltage dividers use the 10 bit ADC sampling rate to determine the current value, nominal resistance and the performance voltage reading from the battery and solar panel.

SMART STREET LAMP OUTLOOK DESIGN

Figure 3 illustrates the installation suggestion of the smart street lamp. The solar panel fixed to the street lamp supplies DC voltage to the storage battery to recharge. Since the rechargeable battery is heat resistance therefore placing the battery on the panel bracket that connects the solar panel does not affects

Figure 3. Physical outlook of smart street light design

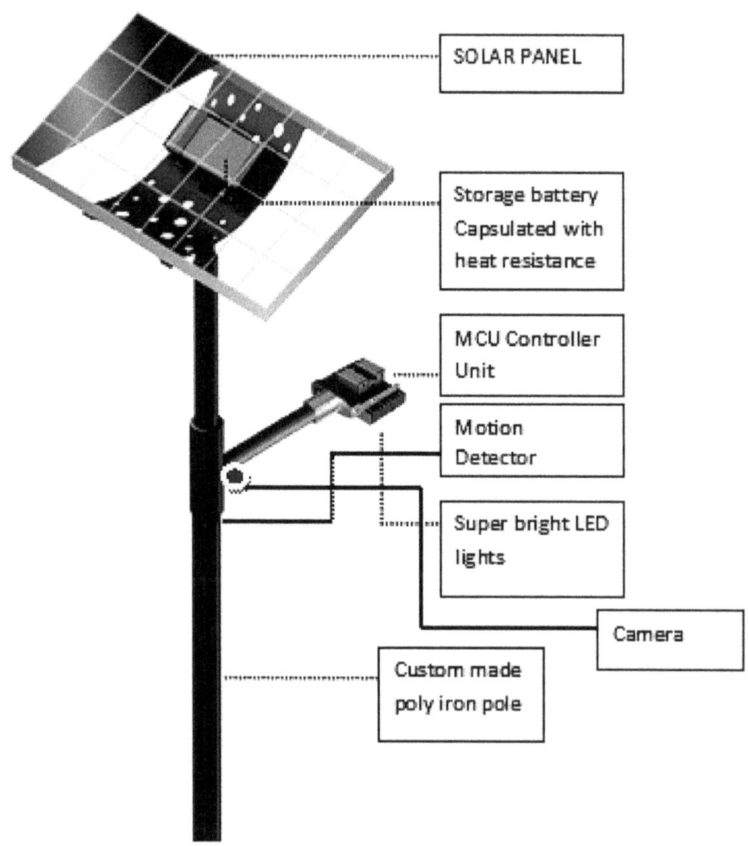

Figure 4. Network configuration of smart street light system

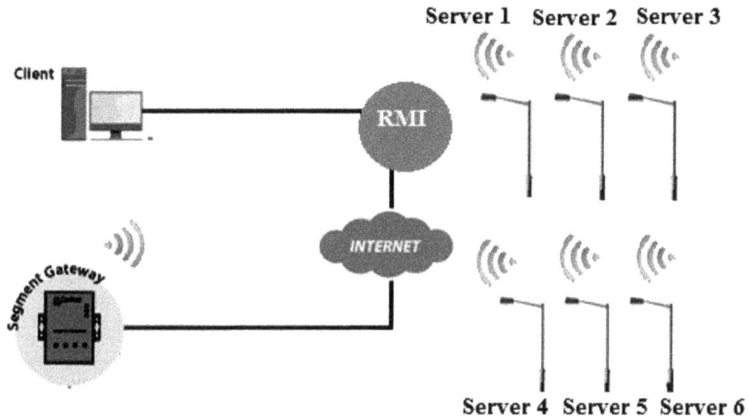

the battery durability. All wires in the street light are concealed in the poly iron rod to avoid damage to the wires from weather condition like rain, mist or hot and wind. The super bright LED is sufficient for achieving the brightness. Weather proof casing will be used for the controller, motion detector and the camera module to avoid damages happen to the hardware components and parts from the unpredicted weather condition.

NETWORK STRUCTURE OF SMART STREET LIGHT SYSTEM

Figure 4 illustrates the client server connection to monitor the health status of each lamp post in a residential area. A Remote Method Invocation (RMI) register register's all the unique street lights that connected in the network structure. Each street light is declared with a separate server with a unique name of the street light. The internet gateway with the wireless enable passes the health information of each street light to the client from the remote method invocation in a JAVA platform. The client will be able to access each server and monitor each street light's health status in a residential area.

SERVICE MAINTENANCE DESIGN

The service maintenance report of the smart street light system is simulated in JAVA platform. To implement Client-Server environment with RMI registry service the following steps are taken as shown in Figure 5.

BRIEF OF REMOTE METHOD INVOCATION

RMI is a distributed system technology that usually implemented in large distributed network system for conveniently receiving information by invocation from a virtual machine to the object methods that run on other virtual machines. This is the best method to handle the individual street light that connected in

Figure 5. Implementation of RMI for smart street light

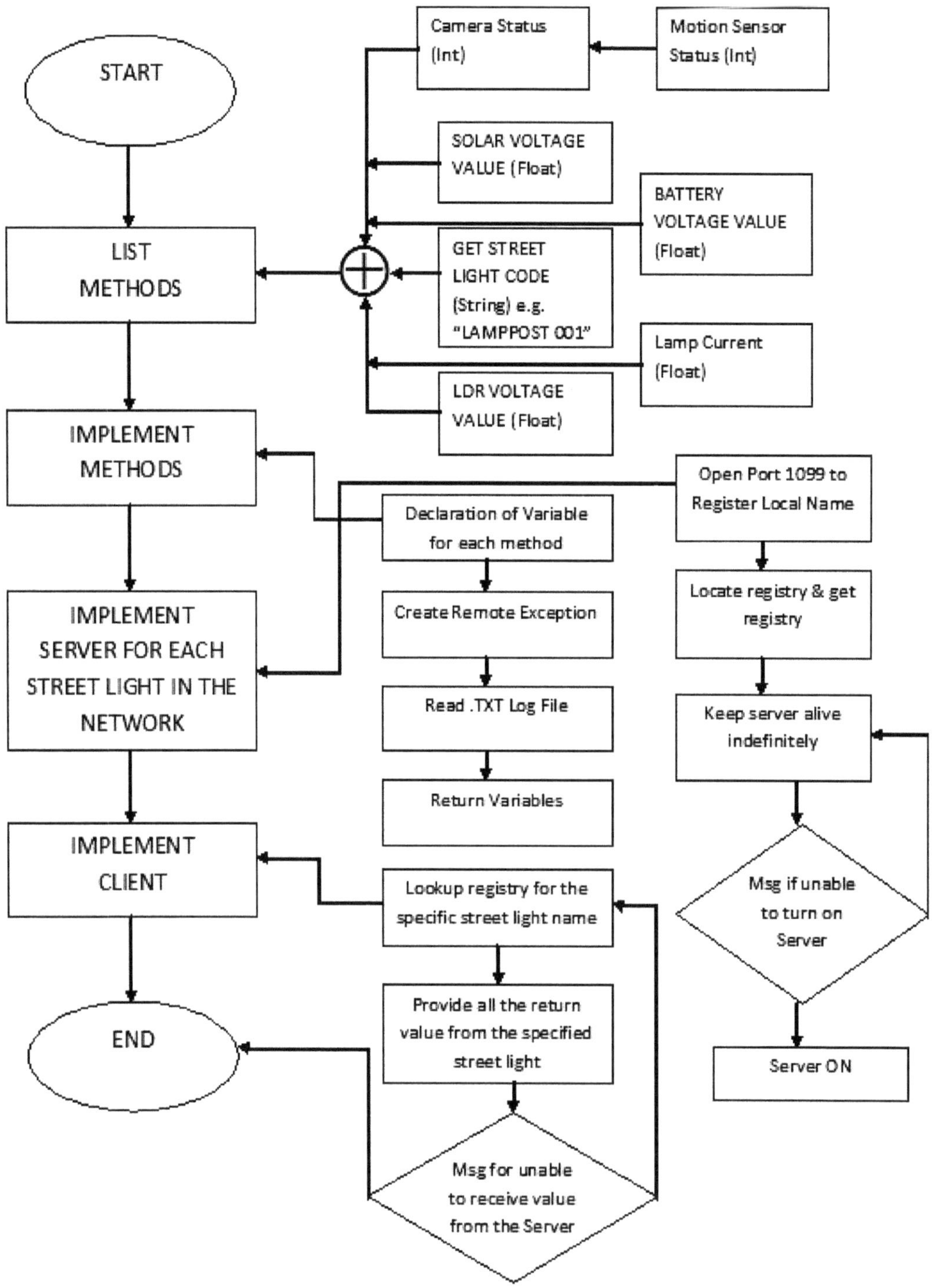

a large distributed network. For instance many street lights are available in residential areas or business centers in urban cities therefore via RMI the object methods are invoked to check the street light health status for further action.

The following are process of establishing RMI:

First an interface must be constructed with the description of the method for allowing client to invoke the methods remotely. Secondly the interface for server is constructed for registering the service with the RMI registry and also to implements security to the JVM. The security is to avoid malicious surprises from the accessibility of remote classes. The third step is to implement the RMI client to call the registry to obtain reference to the remote object. To execute RMI, the RMI registry and servers are turn on and then followed by the client.

In summary the major process to implement RMI is to list the object methods and develop the method's requirements. The server requires RMI registry port accessibility, which is port 1099. Prior invocation from the RMI registry to the object methods the individual server street light will release the information that required for maintenance purpose in condition the servers must be kept alive all time. Once the client connects to RMI registry a lookup table is accessed and scanned to locate the required server for the client to obtain information on the street light health check. Error prompt is set in the program to have feedback of errors in the establishment of RMI in a Client-Server mechanism.

RESULT

The following are result obtained from the Netbean JAVA platform that simulates the concepts of RMI in a Client-Server environment. The dummy data are created and stored in a folder of the same path the project file is created for this simulation project (see Table 1).

The above dummy data is created to resemble actual result that will be collected in log file from a smart street light that connected in the distributed network. For each dedicated server a smart street light unique code is given. The hardware measures the solar panel voltage to determine the ability of the solar panel that charging the rechargeable battery. The battery voltage is measure to determine the sustainability of the battery. The status of camera and motion detector is notified in an integer representation. Logic 1 is notifying that the camera is capturing image because there is similar trigger of logic 1 from the motion sensor to alarm an intruder been detected nearby the smart street light. Both the motion detector and camera has the logic state because the camera will only capture image once there is motion

Table 1. Dummy data to resemble as from the controller unit

Data	Server 1	Server 2	Server 3
Street Light Code	Lamppost 001	Lamppost 002	Lamppost 003
Solar Panel Voltage	12.5V	12.45V	11.45V
Battery Voltage	14.2V	13.8V	13.2V
Camera Mode	1 (HIGH)	1 (HIGH)	0 (LOW)
LDR Voltage	2.56V	1.25V	1.2V
Lamp Current	0.25A	1.3A	1.25A
Motion Sensor Status	1 (HIGH)	1 (HIGH)	0 (LOW)

detection. A zero integer determines that the area is safe and no intrusion happened. The LDR voltage is further processed to determine the light intensity around the area. For instance the 2.56V determines that the light intensity surrounding the lamppost 001 is brighter than the position of lamppost 002 and 003. The lamp current is measured to determine the endurance of LED lamp. If the lamp current reading is higher than 1.35A then the street light should have malfunctioning LED lights.

PROGRAMMING OUTPUT

Figures 6, 7, 8, 9, 10, 11, 12, 13, 14, and 15 show that 3 different smart street lights are used with 3 different servers. Each server is turned on successfully and ready for the client to access via RMI registry. Figure 13 to Figure 15 shows that the client successfully established the connection with the RMI's registry unit and obtained the stored output result of each smart street light's health check information.

DISCUSSION

The proposal is designed to develop a prototype of Smart Street Light System (SSL) for urban places. The SSL design is powered with solar renewable energy to reduce the pollution of CO_2 or SO_2 components that emitted in the air. Proposing of renewable energy resource is the novel in this research proposal but

Figure 6. Switching on smart street light server 1

Figure 7. Status of server 1

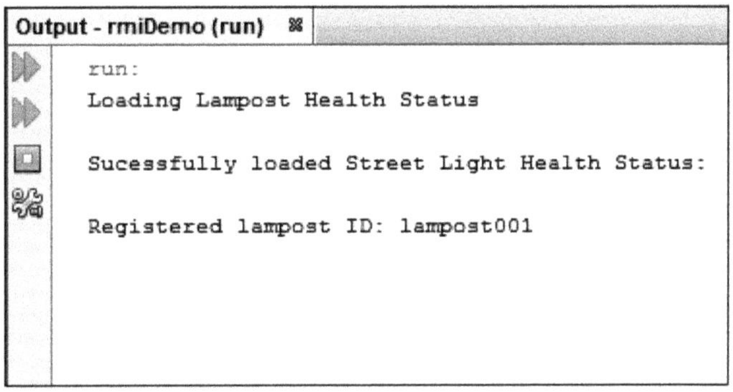

Figure 8. Switching on smart street light server 2

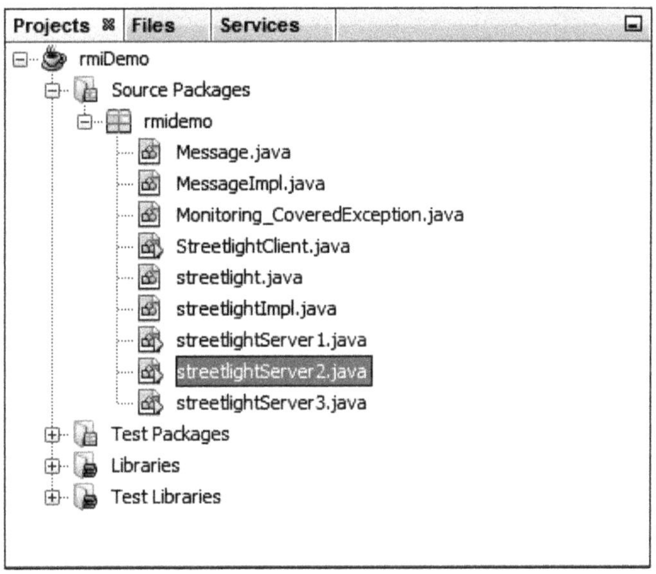

Figure 9. Status of server 2

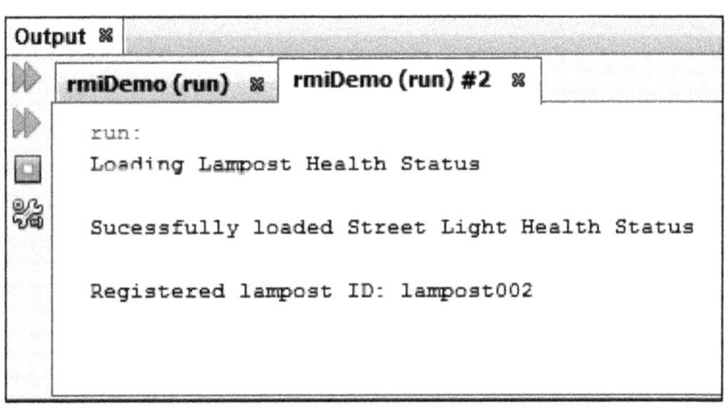

Figure 10. Switching on smart street light server 3

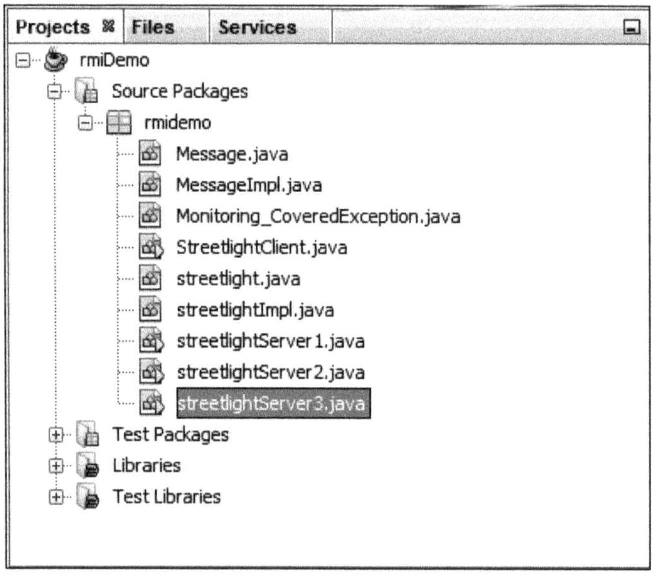

Figure 11. Status of server 3

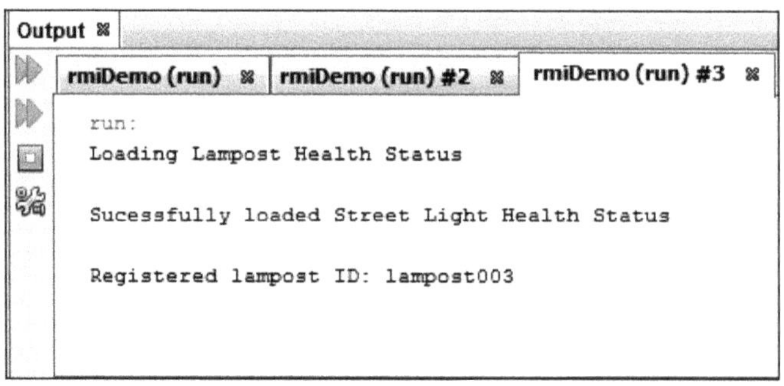

appropriate maintenance scheme must be implemented to maintain sustainability of energy supply to the SSL. The operating mechanism of smart street light system saves energy because the light sensor is set to the LDR nominal resistance with tolerance of 10% to trigger relay circuit for turning on the LED street lamp. Every zone or place has different light intensity therefore it is impractical to set a fixed value for each lamppost. Usually some street lights were found abundantly unattended by the operator to turn it off during day time. The electricity utility board had noted approximately billions spent yearly for having a street light facility throughout the country. Possibly, reducing 20% to 50% of this cost is every helpful.

The rechargeable storage battery is the dependent power used in SSL for night or during darker weather. Measuring the battery capacity on continuous mode will help in managing and servicing a rechargeable battery. The storage battery has the capacity to power SSL for at least two years in continuous if the battery is charged when the battery capacity drops just below than 50 to 60% of overall capacity. This method keeps the battery healthier and long lasting. In this proposal the SSL system comes with a battery charge to charge continuously for 4 to 5 hours daily from the solar power, so that the battery is maintained and sustained.

Figure 12. Switching on smart street light client

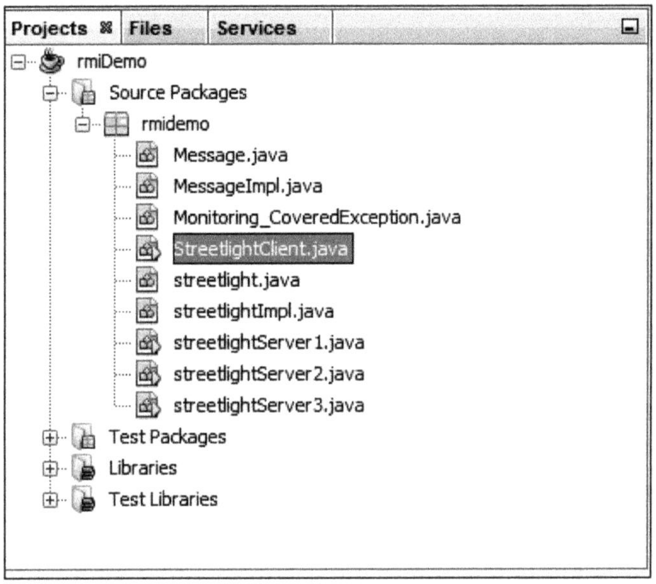

Figure 13. Received record from server 1 of the smart street light

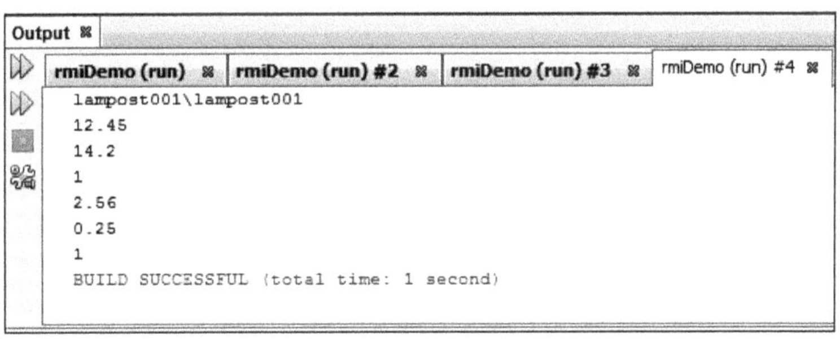

Figure 14. Received record from server 2 of the smart street light

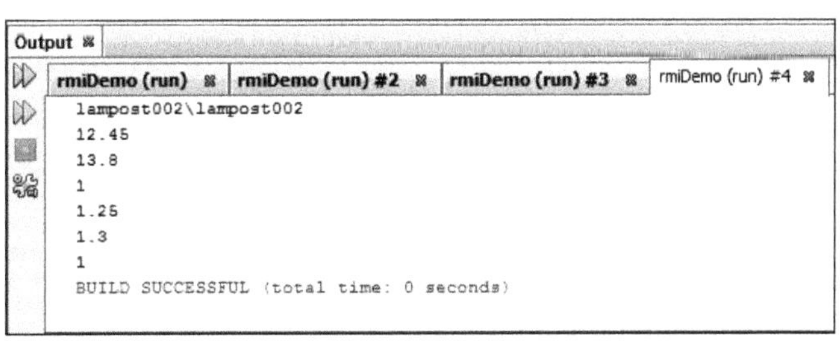

Figure 15. Received record from server 3 of the smart street light

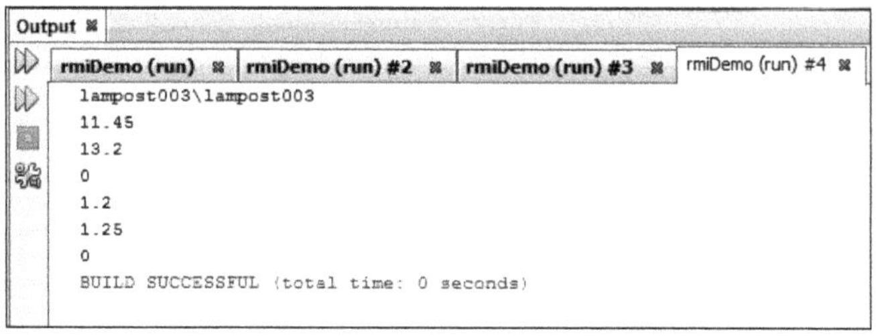

The use of motion detector and camera module simultaneously will help monitor the surrounding area. The trigger from the motion detector will initiate the camera to capture the surrounding and keeps the image in the SD Card, which is an external storage in the microcontroller. The motion detector and camera module rotates around the neck of the SSL pole to trigger and capture any suspicion event around the street light area. The motion detector and the camera are consistently monitored by the maintenance system to ensure residents are protected from theft and intruders.

An 8 bit ATMega microcontroller inside Ardiuno development package with the WiFly shield that can fit the WiFly (IEEE802.11) is sufficient for building the communication system between the smart street light systems. The LDR sensor in the SSL is expected to response once the light intensity at each lamppost has been scanned. The microcontroller GPIO connects to a relay to switch the power from the solar powered to battery. The LDR's voltage drop is measured and sampled in intervals to determine the surrounding light intensity. This value is then sent to WiFly to the remote server to show the status of each lamppost. The current sensor measures the current flow in LED lamp on an interval basis by the current sensor connected to the microcontroller's analogue to digital converter (ADC) input to sample total current flowing to the LED lamp. Once a faulty LED detected, an alert is send from the respective lamppost to the remote monitoring server. The current measurement will ensures that the street light is functioning for 24x7 without issues and faulty LEDs are detectable and repaired in a short time.

The WiFly is configured into multi-hop configuration in order to associate between WiFly in each lamppost till reaches the coordinator WiFly. WiFly is the best transmission device for this application because no RF license is required to operate WiFly module, installation cost is low and easy to replace if any fault encountered. The WiFly has network layer package with security feature therefore data are encrypted before transmission. The only disadvantage of WiFly is the distraction of the geographical factor like tall buildings and trees, which obstructs transmission link between the WiFly's. Appropriate selection of wireless routing protocol such as OSLR or BATMAN will form good transmission between the WiFly's in a MESH network configuration.

To ensure quality of services (QoS) of this system, maintenance health check on certain crucial part of the SSL is obtained to maintain sustainability of solar voltage, battery voltage, the regulation of LDR behavior, status of motion sensor, status of camera and current flow in the LED lamp. The system log file provides the health track record of the lamppost performance in 24x7 hours.

The SSL works on a soft real time processing because for instant voltage drop measurement of the LDR in the lamppost and the packet data transmitted to the coordinator is not a crucial if the transmission delay is not more than a few seconds. Since the aim of SSL system design is to achieve high quality of service, than the real time processing must be functioning like the hard real time processing. The current sensor, voltage monitoring, camera status, motion status and the LDR sensor must be able to process the detection and send in a stipulate time for processing the data for performance analysis of each lamppost that installed. Further to save energy and to handle efficient switching mechanism between the power supply modes the microcontroller must handle data acquisition effectively without delay.

CONCLUSION

The proposal here is to design a smart street light system (SSL) which uses solar power supply to operate the controller, the communication module and to charge storage battery during day time. The charged battery will be used to operate at night. Therefore the system is independent from electric power source. An automation control is required to establish the functions of this system. The software at remote server is used for logging the light intensity obtained from the lamppost via the WiFly module. The bidirectional communication establishes between the remote server and the lamppost enables the maintenance effectiveness of the smart street light. The RMI will ease the maintenance process of the smart street light over remote monitoring. The system can be improved with employing an image processing method to extract image for evaluation purpose.

REFERENCES

Bayar, T. (2011). *Malaysian Explores Its Renewable Options.* Retrieved from http://www.renewableenergyworld.com/rea/news/article/2011/09/malaysia-explores-its-renewables-options

Bettmann, O. (1974). *The Good Old Days They Were Terrible*. Academic Press.

Bullough. (n.d.). Several Views of Metal Hailde and High Pressure Sodium Lighting for Outdoor Application. *Lighting Research and Technology*.

Caponetto, R. (2008). Power Consumption reduction in a remote controlled street lighting system. *Power Electronics, Electrical Drives, Automation and Motion*, 428-433.

De Dominicis, C. M., F. A. (2011). On the development of a wireless self localizing streetlight monitoring system. *Sensors Application Symposium IEEE*. doi:10.1109/SAS.2011.5739806

Galperti, C. A. (2008). An Adaptive System for Optimal Solar Energy Harvesting in Wireless Sensor Network Nodes. *IEEE Transactions on Circuits and Systems*, *5*(6), 1742–1750.

GSM Based Autonomous Street Illumination System for Efficient Power Management. (2013). *Engineering Trends and Technology*, 54-60.

Jherniak. (2012, September 27). *Reduce Electricity Costs on Street Lights*. Retrieved February 12, 2014, from Lighting Controller: www.egenergy.com

Leonowicz, F. L. (n.d.). *Intelligent Wireless Street Lighting System.* Electronic Engineering Department, University Roma Tre.

Liu, Y. C. (2009). Distributed Intelligent City Street Lamp Monitoring and Control System Based on Wireless Communication Chip nR401. In *Network Security, Wireless Communications and Trusted Computing* (pp. 278-281). NSWCTC.

Mohamaddoust, R., Haghighat, A. T., Motahari Sharif, M. J., & Capanni, N. (2011). A Novel Design of An Automatic Lighting Control System for a Wireless Sensor Network with Increased Sensor Lifetime and Reduced Sensor Numbers. *Sensors (Basel, Switzerland)*, *11*(12), 8933–8952. doi:10.3390/s110908933 PMID:22164114

Riener, R. M. (2011). An energy efficient pedestrian aware Smart Street Lighting. *International Journal of Pervasive*, 147-161.

Scott, S. (1904). *History of the Moorish Empire in Europe*. Academic Press.

SDF Solar Aware Distributed Flow in Wireless Sensor Networks. (n.d.). Darmstadt: Technische Universitat Darmstadt.

Search Mobile Computing. (2006, April 17). Retrieved Jun 3, 2013, from Definition of Wireless: http://searchmobilecomputing.techtarget.com/definition/wireless

Solar-Aware Clustering in Wireless Sensor Networks. (n.d.). Swedish Institute of Computer Science (SICS).

Wireless Streetlight Control System. (2013). *International Journal of Computer Applications*, 1-7.

Chapter 7

Harvesting Energy from Microbial Fuel Cells:
Powering Wireless Sensor Networks Operating in Wastewater Treatment Plants

Pedro Serra
University of Beira Interior, Portugal

Antonio Vitoria Espirito-Santo
University of Beira Interior, Portugal

ABSTRACT

Microbial Fuel Cells (MFC) are the main topic of this chapter. Different types of electrochemical devices are presented and their typical power output is compared with other energy sources, providing a framework for the uses and applications of MFC technology. Following an historical approach of how this technology came to be, a more detailed description of some aspects of a typical microbial fuel cell is then brought forward. The energy harvesting concept, its use on low power wireless systems and maximum power point tracking (MPPT) techniques are presented and described. Wastewater treatment plants are a kind of infrastructure where this technology could be applied with a major success to power wireless sensing networks. An experimental setup, develop to improve the use of MFC in waste water treatment plants is presented. This chapter also provides a review on research trends for microbial fuel cells and maximum power point tracking algorithms, therefore, pointing current researches on this technology.

INTRODUCTION

Wastewater treatment processes for agglomerates with less than 2000 population equivalents, especially the ones located on the countryside and in areas with a low population density, have used low cost solutions based on filtration systems (e.g. constructed wetlands and sand filtration beds). Low power wireless sensors networks are becoming a very interesting solution for monitoring the operation of wastewater treatment plants. Their biggest limitation is related with power: the batteries usually applied

DOI: 10.4018/978-1-4666-9792-8.ch007

have a short lifetime, soon needing replacement. This increases operation and maintenance costs and decreases product's appeal gained with the wireless feature. Some bibliographic records point to 74% of all the consumed power of a sensor node being directed at the transceiver module (Huang, et al., 2012). Given this, one of three challenges can be tackled: to decrease the power consumption of this module; to increase the effective rate of transmitted data, decreasing the time for which the module has to be turned on; or to come up with sustainable energy alternatives that can account for all the power needs of these circuits (Nordic Semiconductor, 2012). Energy harvesting, power harvesting or even energy scavenging, is a suitable solution for powering remote devices where no conventional power source is available. For most common remote sensors, power in the range of miliwatts is enough (Raju, 2008).

This chapter aims to provide an overview description of a self-sustaining energy harvesting Microbial Fuel Cell (MFC), a promising solution to this challenge where organic substrates are available.

Firstly, the energy harvesting concept, its use on low power wireless systems and maximum power point tracking (MPPT) techniques will be explored and briefly described. After that, different types of electrochemical devices are presented and their typical power output is compared with other energy sources, providing a framework for the uses and applications of the MFC technology. Following an historical approach of this technology, a description of some aspects of a typical microbial fuel cell is discussed. The chapter closes with the report of an experimental setup and the retrieved data for its operation, while pointing to current and future research trends for microbial fuel cells.

BACKGROUND

Harvesting Energy for Energetically Independent Devices

Throughout history, mankind seems to get more inventive, focused and productive as it gets more limited, be it on tools, food, money, time or space. Currently, particularly due to critical budget constraints, the scientific community is as committed to making brand new discoveries as it is with perfecting already existing processes, materials and methods. Therefore, process monitoring is of the greatest importance. The field of instrumentation and measuring devices, as others, is experiencing an exciting growth (Dewan, Ay, Karim, & Beyenal, 2014). Several very different knowledge domains are getting together and confronting needs and answers. Instrumentation devices are useful in sports - evaluating performances, achievements and material conditions -, in cooking, to improve comfort at home, in medicine, in media devices – like cellphones, tablets or smart watches -, or in industry. Instrumentation devices allow us to keep track of changes and analyze data. As with any other electric device, instrumentation modules require power. And power availability ultimately limits their usefulness: a sensor is no good if it needs to be connected to the main power grid or if it's powered by a battery, running only for a couple of hours. Versatility and reliability are the main keywords for any electrical devices, instrumentation included, irrespective of their application area.

Instrumentation devices can be powered by connecting them to the main power grid, with a suitable power transformer, or by using batteries. The connection to the main power grid severely hinders the versatility of the devices, namely their ability to be used at different locations. Even though wires are the most reliable, faster and safest way to power anything of electrical nature, the connectors are a weak point, susceptible to moisture and dust and the wiring always limits the portability of the device (Li Huang et al., 2010; Smith, 2011; "Ultra low power wireless connectivity technology backgrounder," n.d.). On the other hand, batteries always need to be recharged and, sooner or later, replaced, which is an increased cost. The solution to this challenge resides on energy harvesting, also called power harvesting or energy scavenging. Energy harvesting has been around for centuries and windmills and watermills are the clearer examples. Though these did not produce a measurable amount of energy, this was directly applied to milling. Nowadays, these have been replaced by wind turbines, hydro-electric generators and solar panels.

At a smaller scale, energy can also be harvested from electromagnetic sources (ambient radio frequency, RF), electromechanical transducers (vibration, piezoelectric) and from temperature, pressure or salinity gradients. Harvesting energy from biological sources is a new challenge being explored: though having a limited bio-electro-chemical knowledge, we are at the point where we can use that insight to recover/redirect "biological" power to our benefit. In spite of virtually inexhaustible, and with no or little adverse environmental effects, this harvested energy is also low, unregulated and intermittent (Dewan et al., 2014; Raju & Grazier, 2008). Retrieved power levels are usually low and are most frequently stored to achieve higher levels: an electrical device powered by an energy harvesting mechanism has to work in short periods of time rather than on a continuous mode. Limiting factors for energy harvesting techniques are the quality of the used transducers and the raw available energy.

Until recently, most of the instrumentation devices were integrated in a single assembly and its respective power levels were higher than those achievable with energy harvesting solutions. Also, wireless communication protocols were not as developed as today (Li Huang et al., 2010). The development of the latter, and the optimization of the energy usage by the above devices, created new, modular designs. With several available network topologies (ring, mesh, star, fully connected, line, tree or bus), further enhancement of these systems is possible by applying energy harvesting mechanisms (Li Huang et al., 2010). These modular designs typically have several peripherals and a main central processing unit (CPU) (Dewan et al., 2014). The peripherals usually collect data and send it to the CPU. By segmenting the system's tasks, each module can be independently powered and the system can be disturbed, having a larger area of application.

Electrochemical Devices as Energy Sources

Electrochemical devices can be partitioned into three categories: batteries, fuel cells and solar cells. An electrochemical device converts chemical or light energy to electrical energy through a chemical reaction: if the device uses incident light to produce electricity then it corresponds to a solar cell; if not, then the device can either be a battery or a fuel cell. Batteries are destined to one lifetime cycle and cannot be recycled: their internal fuel cannot be replaced and, once spent, the battery has no more use, making them more pollutant than fuel cells. Fuel cells, on the other hand, can have their fuel replaced, being able to run for a very long time (provided that no other element is damaged). Fuel cells and batteries can be inorganic or organic. For organic types, the devices get the denomination of bio-electrochemical systems (BES) and get subdivided into three types:

1. Cells using an organic material to generate a byproduct, like hydrogen, which is then used as fuel for a conventional, for instance hydrogen/oxygen, fuel cell;
2. Cells using an organic material and enzymes or whole microorganisms to produce electricity;
3. Cells using photochemically active systems and biological partners to convert sunlight to electricity.

Microbial fuel cells (MFCs) and enzymatic fuel cells (EFCs) are two examples of the second category: the latter having catalysts (reaction facilitators) act on isolated and purified enzymes and the first on microbes (Bullen, Arnot, Lakeman, & Walsh, 2006; F. Davis & Higson, 2007; Lovley, 2006; Korneel Rabaey et al., 2007).

The development of these devices came as an alternative to fossil fuels, complementing the clean energy alternatives. These alternatives are of dire need, both to limit the green-house effect and particularly for the developing countries. In India, for instance, it is quite frequent to find people in the rural areas, with limited access to western common energy sources, burning organic waste to have a flame source for cooking (Karmakar, Kundu, & Kundu, 2010). Bio-electrochemical fuel systems were, in 2004, producing power anywhere between 10^{-7} [W] and 10^{-3} [W]. Inorganic fuel cells could output, at the time, power in the ranges of 1 to 10^{9} [W] (Bullen et al., 2006). Since then there has been a significant increase in the power magnitude achievable with a BES, though that is still far from enough to even consider these devices as reliable energy sources for everyday uses. They currently have enough power, however, to feed chemical sensors (usually enzymatic batteries or fuel cells), to power small electronics devices for short instances of time or to trickle-charge capacitors in some particular applications (Bullen et al., 2006; Lovley, 2006). Figure 1 places BES systems and their applications in the overall energy needs panorama.

Comparatively with other organic recycling energy producing technologies, MFCs have some key aspects that make them more advantageous: their energy conversion rate is high; they can be designed to be efficient in several temperatures, being considerably tolerant; the gaseous products of its functioning do not require any treatment; the cathode chamber may not require aeration, on certain conditions, and their principle of operation is very versatile, making it possible to be used with wastewater or with sediments (K Rabaey & Verstraete, 2005).

Figure 1. General overview of the energy produced with each electrochemical energy source. The energy needs of ULP (Ultra low power) and LP (Low power) sensors have also been marked.

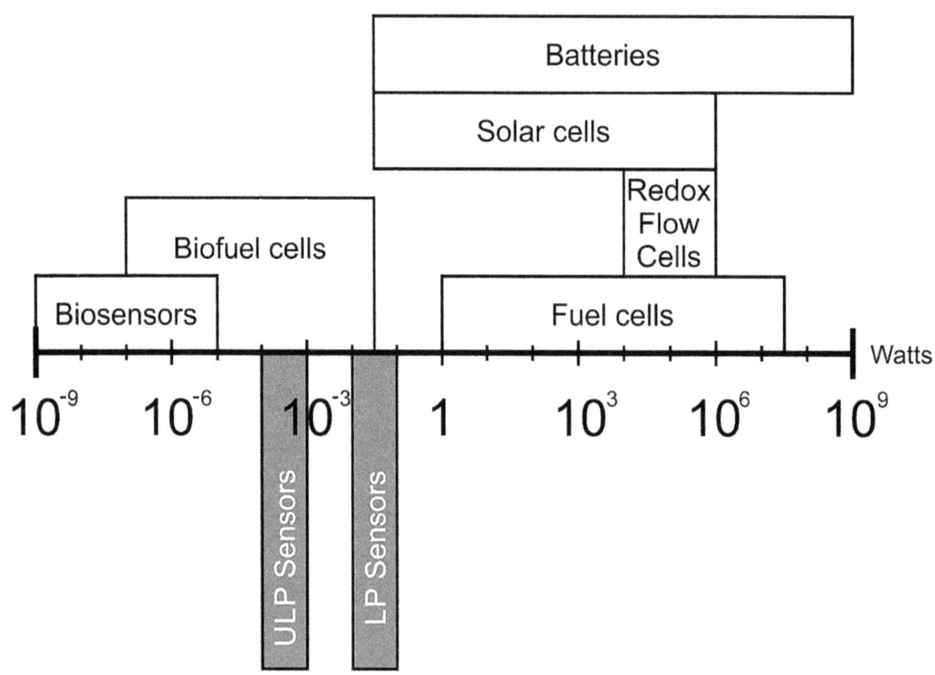

Microbial Fuel Cells and Wastewater Management Systems

The use of microbial fuel cells (MFC) with wastewater substrates is currently of great interest. Having the ability to clean wastewater and, at the same time, producing power or retrieving added value compounds from these residues is now turning heads and gathering attention. Particularly due to the increasing global energy needs (Aelterman, Rabaey, Clauwaert, & Verstraete, 2006; Angenent, Karim, Al-Dahhan, Wrenn, & Domíguez-Espinosa, 2004).

One of the first studies to report the use of microbial fuel cells and real wastewater substrate is from 2004 (H. Liu, Ramnarayanan, & Logan, 2004). Before that, laboratory experiments would use synthetic wastewater or simpler substrates (see the "Substrates" subsection under "Working Principle"). The data retrieved with these trials pointed to estimate power production levels capable of powering wastewater substations, like aerobic trickling filters (He, Minteer, & Angenent, 2005). When using real wastewater, though, these energy achievements would be lower. A power decrease of about 30% was even reported when comparing the functioning of an air-cathode MFC with real and synthetic wastewater (H. Liu & Logan, 2004). The main limiting factors for electricity generation with MFC on real wastewater applications are related with inadequate microbial communities and methanogenesis. There are some researchers who point to the need of a pre-fermentation stage, on the substrate, so as to guarantee that all the fermentable compounds (which give rise to methanogenesis) are converted to non-fermentable compounds (Rozendal, Hamelers, Rabaey, Keller, & Buisman, 2008). All these challenges are under investigation so as to be overcome. It is expected that an adequately dimensioned, continuous flow, MFC behaves effectively (Zhuang et al., 2012). It may even not be a large MFC system but rather a modular type solution.

Nevertheless the advantages on using an MFC on wastewater treatment plants are undeniable: an MFC allows recovering energy, limits energy input to the wastewater treatment system and decreases the sludge production levels. No other current technology can achieve such goals and characteristics. Furthermore, MFC technology (and, on a broader sense, BES) is also capable of biologically producing methane, hydrogen and other chemical compounds of interest. The byproducts of MFC are non-pollutant and single MFCs can be stacked to increase the power production capabilities of this technology (Aelterman et al., 2006; Angenent et al., 2004). MFC have also been compared with anaerobic digesters (AD). Both technologies have the same inputs (wastewater) and outputs (electricity). Even so, the AD's final product is biogas and, for electricity production, this byproduct needs further processing. Also, AD power requirements easily amount to 45-75% of the whole wastewater power plant energy costs. Converting an aeration tank to an MFC equivalent would eliminate these energy costs and could also lead to 10-20% more power production (a total energy balance of about 55-95% more) (Arends & Verstraete, 2012; Wang & Ren, 2013).

MICROBIAL FUEL CELLS

Historical Overview

If the first connection between biology and electricity was made by Galvani, the first fuel cell concept is Grove's responsibility. While Galvani, in 1790, demonstrated that an electrical current can influence a biological response – by the famous frog leg's twitching experiment – Grove, by 1839, was able to force the combination of hydrogen and oxygen to produce water and electricity, reversing electrolysis (Bullen et al., 2006). It was not until early in the 20th century that electro-physiology truly gained momentum. Researchers in Germany and in the US were leading the way on findings related with the association, unclear at the time, between chemical changes and electrical changes on vegetable physiological processes. Haacke and Klein, with investigations 6 years apart, showed the relationship between electrical currents in plants and their physiological processes, being also able to modulate the electrical current characteristics by changing particular aspects of the plant's biological phenomena. The Americans, by Waller, pointed to the possibility that both vegetable and animal protoplasm could give rise to an electrical response. By 1910, a British investigator named Michael Cresse Potter made several experiments so as to determine if one of Waller's conceptions had a more inclusive meaning. According to Waller, in biology, a dissociation process would give rise to an electrical current whereas an association process would need an electrical current. Potter set out to find if there were electrical effects on fermentation or putrefaction by bacteria and if they, as a dissociation process, had the announced effects. He developed a device using a glass jar containing a porous cylinder, a galvanometer, a condenser and a Morse key (Figure 2). Using some copper wires he made the connections between the elements and chose to use platinum for his electrodes. As for the fermentation process he elected the fermentation of yeast upon glucose, mostly because it was a well-known process and had few conditional restraints. Having made all the choices, he poured a glucose solution to the cylinder and added a pre-planned amount of yeast to the outer glass jar. After several studies, made so as to remove any possible sources of mistakes or measurement inaccuracies, he observed a gradually increasing voltage (later on found to be dependent on the concentration of the glucose solution, the temperature and the quantity of yeast added) (Potter, 1911; Rachinski, Carubelli, Mangoni, & Mangrich, 2010).

Figure 2. Schematic representing the experimental set up for Potter's work: A) The cell; B) Mercury cups to facility – improve connections with different cells; C) Morse key; D) Condenser; E) Galvanometer
Adapted from Potter, 1911

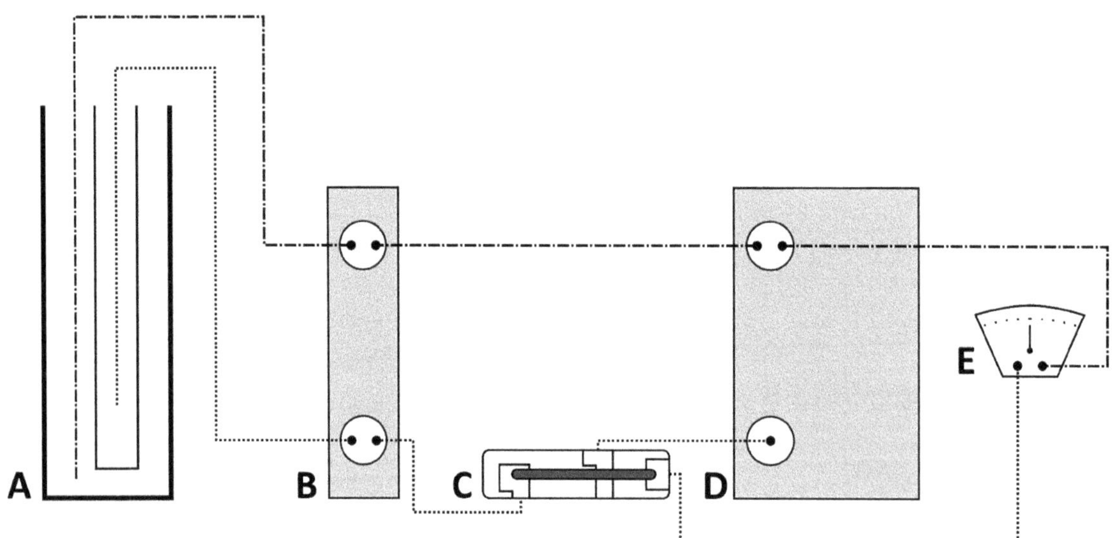

This work was highly underestimated and needed bigger results and a higher impact to get the attention it deserved. By 1931, another researcher brought this work to the spotlight: Barnet Cohen created several fuel cells and connected them in series to evaluate their behavior and electrical output. He was able to produce a 35 [V] potential but only a 2 [mA] current. Even so, his work drew the attention of at least one major player in the energy field: NASA. By 1960, one year after the creation of Project Mercury and the X-15 rocket plane, NASA clearly stated its objective of developing a technology that could generate electricity while organic waste produced by the astronauts was decomposed. Recently, in 2012, they reaffirmed their intentions (Wired UK, 2012). Rohrback and his team made another contribution in 1962: they reported the use of the bacteria *Clostridium butyricum* as a catalyst for producing hydrogen trough glucose fermentation. The cell's hydrogen could then be used, as reported by Delduca et al, in an oxygen/hydrogen fuel cell. From then onwards, several efforts were developed so as to understand what factors would have an effect on MFCs electrical output. In 1966 the attention was on the cell's substrate: rice chaff was used in microbial fuel cells connected in series to produce 40 [mA] at 6 [V], a significant improvement, current-wise, from Cohen's work.

Then, ten years later, focus was on the applications of this technology: efforts were developed to create glucose powered fuel cells for use within heart pacemakers: even though only inorganic electrodes were used, the devices were very susceptible to poisoning by aminoacids, although *in vivo* tests showed a good performance for up to 10 days. The development of photovoltaic cells, a more technically feasible technology, put a brake on the research work in BES. Though the oil crisis stroke in the 80s and renewed the interest in MFCs, true advances and revolutions were only seen around the 90s and the beginning of the 21st century (between 1980 and 1990 the only significant contributions were more theoretical and even so, over enzymatic fuel cells: researchers like Wingard, and Aston and Turner were head names in the field at the time). Researchers like Benetto (who, until his retirement, laid the core funding for the understanding of MFCs' functioning) were responsible for increasing the current density and power output by adding recently investigated electron mediators. By 1991 there were already attempts to apply MFCs to domestic waste water treatment and concomitant energy production (Bullen et al., 2006; F. Davis & Higson, 2007; Debabov, 2008; Du, Li, & Gu, 2007; Karube, Matsunaga, Tsuru, & Suzuki, 1977; Rachinski et al., 2010). It was not until 1999 that the use of electron mediators was set aside and its use practically discarded: a team led by Kim B. H. reported the development of a mediator-less fuel cell and their achievements were so remarkable that this was considered a landmark for MFC's development. If, until then, only two generations of MFCs devices could be identified – generation I (Gen-I) with synthetic mediators and generation II (Gen-II) with naturally occurring mediators – this discovery gave rise to the third generation (Gen-III): mediator-less MFCs (Ieropoulos, Greenman, Melhuish, & Hart, 2005; Logan, 2008). After approximately 10 years, by 2009, MFCs' power designation shifted from being normalized relatively to electrode surface to being normalized to power per volume. This change was a natural result due to the power outputs reported by several researchers: changes like using oxygen at the cathode resulted in around 115 W/m^3 power increase. As years have passed, attention was redirected from exploring materials and factors limiting power on an area basis to the same factors influence on the overall volumetric power achievable. It is expected, in the next couple of years, to achieve power levels similar to what is produced with anaerobic digesters, making this technology strong enough to be more diversely applied.

Figure 3. Microbial fuel cell with acetate as a substrate. With the substrate oxidation, electrons flow from the anode to the cathode through the external circuit. Protons rising from the oxidation reaction are then selectively chosen to go to the cathode chamber through a proton exchange membrane.

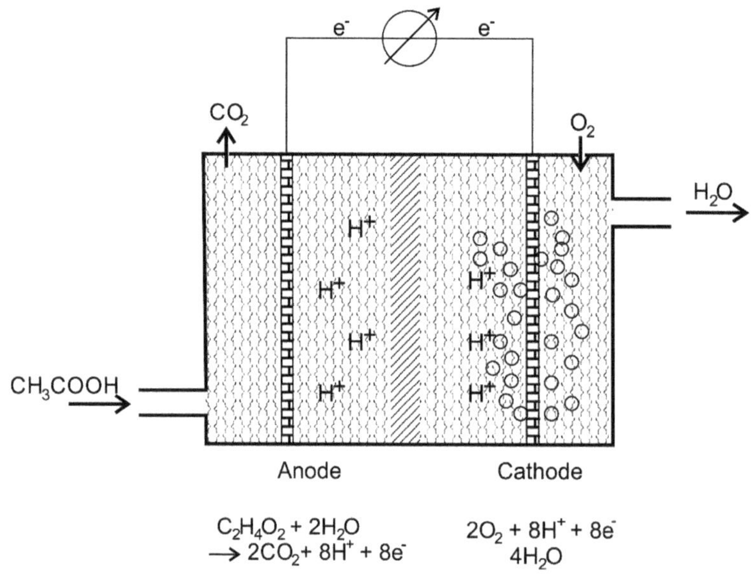

Working Principle

Microbial fuel cells work with microbial beings. But exactly how do they manage to do so? What kind of reactions fuel these devices and how can they be sustained?

Microbial fuel cells are characterized by five major components: the anode, the cathode, the substrate, the membrane and the bacteria. The anode experiences an oxidation reaction (the loss of electrons) and the cathode a reduction reaction (the gain of electrons). These reactions are possible because of the substrate and an external circuitry, responsible for maintaining the conditions for electrons and ions to flow. In a typical fuel cell, electrons produced by the oxidation reaction travel to the cathode, through the external circuit, where they reduce another component. For microbial fuel cells, bacteria are responsible for oxidizing an electron donor and oxygen is frequently the acceptor at the cathode. The membrane keeps both chambers separated and is usually proton permeable, guarantying the presence of protons in the cathode chamber where, with the electrons and oxygen, water is produced (Bullen et al., 2006; F. Davis & Higson, 2007; Du et al., 2007; K Rabaey & Verstraete, 2005). Figure 3 presents a schematic of a MFC with acetate as a substrate.

Electrodes

All MFCs' components are chosen to provide the highest power output possible. Electrodes are no exception. Their characteristics – geometry, materials and chemical properties – are selected to establish the best cost-efficiency relationship. Both anodes and cathodes should be conductive, biocompatible and chemically stable, particularly in contact with the substrate:

1. Non or mildly-conductive materials will limit electron transfer efficiency;
2. Non-biocompatible electrodes will not allow bacterial proliferation;
3. If either the anode or the cathode has a corrosive nature in the presence of the substrate, then the decomposition by-products can reveal themselves as toxic, therefore killing the microorganisms sustaining the MFC.

For materials and geometry, even though the anode and the cathode share some similarities, a more detailed description is presented.

Early work, by Rao and his team in 1976, reported the use of inorganic electrodes made of platinum/nickel combined with polyacrylic acid/polyvinyl alcohol. Even though these managed to achieve remarkable power output levels (at the time), they were also extremely sensitive to the negative influence of aminoacids (F. Davis & Higson, 2007).

Since 1993, anode materials' have changed. Allen and Bennetto started off with two choices: graphite felt and reticulated vitreous carbon (RVC). By 1996, Cooeny et al used an activated charcoal anode. Again, RVC and, anew, carbon felt were used between 1999 and 2000 by Kim et al. More recently, in 2000, 2002 and 2003, graphite has been the choice for several researchers: Chaudhuri and Lovley, Park and Zeikus, Rabaey et al, and Schroder et al. Other BES fuel cells, specifically EFCs, have used graphite as well, platinum and platinum black, carbon paper and carbon cloth (Bullen et al., 2006). Unpublished work, described by Logan on his book, was developed using conductive polymers as the anode material. Early results pointed to concerns with the stability of these materials and their erratic and inconsistent performance. The integral use of metal or metal coatings was also investigated. However, the results were very far of regarding expectations and needs. Other lines of research are investigating, and obtaining significant results, non-metallic treatments and modifications to the anode material. Bounding mediators to graphite electrodes and treating carbon cloth with ammonium gas preparations are a few examples (Logan, 2008).

The most adequate anode materials are non-fouling, need to have high-specific surface area (area per volume) and high porosity. Conductivity wise, copper is the best choice, followed by graphite, carbon and conductive polymer sheets (Logan, 2008).

For cathodes, materials aren't very different. In fact, the difference typically resides in having a catalyst (platinum or transition-metal components for oxygen reduction). Choosing a cathode material is of the greatest importance and difficulty in an MFC: a cathode should be able to facilitate a reaction simultaneously with electrons, protons and oxygen. Research started with pentachlorophenol (PCP) cathodes in Allen and Bennetto's team by 1993. 6 years later, Kim et al applied a RVC cathode and by 2000 platinum and graphite cathodes were used by Kim as well as by Park and Zeikus. From then onwards, between 2002 and 2004, graphite became a natural choice for several researchers: Kim, Tender, Park and Zeikus, Niessen, Rabaey, Chaudhuri and Lovley, and Schroder (Logan, 2008; Logan et al., 2006).

Graphite and carbon are usually the first choices for either electrode. Their cost-efficiency relationship is one of the best, they can be easily shaped and they are durable. There are even reports of high-school students using them as off the shelf materials to build MFCs' electrodes for science fairs. In the face of their different configurations, RVC and carbon foams hold the best records for carbon electrodes: they are thick, porous and have excellent conductivity values. Graphite materials seem to have a better behavior for fiber and brushes configurations, resulting in the highest specific surface areas and porosities. They can also be shaped into rods (low internal porosity), sheets (not porous), felts and granules. Felts and rods result in approximately the same current production (normalized to the surface area) and granules need to be in contact with each other the guarantee a high conductivity (Bullen et al., 2006; Du et al., 2007; Logan & Regan, 2006; Logan, 2008; Logan et al., 2006).

Membrane

There are several different reasons for using a membrane in an MFC. Membranes are typically associated with two-chambered MFCs and can be used specifically to conduct protons between the electrodes and to separate the anodic from the cathodic chamber, protecting the anode from oxygen exposure. As a physical barrier, the membrane will stop the transfer of other chemical species between chambers and fuel crossover. They can also be used in single-chamber MFCs to isolate the cathode from the catalyst. They can be cation exchange membranes (CEMs), anion exchange membranes (AEMs) or bipolar membranes. The schematics for a CEM and an AEM are shown in Figure 4.

CEMs have extensively been used in several studies. Nafion and Ultrex are two commercially different CEMs, usually referred to as proton exchange membranes (PEMs): while Nafion membranes are prone to biofouling by ammonium they also have the lowest resistance of the two. Ultrex is cheaper and a typical choice for studies where ferricyanide is used as a catholyte.

AEMs, contrary to CEMs, transport pH buffers (phosphate ions). In 2007, Kim et al reported a study where the use of an AEM (AMI-7001) resulted in a more stable pH value though the cathode's pH had a higher increase than what had happened with a Nafion membrane (J. R. Kim, Cheng, Oh, & Logan, 2007).

Bipolar membranes are also available for use with MFCs: they consist of an anion and a cation membrane joined in series. Instead of proton flow, a voltage is developed across the membrane and that results in the transport of anions (OH^-) to the anode and cations (H^+) to the cathode. Comparatively to other membranes, this type has a higher internal resistance and decreases the cell power output.

Other membranes are also accessible: salt bridges – though producing at least one order of magnitude less power than CEMs are an option for first studies; and kaolin septum – with no disadvantages, apart from fouling.

Figure 4. On the left (A), anion exchange membrane (AEM). On the right (B), cathode exchange membrane (CEM). Negative ions are conceptually represented by chloride ions (Cl^-) and positive ions by sodium ions (Na^+).

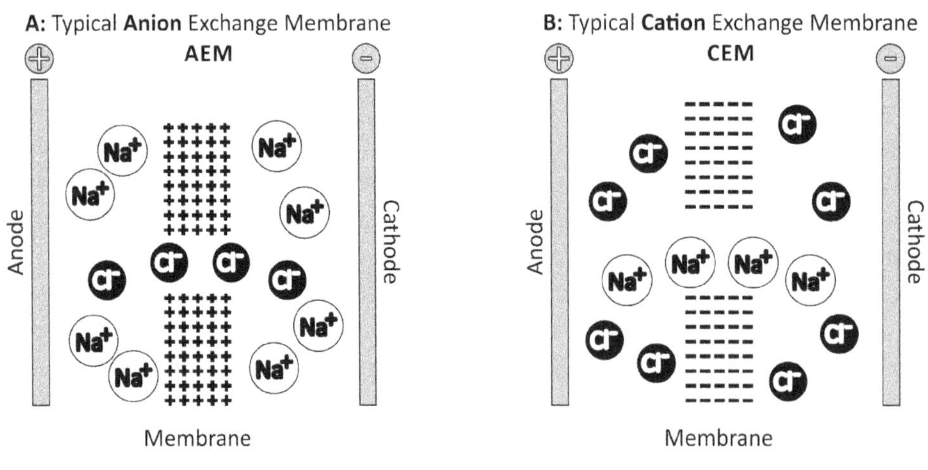

Anyhow, membranes counterbalance their need with some disadvantages: increased cost and decreased system performance (comparing with membrane-less studies) due to an increase in internal resistance (Du et al., 2007; Logan & Regan, 2006; Logan et al., 2006; K Rabaey & Verstraete, 2005).

Substrate

Substrates can also be called fuel sources. They are the nutrients for the bacterial population and, ultimately, their energy source. The fuel source, or substrate, does not have to have a dynamic delivery process to the MFC anodic chamber: fuel should, at the very least, be stored and supplied from outside of the reaction chambers (to satisfy the requirements of a fuel cell). The continuous replenishment of fuel is a recent technique that is applied in order to mimic real systems. Substrates are not electrochemically active but have great energy stored in the connections between their different compounds. Substrate characteristics have been found to have significant impact on the fuel cell's electricity generation process. Several studies have been conducted to clarify the latter relationship: according to their findings, a complex substrate builds a diverse and electrochemically active microbial community and the use of a simple substrate results in a faster and easier decomposition process, improving the electric output of the system (Bullen et al., 2006; Lovley, 2006; Pant, Van Bogaert, Diels, & Vanbroekhoven, 2010).

Early studies began by using simple substrates to evaluate MFCs. Everything was quite new so there was a need to have the fewest variables as possible. Acetate and glucose were the first choices. Acetate is the immediate energy source for the citric acid cycle (more commonly referred to as Krebs cycle) and leaves, at room temperature, no possibility for other microbial respiratory process to occur. It is also the final product for several other metabolic pathways. When compared to the use of protein-rich wastewater, acetate-based MFCs achieved more than two times the maximum electrical power than with the mentioned wastewater. Glucose did not lead to as impressive outputs: comparing acetate and glucose energy conversion efficiency, it was found that acetate had a 42% conversion efficiency rate and glucose only 3% (Pant et al., 2010). These results seem to be justified by the fact that glucose is a substrate usable by other metabolic energy retrieving processes, like fermentation and methanogenesis. These processes are not capable of positively contributing to the MFC. Glucose was responsible, however, for the development of a more diverse microbial culture. The use of lignocellulosic biomass is representative of the complex substrates already mentioned. They cannot be directly used by the bacterial colonies unless converted to monosaccharides or other simpler compounds. It has been proven that basic cellulosic components, monosaccharides, are good starting points for electricity generation in MFCs (Logan & Regan, 2006; Pant et al., 2010).

The microbial community working with this raw material has, however, to have cellulolytic and exoelectrongenic characteristics. Comparatively with glucose, this fuel type seems to, however, have a lower power output and no microorganisms have been found to be capable of digesting pentoses, one of the main sugar types resulting from cellulose breakdown. As lignocellulosic biomass is a result of residual plant material, MFCs using this substrate are not effective. The application of MFCs technology in wastewater would result in significant financial gains: not only would the wastewater be bioremediated but that process would concomitantly be electrically sustained (Pant et al., 2010). Synthetic wastewater has a well-defined composition and is an emblematic choice to check the influence of its composition on the performance of a MFC. Rodrigo et al found that for wastewater with the same pollutants (organic glucose and peptone), the slower the MFC's feeding rate, the higher was the electrical output, probably because of the pilling up of intermediates (Rodrigo, Cañizares, García, Linares, & Lobato, 2009). Other wastewater types have been evaluated: brewery wastewater, with high organic matter content (almost ten times above domestic wastewater) - low strength, low concentration of inhibitory substances - starch processing wastewater (also with a high organic content), dye wastewater (a choice driven by the importance of their removal from water resources), paper recycling wastewater, biorefineries wastewater and swine wastewater. Non-conventional substrates, like cellulose and chitin have also been studied. These compounds are considered non-conventional because of their particulate nature. It was found that for solid, or particulate, substrates like these, a low hydrolysis (cleavage of chemical bounds by the addition of water) rate limits power production. Landfill leachates were scoped by some studies in 2009 but no power output was reported. The concept was validated but further studies are needed to understand their full potential (Pant et al., 2010).

The catalysts for all these organic fuels must be of organic nature as well. These catalysts, in a MFC, are whole organisms and are suitable to operate on different working conditions while mildly withstanding poisoning and, because of their full enzymatic pathways, successfully converting the substrates to carbon dioxide and water (Bullen et al., 2006).

Bacteria

The bacterial population on an MFC is also a critical factor determining its power output. The basic structural, functional and biological unit of all living organisms is the cell. They are usually called the "building blocks of life" and can either be eukaryotic or prokaryotic, according to their elements. A eukaryotic cell has a nuclear membrane (delimiting the nucleus, where the genetic material is), its ultimate characteristic; a prokaryotic cell does not – in fact, its genetic material (DNA) is unbound in the cytoplasm of the cells. A bacterial organism is made of prokaryotic cells.

Prokaryotic organisms, as well as eukaryotic, have different processes from which they can obtain energy and nutrients. The overall set of reactions that occur in living organisms, and that are required for energy generation and its use to synthesize cell material from small molecules in the environment, is called metabolism – from the Greek word "change" (μεταβολή). If the metabolism has an energy-generating component, it is called catabolism; if it has an energy-consuming, biosynthetic component, it is called anabolism.

Because prokaryotic organisms have been around for longer that eukaryotic ones, their metabolic pathways are incredibly adaptive: according to their living conditions so is their metabolism. So, the same prokaryotic organism, living in two different environments, can exhibit different metabolic pathways.

There are two major metabolic pathways: *Heterotrophy* and *Autotrophy*. The separation between both lies in the use of an organic compound as a source of carbon and energy – *Heterotrophy*; or carbon dioxide's direct fixation – *Autotrophy* (the fixation is made by a compound known as RUBP Carboxylase). A further categorization of these two major catabolic pathways is presented.

For *Heterotrophy*, Fermentation and Respiration are two energy producing alternatives: Fermentation occurs when energy is derived from the partial oxidation of an organic compound using organic intermediates as electron donors and electron acceptors. No outside electron acceptors are involved and no membrane or electron transportation system is required; for Respiration, there is a complete oxidation of the substrates by an outside electron acceptor.

As *Autotrophy* is concerned, there are also two major catabolic alternatives: Lithotrophy implies the use of an inorganic compound (no carbon and no hydrogen in its constitution) as a source of energy and can be used by hydrogenic bacteria, methanogens (which transfer electrons from H_2 to CO_2 while reducing

Figure 5. Major metabolic pathways. Heterotrophic organisms obtain carbon from organic compounds; Autotrophic organisms make direct carbon fixation.

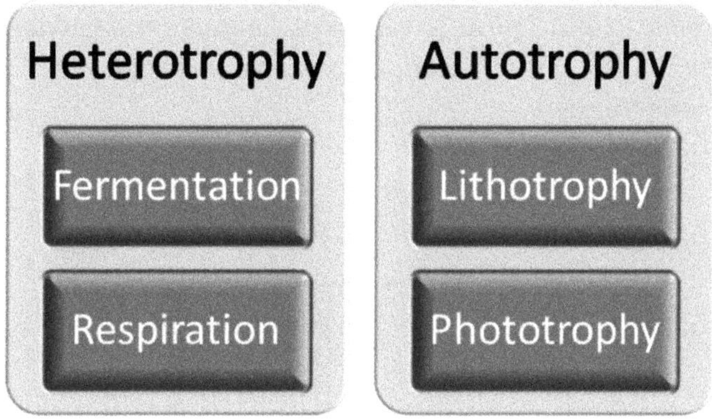

it to methane, CH_4), carboxydobacteria ($CO \rightarrow CO_2$), nitrifying bacteria ($NH_3 \rightarrow NO_3$), sulfur oxidizers and iron bacteria; Phototrophic organisms take advantage of light and use it as a source of energy for growth. Phototrophic metabolisms can also be divided into three major types: oxygenic photosynthesis, anoxygenic photosynthesis and nonphotosynthetic photophosphorylation.

All these metabolism types can be grouped into three major categories, according to the final electron acceptor: aerobic for using molecular oxygen; anoxic for using non-molecular oxygen; and anaerobic when there is no use of oxygen (single or molecular) (Todar, 2008). Figure 5 synthesizes all these concepts.

For a MFC to work, bacteria, while processing the substrate, should be able to transfer the electrons produced by their catabolic reactions to the anode. Exoelectrogens, electricigens, electrochemically active or anode respiring bacteria are all common terms applied to describe bacteria that are capable of doing just that. They can do so resorting to outer membrane components (cytochromes: inner-membrane, periplasmic and outer-membrane redox proteins), excreted mediators (also known as electron shuttles) or synthesizing appendages capable of transferring electrical current (nanowires or pili). Figure 6 makes a clearer explanation of how these processes affect electron production and transfer.

Having this ability doesn't necessarily mean that it'll be used. And that's where the metabolism choice takes part. For an MFC, the best achievable results will appear when the electron output is at its highest; different catabolic reactions give rise to different net electron production. Choosing conditions that favor an aerobic metabolic path (or a bacterial strain) will be of no use: if the final electron acceptor is molecular oxygen, than the electrode will not be used (no power output); if we also consider that the substrate used will be organic, we can accurately determine that the bacterial strains to use in an MFC anode chamber must have an anaerobic heterotrophic metabolism, preferably anaerobic respiration (Logan, 2008, 2009; Lovley, 2006; Rozendal et al., 2008).

Potter's early investigations used living cultures of *Escherichia Coli* and *Saccharomyces*. The first problems encountered with these species had to do with the low power output because no electron mediators had been used. Since then, several studies have been conducted and presented and a clearer picture seems to be rising, matching exoelectrogens and the cell's power output (Du et al., 2007).

For *Escherichia Coli*, *Pseudomonas*, *Proteus* and *Bacillus* species, the use of artificial mediators cannot be ignored. If it weren't for them, the power output of MFCs with these bacterial populations would be neglectable since these strains cannot effectively transfer electrons at key metabolic points.

Figure 6. Different types of electron transfer methods. Glucose has been used as an example substrate. In a) an example of an indirect microbial fuel cell, with hydrogen (metabolic end product) used for electron production at the anode; In b) a mediator-driven microbial fuel cell is shown: it'll be the mediator, converted from its reduced to oxidized form, to transport the electrons resulting from the bacterial metabolism to the anode; finally, in c) the schematic displays an air-cathode MFC with electron transport proteins making the electron transfers
Adapted from Lovley, 2006

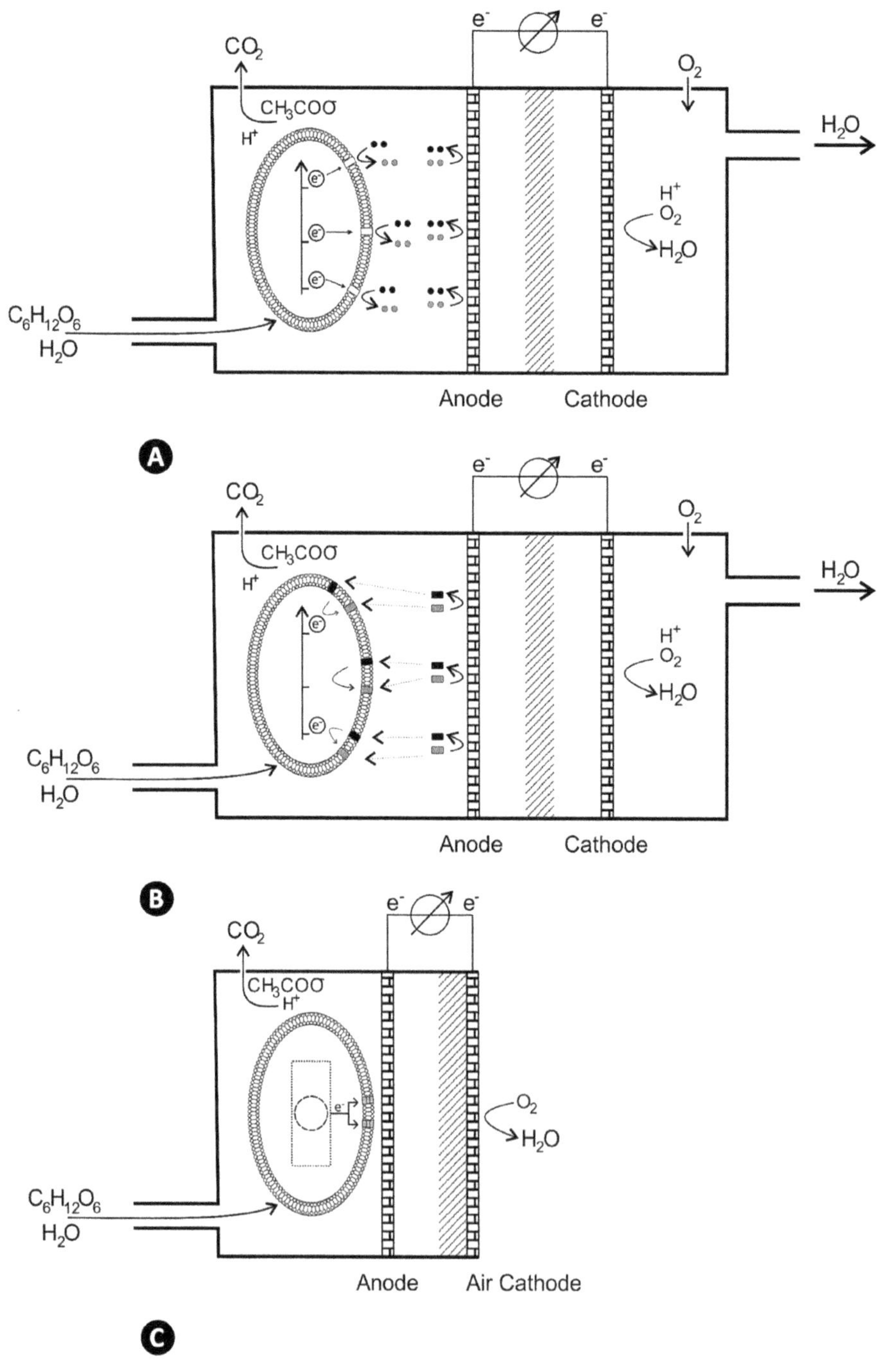

Typical electron shuttles include thionine, benzylviologen, 2,6-dichorophenolindophenol, 2-hydroxy-1,4-naph-thoquinone and various phenazines, phenothiazines, phenoxoazines, iron chelates and neutral red (Karmakar et al., 2010; Lovley, 2006).

The use of artificial shuttles, or mediators, as several shortcomings: firstly, even though some electrons are captured by the shuttles, some others are kept in chemical connections and continue to be processed in parallel metabolic paths (fermentation); secondly, the use of artificial shuttles will increase the difficulties of maintaining a MFC: because the fuel will be recirculated, there is a need to keep adding electron shuttles. Besides, many of the electron shuttles used are pollutants and cannot be used because of it (Lovley, 2006).

There are also other species that do not require artificial mediators, they biosynthesize them. This has energetic costs and is usually not a good alternative as it deduces the net electron production of the MFC. The *Pseudomonas aeruginosa* strain, and the *Lactobacillus* and *Enterococcus* species are examples of bacteria capable of producing its own mediators. Experiments with similar bacterial strains for a one-time replacement of the substrate resulted in a decreased power output of about 50%, which seems to point that species with these characteristics will not but a popular choice.

For whichever form of bacteria previously mentioned, studies report that they also do not fully oxidize the substrate they are in.

Only one type of exoelectrogens remains: bacteria which are able to "directly" transfer electrons to the electrode. *Shewanella putrefaciens* have outer-membrane components, named cytochromes, which are vital in this process. These cytochromes make the catabolic electrons available at the cell's membrane and they end up reacting with other substrate components which can directly be oxidized by the anode. Other strains, like *Aeromonas hydrophila* and some *Clostridium* and *Geobacter* species have also been reported to have the ability to directly interact with the electrode. However, they seem to do so using some biosynthesized conductived appendages - nanowires or pili – and forming a biofilm in the anodes surface.

Other than considering the bacteria-electrode interface, other studies have been developed so as to gain an understanding of how power production depends on single or multi strain/species of bacteria at the anode (Bullen et al., 2006; Debabov, 2008; Du et al., 2007; Logan, 2008, 2009; Lovley, 2006; Korneel Rabaey et al., 2007).

Architecture

This section will address construction characteristics disregarding electrodes, membranes, bacteria or substrate choices as these have been previously addressed.

Bennetto and his team were of the first investigators to report a low cost MFC device. This construction was mostly used for demonstrations and proof of concept studies. In its simplest form, the container was a large glass tube. The electrodes were recycled from old batteries and the connections made using an ordinary copper wire. A dialysis membrane, in a shape similar to a bag, was applied to the anode to separate it from the cathode. The anodic bag had the microorganisms, a mediator and a small amount of glucose.

In spite of this cell not producing useful amounts of energy it did, and still does, establish the building grounds for an MFC (Bennetto, 1990). One year before, in 1989, Sell et al reported the first air-cathode MFC (Sell, Kramer, & Kreysa, 1989). The development of different designs spread through time and pinpointing when and which architecture came to be is a rather unfair task. Currently, just as Figure 7 shows, MFC designs can be organized by number of chambers - single and two-chambered MFCs; geometry of electrodes - air-cathode MFCs, MFCs with soluble catholytes or poised potentials and tubular MFCs; or number of cells – stacked constructions (Bennetto, 1990; F. Davis & Higson, 2007; Debabov, 2008; Du et al., 2007; Logan, 2008).

Two-chamber MFCs are the typical construction for laboratory studies. Bottle reactors, usually in a H-shape, and cube reactors as reported by Bennetto, are a classical two-chamber MFC example. They are easily autoclaved and sterilized and can be run in batch or continuous mode. They are useful to determine whether or not power generation is possible and can be used to establish a timeline relationship for the development of microbial communities. They are not adequate to improve or discover which factors affect power output though they are perfectly suitable to be used in a commercially viable MFC design (Bennetto, 1990; Du et al., 2007). More recent two-chamber devices are: bottle type - You et al, in 2006, described a bushing MFC: a plastic bottle with 4 bars had a CEM hot pressed in them and a cathode concentrically placed inside the bottle; Tubular devices – built to run in continuous flow mode, the substrate in these designs flows from the anode to the cathode, either in a bottom to top configuration or in an outer to inner path (for cylindrical designs). For vertical constructions, up-flow, the first devices had their chambers separated by glass wool and glass beds. Rabaey and his team, in 2005, presented a tubular device made using a granular graphite anode delimited by a PEM sleeve and a sacrificial cathode. Up-flow MFCs are mainly used in wastewater treatment since the energy needs for fluid recirculation

Figure 7. The three major types of architectures for MFCs. In a) single cell (air-chatode) geometry; b) represents the dual chamber disposition and c) refers to the tubular devices
Adapted from Du et al., 2007 and Logan et al., 2006

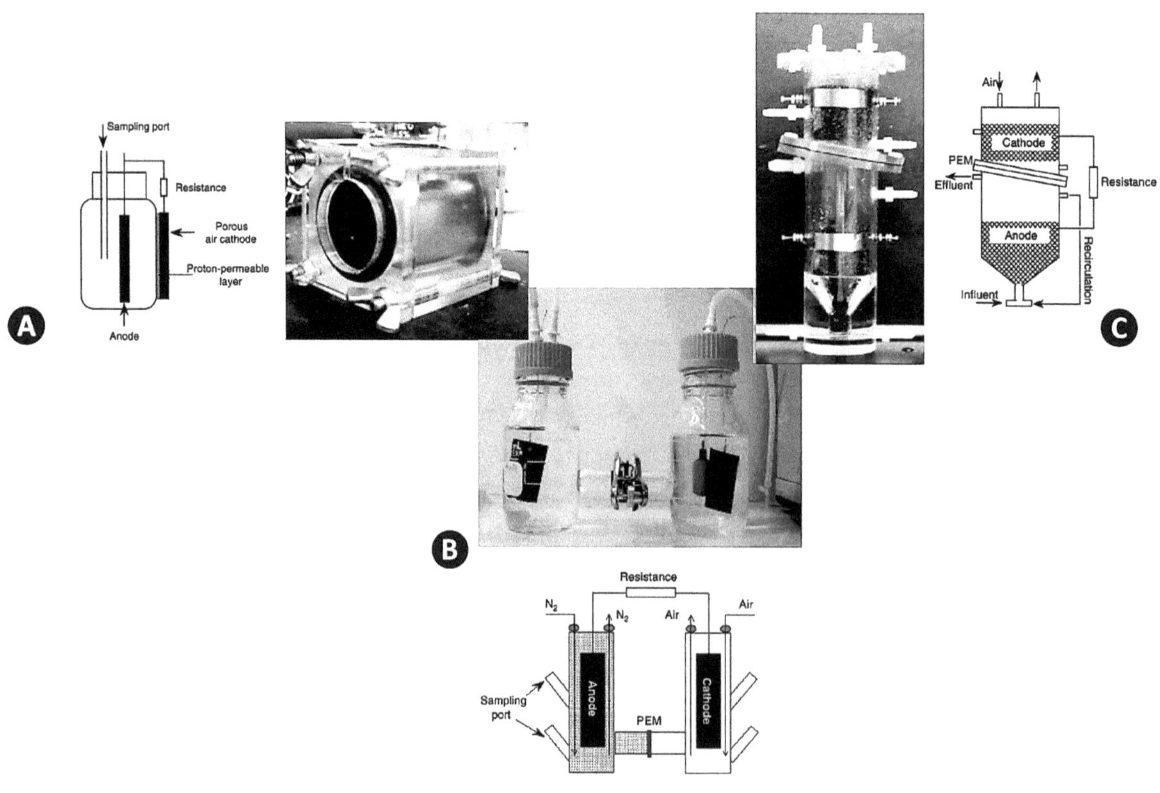

are greater than the power output (F. Davis & Higson, 2007; Debabov, 2008; Du et al., 2007; Logan & Regan, 2006; Logan, 2008).

Single-compartment MFCs have simpler and cheaper designs. The cathode is frequently exposed to air leaving behind the need to aerate its chamber. Bottle type devices are also the simplest examples. The most typical constructions consist of a single laboratory bottle with a side-arm where the cathode (generally a membrane) is placed. The side arm has a larger diameter than usual to increase oxygen exposure. Cube reactors are also a reality for single-chamber designs: the geometry reported by Liu and Logan, in 2004, described the construction from a 4 cm block of acrylic drilled to produce a cylindrical chamber with a radius of 3 cm. The reactor was oxygen free and sealed to prevent any leaks. Ringeisen's team, in 2006, proved the scalability of the design having built a bigger reactor. Other constructions have been reported. Of particular interest is a flat plate device (built to mimic a hydrogen fuel cell in hopes to achieve similar power outputs) built by Min and Logan. Their main objective was to come up with a construction of decreased space between the electrodes. The two plates were 15 cm wide and 15 cm long with only 2 cm of height. The characteristic serpentine flow path had only 0.7 cm wide and 0.4 cm deep.

Stacked MFCs can be seen as another type of constructions. They can be described has the connection, in parallel or series, of individual MFCs which can, in turn, be of any type. These devices are most common in applications where a single device can't power the application for what it was devised (Logan, 2008).

MAXIMUM POWER POINT TRACKING

The Theorem

The efficiency of a system is a measure of how well the output reflects the input. Computed by the ratio between two equal parameters – one referring to the input, another to the output -, its value can be anywhere between 0 and 1 (or, more commonly, 0 and 100%). Since there is still no way to remove a system's losses, an efficiency of 100% is still not attainable. So, in order to achieve higher efficiencies, every other parameter of the system has to be improved. Let's consider an electrical system, simplified to its Thévenin equivalent as in Figure 8.

Considering the system's losses have been optimized to the fullest, the efficiency of the system can be expressed by:

Figure 8. Thévenin equivalent circuit with a resistive load (R_L)

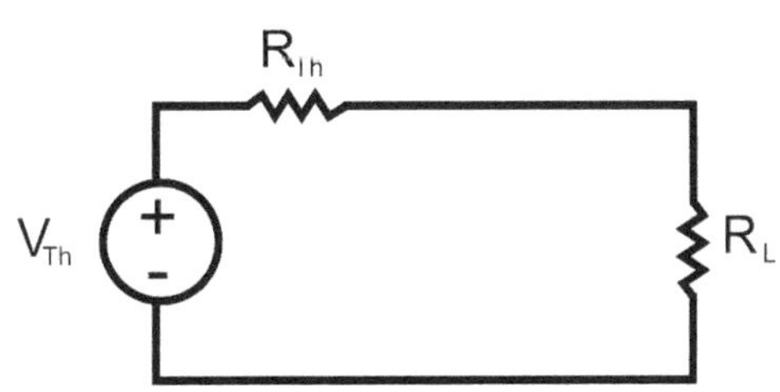

$$P_{eff} = \frac{P_{Out}}{P_{Total}} = \frac{i^2 R_L}{i^2 R_L + i^2 R_{Th}} = \frac{\left(\frac{V_{Th}}{R_{Th}+R_L}\right)^2 R_L}{\left(\frac{V_{Th}}{R_{Th}+R_L}\right)^2 R_L + \left(\frac{V_{Th}}{R_{Th}+R_L}\right)^2 R_{Th}} = \frac{R_L}{R_L + R_{Th}} \quad (1)$$

According to the ratio above, efficiency increases with the increase of R_L values relatively to those of R_{Th}. That difference would, however, limit the power actually delivered to the load. Proper dimensioning of the resistive load value will guarantee that P_{Out} is maximized:

$$P_{R_L} = i^2 R_L = \left(\frac{V_{Th}}{R_{Th}+R_L}\right) R_L \quad (2)$$

Since V_{Th} and R_{Th} are fixed, the power will ultimately be determined by R_L:

$$\frac{dP}{dR_L} = V_{Th}\left(\frac{\left(R_{Th}+R_L\right)^2 - 2R_L\left(R_{Th}+R_L\right)}{\left(R_{Th}+R_L\right)^4}\right) \quad (3)$$

Power is maximized for:

$$\frac{dP}{dR_L} = 0 \Leftrightarrow \left(R_{Th}+R_L\right)^2 = 2R_L\left(R_{Th}+R_L\right) \Leftrightarrow R_L = R_{TH} \underset{\left(\frac{V_{Th}}{R_{Th}+R_L}\right)R_L}{\Rightarrow} P_{R_L} = \frac{V_{Th}^{\ 2}}{4R_L} \quad (4)$$

With $R_L = R_{Th}$, $P_{eff} = 50\%$. So, in order to achieve the maximum power transfer between source and load, the efficiency of the system has to be of 50% (Mei, Shan, Liu, & Guerrero, 2011; Nilsson & Riedel, 2001).

Application Comparison and Review

As reported on [5], studies of power harvested with microbial fuel cells are usually conducted with polarization curves. Retrieving such data can be accomplished with a potentiometer (analogue or digital) or a potentiostat [5], [20]. However, their only use is for measuring the MFC's power, not to enhance it. The improvement of the power production levels on MFC's can be accomplished by optimizing its materials and architectures and by enhancing the energy collection system: hardware and software-wise. To enhance the harnessed power, other energy harvesting technologies, like photovoltaic systems, make use of maximum power point tracking algorithms (MPPT). These methods allow for the continuous tracking of a device's IV curve and for actively dimension the external load to it applied, in order to increase its maximum achievable power. Literature records for the application and study of such algorithms to MFCs fades in comparison to that of photovoltaic systems. For that reason, and in order to understand how to develop and apply such algorithms to MFCs, a thorough analysis of MPPT algorithms on photovoltaic systems is presented.

Maximum Power Point Tracking on Photovoltaic Systems

The use of the maximum power transfer theorem has significant expression with photovoltaic (PV) systems. Photovoltaic systems are subjected to varying conditions throughout the day: from sunrise to sundown, the irradiance and temperature levels of the photovoltaic panels change. Since PV systems have a limited lifespan and high initial cost (not only with the panel itself but also with the energy conditioning equipment), retrieving the maximum amount of power for as long as possible reduces the net cost of the system (Jain & Agarwal, 2007; Kamarzaman & Tan, 2014). For this, two solutions can be implemented: mechanical tracking of the sun and electrical tracking of the maximum power point. Mechanical tracking of the sun has a limited use because it is only useful for good irradiance conditions. For itself, this solution can't get the system to operate at the best possible efficiency. Electrical tracking, achievable with a power conditioning stage, can respond to different environmental conditions, improving the system's versatility (Jain & Agarwal, 2007). Electrical tracking corresponds to the application of maximum power point tracking algorithms.

To find the maximum power point (MPP) for a PV cell, without an algorithm, elaborate mathematical computations are needed and the solution has to be found with the help of numerical methods. If we start with a theoretical model of a PV cell, the IV behavior is given by (Enrique, Durán, Sidrach-de-Cardona, & Andújar, 2007):

$$I = n_P \left[I_L - I_S \left(e^{q\left(\frac{V}{n_S} + \frac{IR_S}{n_P} \middle/ AKT\right)} - 1 \right) - \frac{\frac{V}{n_S} + \frac{IR_S}{n_P}}{R_P} \right] \qquad (5)$$

And power by:

$$P = V \cdot n_P \left[I_L - I_S \left(e^{q\left(\frac{V}{n_S} + \frac{IR_S}{n_P} \middle/ AKT\right)} - 1 \right) - \frac{\frac{V}{n_S} + \frac{IR_S}{n_P}}{R_P} \right] \qquad (6)$$

If we intend to maximize the above parameter:

$$\frac{dP}{dV} = 0 \qquad (7)$$

Just by direct observation, the complexity of the above equations becomes clear. So, and in order to have a way to effectively apply the theorem to real systems, simplified models have been developed. To date, three models have been presented: a generic model, with a single diode, and two resistors (one in series and the other a shunt); the two diodes model, more accurate than the generic model but needing more parameters to compute; and the simplified model, which is in fact the two diodes model but with

Figure 9. At left: simplified model of a PV cell; at right: characteristic I-V curve

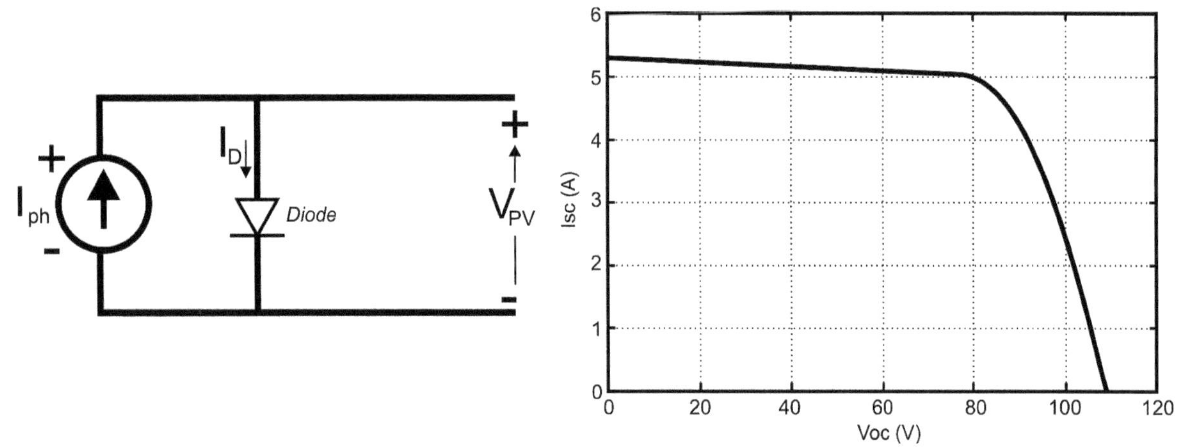

fewer parameters to compute (Chan & Phang, 1987; Ishaque, Salam, & Taheri, 2011; Kamarzaman & Tan, 2014; Rodrigues & Melício, 2011; Tsai, Tu, & Su, 2008).

According to the simplified model, represented on Figure 9, the IV characteristic of a PV cell can be described by:

$$I_{PV} = I_{Ph} - I_0\left(e^{V_{PV}/V_T A} - 1\right) : I_{Ph} = \lambda\left(I_{SC} + K_1\left(T - T_{Ref}\right)\right) \quad (8)$$

Where:

$I_0 \equiv$ cell's saturation of dark current;
$T \equiv$ temperature of the p-n junction in Kelvin;
$I_{SC} \equiv$ cell's short-circuit current at standard test condition (1000 W/m^2 and 25°C);
$K_1 \equiv$ coefficient of the cell's short-circuit current;
$T_{\mathrm{Re}f} \equiv$ cell's reference temperature;
$\lambda \equiv$ solar irradiance in W/m^2.

With these simplifications, applying maximum power point tracking (MPPT) algorithms is faster and more efficient (Kamarzaman & Tan, 2014).

Maximum Power Point Tracking Algorithms

The energy levels produced by PV cells or arrays are not constant and have a fluctuant nature. Even though applying MPPT algorithms makes the system work at maximum power, that point changes with environmental conditions. In order for a battery to be charged, or to deliver the power to the power grid, power conditioning stages are essential (Salas, Olías, Barrado, & Lázaro, 2006). They are responsible for stepping voltage up or down (boost or buck converter), creating AC or DC signals as the final application requires. It is also through them that MPPT algorithms are applied.

Figure 10. Block diagram for a typical topology in PV systems

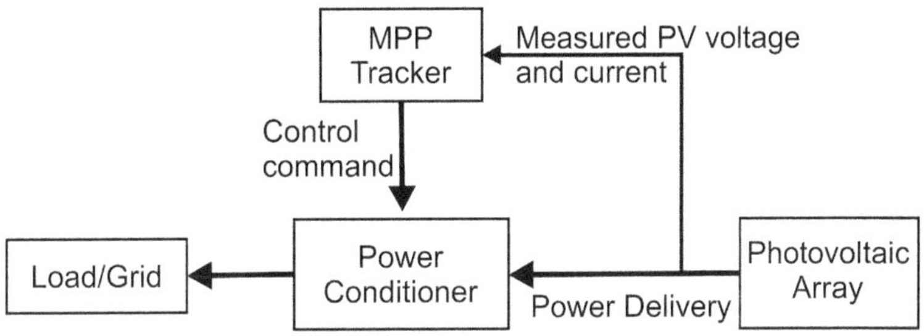

DC/DC converters are the main element in the power conditioning stage. They may be followed by DC/AC converters if power is to be delivered to the grid. For MPPT application, a MPPT controller is applied to the converter. It's inputs are collected from the PV cell/array and the controller acts on the converter which influences the cell's/array's operating point (Figure 10) (Balato & Vitelli, 2014; Carannante, Fraddanno, Pagano, & Piegari, 2009; Enrique et al., 2007; Jiang, 2013; Muthuramalingam & Manoharan, 2014; Petrone, Spagnuolo, & Vitelli, 2012; Salas et al., 2006; Xiao & Elnosh, 2011; Xu, Mu, & Jin, 2014; Zhou, Holland, & Igic, 2014).

DC/DC converters take a fluctuating, non-periodical, PV output voltage and, due to the relationship between its conducting (ON) and non-conducting (OFF) regimes – duty ratio – output an averaged voltage. This technique is called pulse width modulation (PWM). Duty ratio, D, is set by the commutation frequency of the transistor in the converter and ultimately leads to new operating points of the system. The DC/DC converter duty ratio is responsible for impedance matching and, therefore, working in the MPP (Alajmi, Ahmed, Finney, & Williams, 2011; Chao, 2014; Enrique et al., 2007; Esram & Chapman, 2007; Jiang, 2013; Kollimalla & Mishra, 2014; Mei et al., 2011; Muthuramalingam & Manoharan, 2014; Rahmani & Seyedmahmoudian, 2013; L. Zhang, Hurley, & Wölfle, 2011).

There are three major DC/DC converter configurations: the buck converter, the boost converter and the buck-boost converter. All three configurations are presented on Figure 11. The buck converter, voltage step-down or current step-up converter, is more efficient than linear regulators since it does not dissipate energy as heat. Its output voltage can be between 0 and its input voltage, V_i. The boost converter, voltage step-up or current step-down converter, can output voltages between V_i and ∞. The buck-boost converter is the more versatile configuration, being able to output voltages between 0 and ∞. Therefore, they are the best solution for MPP tracking (Enrique et al., 2007).

There are over 10 different MPP tracking algorithms. Their classification can be made according to several parameters and differs from researcher to researcher. Table 1 sums up different classifications for different researchers regarding the same algorithms. There are 18 major algorithm groups. Each group is represented on Table 1 with the following correspondences:

1. Uniform irradiance condition (Kamarzaman & Tan, 2014);
2. Partial shading condition(Kamarzaman & Tan, 2014);
3. Low-power methods (Lopez-Lapena et al., 2010);
4. Analogue (Esram & Chapman, 2007);
5. Digital (Esram & Chapman, 2007);
6. Fast and continuous tracking (Esram & Chapman, 2007);

Figure 11. DC/DC converter topologies: a) buck converter; b) boost converter; c) buck-boost converter. Each plot (from d) to f)) represents the resistance ratio vs duty ratio

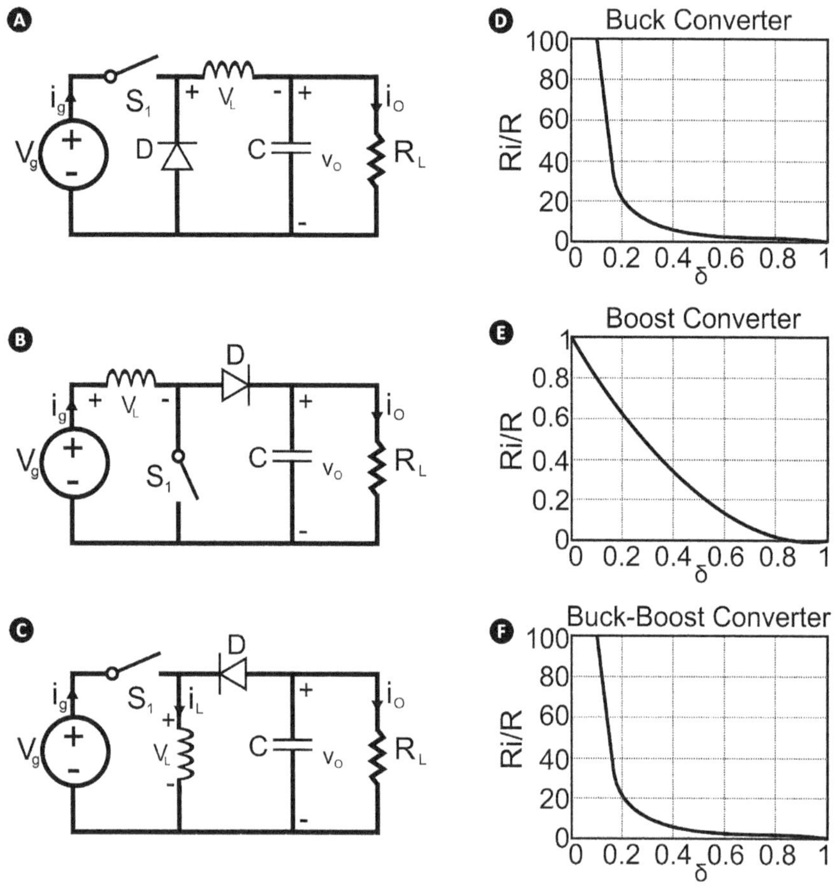

Table 1. Classification of the currently available MPPT algorithms

	1	2	3	4	5	6	7	8	9	10	11	12	13	14	15	16	17	18
Perturb and Observe	X		X			X				X					X		X	
Incremental Conductance	X					X	X			X					X		X	
Fractional V_{OC}	X		X	X				X	X		X				X		X	
Fractional I_{SC}	X			X				X	X			X				X		X
Ripple Correlation Control	X			X		X						X				X		X
Current Sweep	X						X		X									
Load Current or Load Voltage Max.	X			X		X												
dP/dV or dP/dI Feedback control	X				X									X	X		X	
Particle swarm optimization		X																
Genetic algorithm		X																
Differential evolution		X																
Artificial Neural Network		X			X	X			X				X					
Fuzzy Logic		X			X	X			X				X					

7. Minimum cost (Esram & Chapman, 2007);
8. Easy implementations (Esram & Chapman, 2007);
9. Periodic tunning (Esram & Chapman, 2007);
10. Heuristic search (Xiao & Elnosh, 2011);
11. Extreme value search (Xiao & Elnosh, 2011);
12. Linear aprox. (Xiao & Elnosh, 2011);
13. Intelligent control (Xiao & Elnosh, 2011);
14. Linear control (Xiao & Elnosh, 2011);
15. Two variable methods (Esram & Chapman, 2007; Salas et al., 2006);
16. One variable method (Esram & Chapman, 2007; Salas et al., 2006);
17. Direct (Esram & Chapman, 2007; Salas et al., 2006);
18. Indirect (Esram & Chapman, 2007; Salas et al., 2006).

Hill Climbing Techniques

Two of the most referenced MPPT algorithms are hill climbing techniques: the perturb and observe algorithm and the incremental conductance algorithm. The technique's name derives from the shape of the PV curve.

The perturb and observe (P&O) algorithms can be further classified according to its perturbation step or perturbation parameter (Rawat & Chandel, 2013).

The perturbation step can be:

- **Fixed:** Classical Perturb and Observe;
- **Adaptive:**
  - **Size Wise:** Adaptive P&O; and
  - **Direction and Size Wise:** Three Point Weight Comparison.

As for the perturbation parameter, options are:

- Reference voltage perturbation;
- Reference current perturbation; and
- Duty ratio perturbation.

The classical P&O algorithm is further explored. Two initial measurements of voltage and current are taken for two separate time instants. The second time instant values' result from a perturbation. The power is calculated and compared for each pair, by subtracting one another. If the difference is positive, that means power in the second instant is higher than for the first and the systems is further perturbed with the parameters of the first perturbation. If the difference is negative, then the perturbation direction is changed. The algorithm imposes a cyclic perturbation operation that ensures the operating point is at the MPP region. Its flow chart is presented on Figure 12 (Esram & Chapman, 2007; Kamarzaman & Tan, 2014; Yali Liu et al., 2014; Rawat & Chandel, 2013). P&O algorithms have been studied and improved according to several considerations. On (Kollimalla & Mishra, 2014), researchers have adapted the algorithm to better answer sudden changes in irradiance levels by perturbing current and using a boost converter. The team on (Carannante et al., 2009) also studied the application of the P&O

Figure 12. Flowchart for the classical perturb and observe method

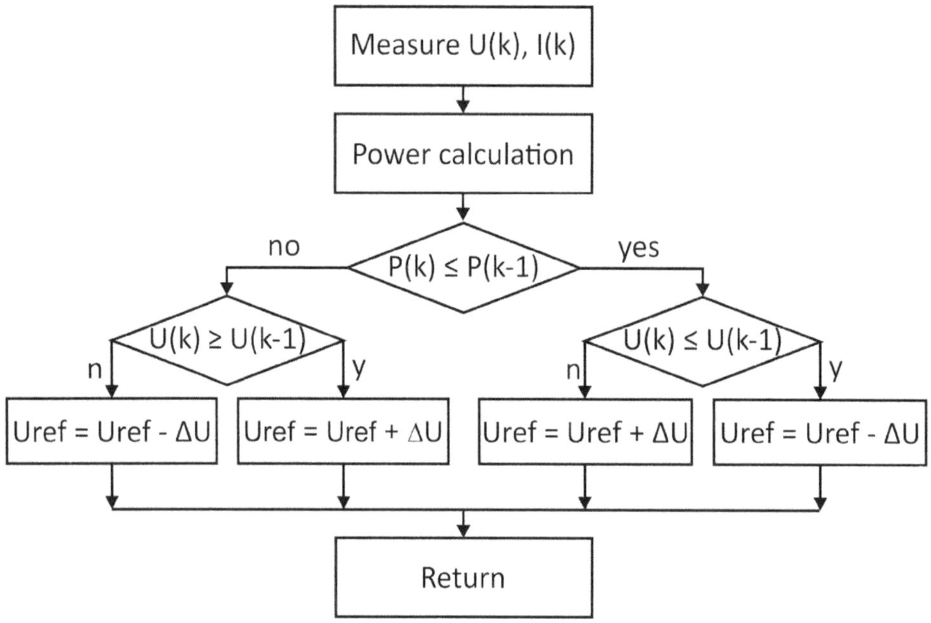

Figure 13. Flowchart for the incremental conductance algorithm

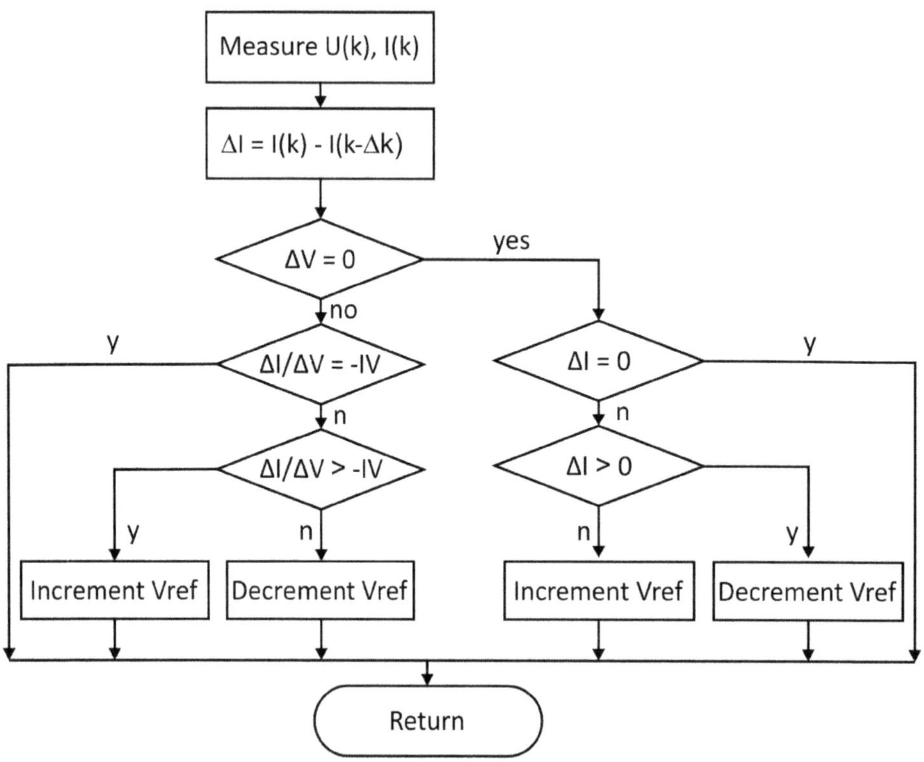

algorithm to variable irradiance conditions and current perturbations. The size and frequency of the P&O perturbations is studied on (Jiang, 2013). The three-point P&O algorithm is addressed with (S.-T. Kim, Bang, Lee, & Park, 2014).

As far as the incremental conductance algorithm goes, according to (Esram & Chapman, 2007), a voltage is imposed on the PV module at every iteration. After that, the incremental change on conductance is measured and compared to the instantaneous conductance for determining the relative position of the operating point regarding the MPP. The method is described as being more efficient than the P&O algorithm because it does not fluctuate over the MPP. Its flowchart is represented on Figure 13.

The incremental conductance method and the perturb and observe method are usually described as separate algorithms. A careful analysis of both, leads to the conclusion that they are, in fact, equivalent. They are studied both from a mathematical stand-point and from an application perspective and results show a deviation of less than 0.15% for their efficiencies (0.13% for dynamic conditions and 0.02% for static operation) (Sera, Mathe, & Kerekes, 2013).

Fractional algorithms

These algorithms are based on the idea that the voltage or current at the MPP is a set fraction of the open circuit voltage or short circuit current at a given instant. As the name implies, the fractional voltage open circuit algorithm measures the open circuit voltage V_{oc} of the PV system and sets its voltage operating point V_{MPP} according to:

$$V_{MPP} = k \times V_{OC} \tag{9}$$

The variable k has a value between 0.71 and 0.78 and is dependent on the PV characteristics. The fractional short circuit current algorithm measures the short circuit current I_{SC} of the PV system and sets its current operating point I_{MPP} according to:

$$I_{MPP} = k \times I_{SC} \tag{10}$$

As it was with the proportionality constant on the open circuit voltage algorithm, here k is also attributed according to the PV characteristics, usually resting between 0.78 and 0.92.

The use of the V_{OC} algorithm is simpler and cheaper than the use of the I_{SC} variant, although not so accurate. Both methods work at the MPP region rather than on the MPP itself. A pilot cell can be used with these systems so as to periodically measure the unloaded voltage and current values of the system. Otherwise, load shedding on the measuring process leads to power losses (Esram & Chapman, 2007; Kamarzaman & Tan, 2014; Yali Liu et al., 2014; Lopez-Lapena, Penella, & Gasulla, 2010).

Load Current or Load Voltage Maximization

A PV system connected to a power converter and subjected to a MPP algorithm makes the converter load work at its maximum power. The reverse, considering a lossless system and according to the maximum power transfer theorem previously approached, is also true: a load working at its maximum power will make the power converter induce a MPP working regime for the PV system. Since there are only three types of loads – voltage-source type, current-source type and mixed types – and considering non-negative impedance types, either voltage or current can be used to drive the system to its MPP, considering appropriate matching between load type and stimulus (Esram & Chapman, 2007; Kamarzaman & Tan, 2014; Salas et al., 2006; Shmilovitz, 2005).

Current Sweep

This algorithm usually takes up to 50 ms to run: it applies a sweep waveform to the PV system, gathers the data for building an IV characteristic curve and finds the MPP with it. Then the system is biased and its operating point is positioned at the MPP. Due to the time the algorithm takes to reposition its operating point, the MPP may shift and the repositioning may not be successful. Besides, and because of its stepped up energy requirements, it is only useful if the power consumption it takes is lower than the retrieved power from the system. This method is particularly useful for circumstances where multiple local maximum points are present. In this circumstance, other algorithms, like those from the hill climbing techniques, are not so accurate (Esram & Chapman, 2007; Kamarzaman & Tan, 2014; Salas et al., 2006).

Ripple Correlation Control

The ON/OFF behavior of the power converter device leads to ripples on the retrieved power. The ripple correlation control algorithm, or RCC, makes use of these ripples to achieve the MPP. With the time derivatives of power related with voltage or current derivatives, the operating point position is found relative to the MPP. Its repositioning is achieved by changing the duty ratio of the converter and getting the power gradient equal to zero. The RCC algorithm is quick and accurate, even in varying irradiance levels. Since it does not require any previous information or knowledge about the PV characteristics, its implementation is easier. The algorithm has analogue implementation and is considered a two variable method (Esram & Chapman, 2007; Kamarzaman & Tan, 2014; Salas et al., 2006).

Stochastic Algorithms

The irregularity and unpredictability of environmental conditions are behind the need to develop and adapt probabilistic and artificial intelligent methods. Fuzzy logic controllers (Alajmi et al., 2011; Rahmani & Seyedmahmoudian, 2013), neural networks (Esram & Chapman, 2007; Kamarzaman & Tan, 2014; Yali Liu et al., 2014; Muthuramalingam & Manoharan, 2014; Salas et al., 2006), particle swarm optimization (Chao, 2014), differential evolution or genetic algorithms (Esram & Chapman, 2007; Kamarzaman & Tan, 2014; Yali Liu et al., 2014; Muthuramalingam & Manoharan, 2014; Salas et al., 2006) are some examples of stochastic MPPT algorithms.

These methods are of digital nature and implementation and mostly inspired on the behavior of biological organisms. They are usually applied in systems where irradiance and temperature values may change rapidly. Apart from fuzzy logic controllers, a training period is needed for the accurate operation of these methods. This training is usually conducted with the actual PV system so that the algorithm can adapt itself to its particular conditions. Due to this, these methods have a higher cost and are less common in large scale applications (Esram & Chapman, 2007; Kamarzaman & Tan, 2014; Xiao & Elnosh, 2011).

Maximum Power Point Tracking on Microbial Fuel Cells

The improvement of power production levels on bioelectrochemical systems (BES) can be accomplished, not only by optimizing materials and architectures, but also by enhancing the energy collection system: both in hardware and in software. Of particular interest are the improvements on microbial fuel cells (MFC) for their power production capabilities.

The most frequent choice for the final destiny of the harvested power is a resistor: with a fixed or floating value (potentiometer), it simultaneously connects the anode to the cathode, without shorting them, and dissipates the harvested energy as heat (Dewan et al., 2014; Liang et al., 2011; Park & Ren, 2012a; Wang, Park, & Ren, 2012). The use of a potentiometer, or, for more accurate results, a potentiostat, is, nonetheless, extremely useful for conducting polarization trials and finding the MFC's internal resistance. According to the maximum power point theorem, the cell's internal resistance can be determined by finding the external resistance value that leads to the highest retrieved power (Molognoni et al., 2014; Park & Ren, 2012a; Wang et al., 2012).

In order for rendering the energy as useful, it has to be stored with capacitor based circuits (passive strategies) or DC/DC converters (active strategies) (Wang et al., 2012). Passive strategies correspond to the use of capacitors or charge pumps. Studies have been conducted so as to explore the benefits of discharging the capacitor to a load or back to the fuel cell. In (Liang et al., 2011) proof was made that a BES power output could be increased if the storing capacitors discharged back to the cell. That, however, converts the BES to a microbial electrolysis cell (MEC), a cell whose sole purpose is to use microorganisms to oxidize organic matter. Still, storing the energy to a capacitor and making it discharge to a load results in higher power than simply using a load (Dewan, Beyenal, & Lewandowski, 2009). The charge and discharge frequency of capacitors was studied on (Ren, Xia, Yuan, Liang, & Huang, 2013) and higher average currents were obtained for higher switching frequencies, though a too much high value may not allow an appropriate charge recover efficiency by the cell.

Though of cheap and easy implementation, these hardware only techniques don't allow for a continuous adaption to the MFC's dynamic behavior. With time, and under the influence of several parameters, the MFC's internal resistance changes; in order to achieve the maximum power transfer, the charging system has to adjust itself to new conditions.

Active devices are capable of doing so. DC/DC converters are of hardware and software nature and their behavior can be made to dynamically fit to new conditions, improving the harvested energy levels (Liang et al., 2011; Meehan, Gao, & Lewandowski, 2011; Park & Ren, 2011, 2012a; Wang et al., 2012).

Conventional DC/DC converters, as the ones used with PV systems and described above, have significant energy losses at low input voltages, like the ones of MFCs. The diodes typically used have a forward voltage drop of 0.4 V and that can be higher than a cell's output voltage (Y. Kim, Hatzell, Hutchinson, & Logan, 2011; Park & Ren, 2012a). A solution, reported in (Boghani et al., 2014; Y. Kim et al., 2011), is the series connection of several cell's. Without proper corrections, however, a voltage reversal phenomenon usually occurs. Voltage reversal happens when anode potential becomes positive and similar to the cathode due to low substrate concentrations (relative to that of an adjacent cell). It takes only one cell with voltage reversal to induce failure on the assembled set. The use of two stage DC/DC converters can address this issue and is further applied to single or multiple MFCs. The first stage, prior to the DC/DC converter itself, is usually a charge pump: this stage stores energy to a level usable by the DC/DC converter. A simpler array of capacitors could be used to replace the charge pump but that would lead to voltage drops and hinder the converter functioning (Meehan et al., 2011). The topologies for the DC/DC converter following the charge pump have also been studied, particularly those who achieve voltage step up: boost topologies, flyback topologies and a oscillator voltage-lifted topology. The latter, developed by Degrenne et al, could be customized but presented lower efficiency levels when compared to the boost or flyback topologies (which achieved similar efficiencies of around 80.6%) (Nicolas Degrenne, Allard, & Buret, 2011).

To achieve an appropriately dimensioned converter, 3 major considerations have to be made:

1. **Inductance:** The inductor acts as an intermediate storage element, affects current generation and current change rate, and sets the MFC terminal voltage;
2. **Duty Cycle:** Establishes the relationship between the energy extraction and the MFC recovery;
3. **Switching Frequency:** the frequency of the converter.

Experimental results from (Park & Ren, 2012b; Wang et al., 2012) show that low inductances and small frequencies lead to high currents and low voltages. On the other hand, small duty cycles induce small currents and higher voltages. Changes on the 3 parameters above affect cathode potentials, while the anode potential remains stable.

To further minimize the power losses of a DC/DC converter, diodes are replaced with transistors. The work by Jae-Do Park, reported sequentially in (Park & Ren, 2011, 2012a, 2012b), describes the adaptations implemented on a DC/DC converter in order to replace the diode with a MOSFET transistor, since an appropriate turn-off time has to be dimensioned to guarantee no reverse power flow happens. These changes allowed for a 73% improvement on the efficiency of the harvesting process.

Another great thing about DC/DC converters is that they allow the implementation of MPPT algorithms, just as with PV systems. MPPT algorithms will guarantee that the fuel cell and the power management system operate at the ideal power transfer point, maximizing the retrieved power. The first work with MFCs and MPPT algorithms, but no DC/DC converters, was by Woodward et al (Molognoni et al., 2014; Woodward, Perrier, Srinivasan, Pinto, & Tartakovsky, 2010). His team investigated the effect of the perturb and observe algorithm, the gradient method and the multiunit method. The gradient method was discarded as a suitable MPPT technique as it failed to optimize the cell to which it was associated. The multiunit method had a faster convergence than the perturb and observe but its tuning is more complex since it is based on the static characteristics of the MFC where it is applied. The perturb and observe algorithm is easily tuned but the frequency of the induced perturbations has to be carefully dimensioned so as to not induce instability in the system (Woodward et al., 2010). The relationship between the start-up time of a MFC and the use of a MPPT algorithm was also studied (Boghani, Kim, Dinsdale, Guwy, & Premier, 2013). Either with or without poised anode potentials, the use of a MPPT algorithm reduced the start-up time by 45% when compared to the use of a fixed load. A fractional open voltage method was also researched but results were inconclusive (N. Degrenne, Buret, Allard, & Bevilacqua, 2012).

A careful study and test of other MPPT algorithms should be conducted since extended benefits can be achieved with other techniques. Due to the dynamic behavior of a MFC, an algorithm should have:

- A wide input range so that it can answer to different characteristics of wastewater in real applications;
- A quick answer so that the cell may work at the MPP for as long as possible;
- Increased stability to not induce major changes in the bacterial community of the cell;
- Limited complexity so that it may be cheap and less power consuming.

SOLUTIONS AND RECOMMENDATIONS

Microbial Fuel Cell for Laboratory Experiments

When compromising with a model reactor for experimental evaluations, two key aspects were found to be of greatest importance:

- The reactor needs to be versatile and cheap: using the same building blocks, the reactor has to guarantee that studies relating power and electrode positioning, geometry and configuration, membrane composition, bacterial communities, catholyte and anolyte and substrate elements can be made;
- Its operation has to be safe and reliable: its architecture and composition has to assure that no leaks, between chambers or lids, will occur.

A first approach would be to use two cubic geometries. The production costs would be lower and the instrumentation positioning would be easier. The electrodes would to be positioned at the edges of the chamber, as it happens with air-cathode architectures. This would limit biofouling of the electrodes even though also limiting the reactor power production capabilities.

In spite of the advantages presented with these choices, a cubic reactor would not be as versatile as needed and there would be some dead areas on the reactor chambers. These areas would allow for bacterial deposition and would hinder the reactor functioning and maintenance.

A cylindrical geometry has been chosen for current lab trials. This geometry solves the issues with the dead areas previously mentioned while allowing for free positioning of the electrodes within the reactor. This is achieved with a crown structure - of diameter size equal to the inner diameter of the cylinder tube - made out of two individual parts. Figure 14 shows the currently laboratory implemented model.

The membrane positioning was made outside of the reactor chamber. Doing it inside the chamber presented a risk of not adequately separating the anode and cathode mediums (diameter differences) and also diminished the reactor total reaction volume.

The total reaction volume for the reactor corresponds to the volume of each chamber minus the volume occupied by the crown assemblies and the lid projections (for adequate fitting). Each crown assembly can be seen as a hollow cylinder with an outer radius of 4.6 cm and an inner radius of 3.6 cm. The volume for each hollow cylinder is, approximately, 12.88 cm^3. Each chamber is composed of a 4.6 cm radius cylinder on 12 cm (12.5 cm minus the lid projection, 0.5 cm), totaling about 763.41 cm^3. The total reaction volume for each chamber is, therefore, 750.53 cm^3 (the volume for each cylinder minus the volume occupied by a crown assembly), or 238.9π cm^3. Considering that the reactor has two chambers, the total reaction volume for the reactor is then 1501.05 cm^3 or 1.5×10^{-3} m.

Electrical Characterization of the MFC

The first data retrieved with the described cell corresponds to an outline of its electrical characterization. Implementing this as the first trial was needed so as to validate the geometry and its power production capabilities. In order to outline the influence of the external resistance on the MFC's power production capabilities, trials were conducted by manually varying a resistance load, applied to the MFC, and mea-

Figure 14. Final prototype. A) shows the prototype model built with a 3D CAD program; B) and C) highlight the crown and tube dimensions respectively

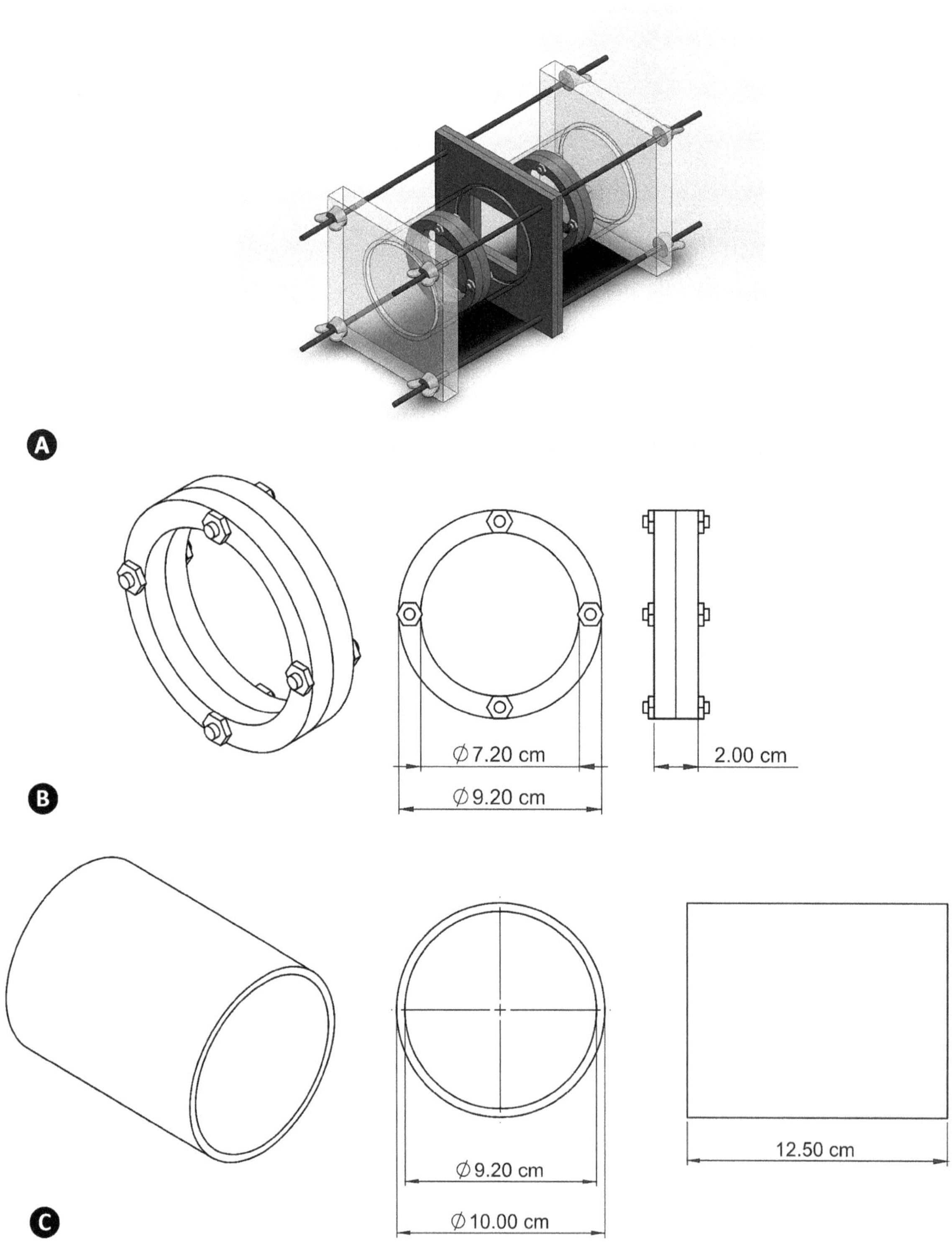

Figure 15. Assembled reactor

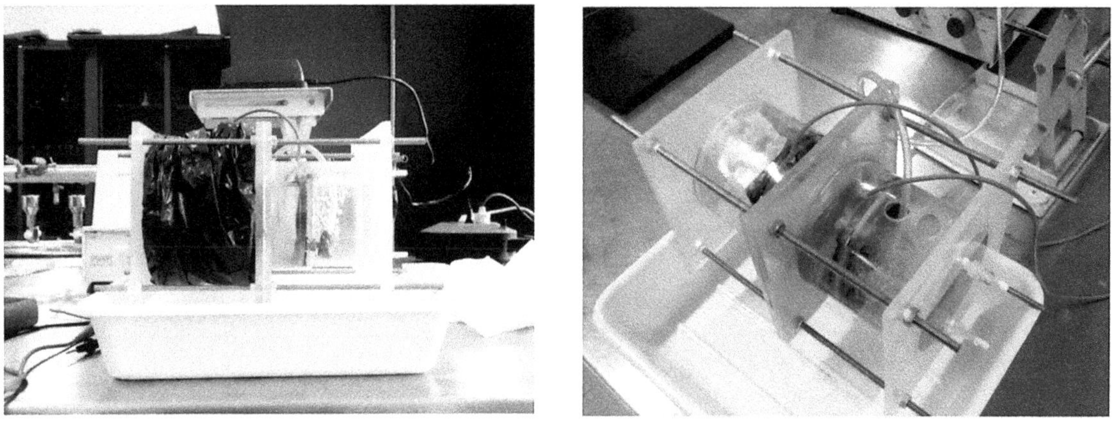

suring its voltage and current. The application of a load is a needed element to improve energy production (as stated in (Boghani et al., 2013)) so the trial would not be accurate if a load was not connected.

The MFC was evaluated for an anaerobic and light protected anode chamber and an aerated and light exposed cathode. The assembled reactor, on Figure 15, corresponds to the conceptual model of Figure 14. Besides the reactor, only an aeration device (air pump) and an external electrical load (a variable resistor) were used.

The substrate used as a bacterial medium was a solution of acetate. A period of roughly 2 days was provided for reactor acclimation. A proton exchange membrane (Ultrex CMI-7000, (Membranes International Inc., n.d.)) was applied between chambers. The electrodes were of a carbon fiber mesh and size equal do the crown inner diameter, 7.2 cm. The experimental setup assembled for polarization trials consists of a potentiometer, varied between 0 and 5 kΩ, and multimeters for evaluating circuit voltages and currents. Open circuit voltages were also measured between each resistor increase. A set of 14 voltage values was retrieved and used for building the polarization curve plotted on Figure 16.

The retrieved data shows that high external resistance values limit current output and voltage production between the module's terminals (plot A). As was stated earlier on, the external resistance also affects the module's internal resistance (plot B). According to plot C, the cells maximum power point occurs approximately at 39 $mW.m^{-3}$ for 220 mV and 0.2 mA (the hill on plot C corresponds directly with the crossing of the curves on plot A). For that power level, the external resistance was of 1.09 kΩ and the estimated internal resistance of 0.96 kΩ, which supports the maximum power point transfer theorem already approached.

Manually varying a resistance load, applied to the MFC, and measuring its voltage and current, turned to be cumbersome, time consuming and not as accurate as needed. The method was only used to validate the power generating capabilities of the cell. Improvements of the method and the development of a fully capable platform for biochemical and electrical characterization of a microbial fuel cell are presented next.

Figure 16. Plots for the experimental data. A) voltage and current curves for increasing external resistance values; B) internal resistance variation according to the voltage and current of the module; C) normalized power curve (for a reactor volume of 1.49 x 10^{-3} m^3)

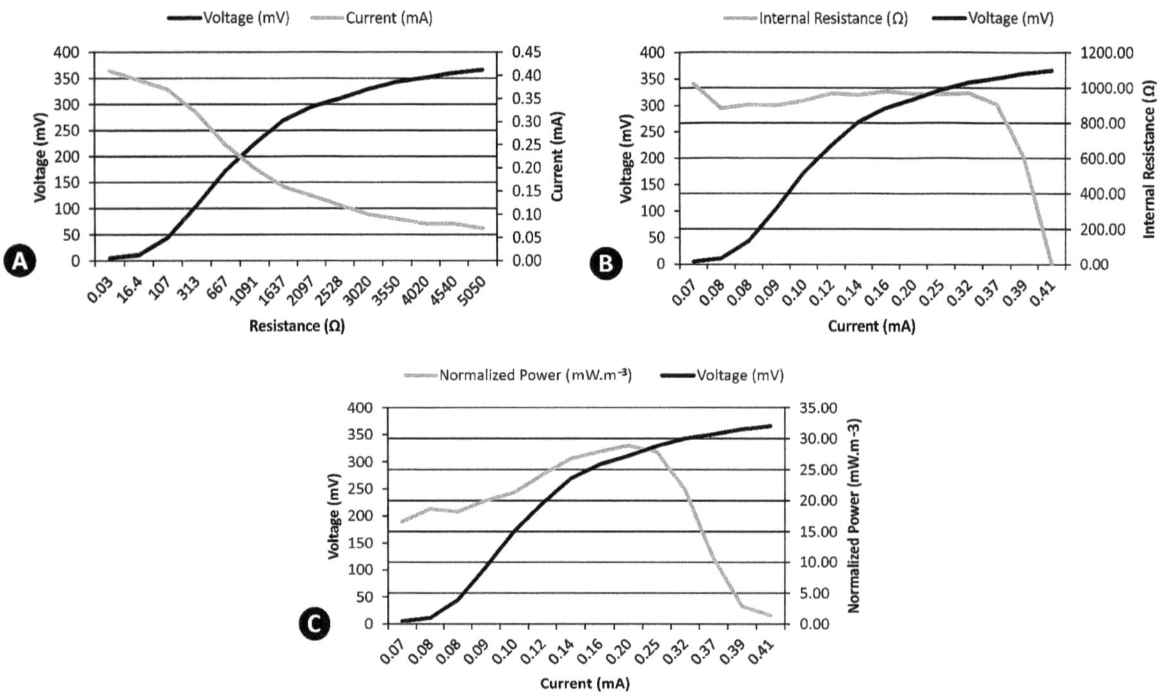

Validating Solutions: MFCs for Wastewater Instrumentation

A detailed study on how MFCs operate is essential and aspects such as setting separate cameras or using single cameras, the geometry of the electrodes and solutions to increase their surface area, the behavior of bacteria for different pollutant load conditions and hydraulic loading rates, the material for manufacturing the electrodes, and influence of external factors such as temperature, the availability of organic matter, and electric loads should be studied in order to adequately develop a MFC for sewage treatment plants beds. This study needs to be aided by an instrumentation platform, gathering sensors like pH, dissolved oxygen (DO), oxidation reduction potential (ORP), temperature sensors and a load varying assembly. The platform will allow for the study of the biochemical changes induced by the implementation of different loads, correlating its influences and allowing a better understanding of the effects of power conditioning strategies on the biochemical characteristics of a MFC. It can also be configured so as to effectively implement MPPT algorithms. With a MPPT algorithm in place, a detailed biochemical study can be conducted and further improvement strategies may become clearer.

The use of exchange membranes should also be studied. The absence of exchange membranes can reduce the cost of the energy capture mechanism, while reducing the need for maintenance, and increasing the lifetime of the device. It will be necessary to identify the most common specie/genera of bacteria living in sewage treatment plants and quantifying their performance will also be needed in order to predict the electricity generation capabilities of the assembly. This task should be carried out through different evaluation procedures in laboratory trials (where the operating conditions are defined and controlled) and at a real scale sewage treatment plant (where the operating conditions are less predictable).

The adaptation of the MFC technology to the sewage treatment plant bed should be achieved without any negative impact on the performance of the bed.

Previous studies of the authors have shown that the operation of an MFC varies throughout the day, either by temperature variations or the availability of substrate. The use of maximum power point (MPPT) techniques capable of automatically adjusting the mechanism for a more efficient energy collection can limit negative influences whilst enhancing the system's energy retrieval. Inadequate loads at the cell's terminals result in decreased output voltages. The use of a boost converter topology can help solving this issue, provided its operation is adjustable. A suitable system should have the ability to acquire the energy necessary for its start-up, while having two different phases of operation. Initially, the system would harvest the energy necessary to start the main control unit (MCU), which would also setup the MPPT mechanism. After stable voltage levels are achieved, the MCU, which performs the management of the energy harvest, would control the second phase of operation. This phase would consist of the tracking of the power harvesting efficiency.

The energy produced, if properly managed and stored, could be used to supply instrumentation devices for monitoring the plant (e.g. physical-chemical probes, pumping devices and water level measurement devices). These findings are important since they contribute to extending the lifetime of the equipment, without using batteries, which will decrease the maintenance costs and minimize the pollution associated with batteries.

The Envisioned System

A distributed system, energetically independent, that makes no use of batteries and is composed of a low power, wireless sensor network can be a proof of concept for the current state-of-the-art technology. A proposed system would be composed of two modules: the energy harvesting module (module I); and the instrumentation and communication devices (module II). The system would use adaptive sensing: while module I continuously stored the energy into a storage element (e.g., solid state batteries or supercapacitors), module II would wake up with certain criteria, would do a pre-defined set of tasks and would go back to sleep.

Having identified the devices to equip the instrumentation platform, time and energy requirements for each wake-up cycle would be very well-known and, therefore, so would be the systems' energy needs.

This behavior would allow operation with low energy demanding. The concept also considers a dimensioning parameter: the dimension refers to the number of two module sets on a given process.

Having two possibilities for dimensioning (system's dimension and module size), adds to the system's versatility and application. The system could, this way, increase its range and aid in covering large, medium or small deployment fields; a terminal set, also equipped with a module I, would collect all data and make it available to a user (on a cloud server or via USB). Figure 17 shows the system's setup.

Instrumentation Platform

Several studies have shown that for MFCs using similar bacterial population and fuel sources, the outputted power can have up to an order of magnitude of difference. The set of all parameters related with the reactor configuration are usually referred to as over-potential or internal resistance (B. Kim, Chang, & Gadd, 2007). Besides internal resistance parameters, the load applied between the electrode terminals also critically conditions the retrieved power levels.

Figure 17. System overview

The best way to study and determine how to enhance a MFC's power output is to evaluate the above parameters in vivo, due to their dynamic characteristics and co-dependent behavior. As there are so many parameters that could be of importance (and, currently, no way to simultaneously measure them all), a choice needs to be made. For the first studies to be conducted with the architecture on Figure 14, three variables have been chosen for over-potential measurements: pH, dissolved oxygen (DO) and oxidation-reduction potential (ORP).

The ORP sensor will allow a continuous estimation of the MFC's power generation capabilities. As with any other fuel cell, the potential difference between electrodes is a major player in the cell's expected electrical output. The data retrieved by this sensor will also allow for the accurate pinpointing of when the fuel source (substrate) has to be replaced: a low absolute ORP value on the anode chamber means that the oxidation potential is low and the MFC's power production capabilities will be limited.

The DO parameter is a regular experimental measurement when working with water derived substrates. DO sensor measurements can be expressed in mg/L or % saturation and they translate the concentration of oxygen in water. This sensor is usually applied at the cathode chamber of an MFC and guarantees a close monitoring of the oxygen values: low oxygen values at the cathode lead to a drop in the cell's electrical output, as the reduction reaction will be severely hindered.

Measuring pH is a necessity due to the requirements of the biological population at the anode: suitable conditions of temperature, substrate and pH must be maintained in order to keep the oxidizing potential at its highest.

All sensors where acquired to Atlas Scientific™, after carefully considering the commercially available offerings. All three sensors have their working principle based on electrochemical batteries: they are composed of two electrodes (an anode and a cathode), with the reference electrode submerged on a known solution and separated by a permeable membrane.

As for the external load evaluation and MPPT algorithm studies, a customized assembly was developed and is described next.

All these peripherals should be configurable so as to address different working conditions and trial runs. The platform should also allow for the storage and display of the retrieved data so that real-time feedback is available as well. This is only achievable with a microprocessor.

pH Sensoring

pH is a measurement of how acidic a solution is: it is a logarithmic measure of hydrogen ions (H^+) concentration per liter of solution and can have values between 0 and 14 (corresponding to hydrogen ion concentrations from 1 to 10^{-14} [mol/dm^3]); an acidic solution has a pH value below 7 (a concentration of H^+ inferior to 10^{-7} [mol/dm^3]) and a basic solution a pH value above 7. If the measured pH value is 7, then the solution is said to be neutral.

A pH sensor works with a pH probe, which has a measuring electrode, a reference electrode and a temperature sensor, and a conditioning circuit.

The measuring circuit can be summed up as a battery where the positive terminal is the measuring electrode and the negative one the reference electrode. The circuit will compare the measuring electrode's potential (which has a direct relationship with the H^+ concentration) with the reference electrode's potential. As described before, this sensor works as an electrochemical battery and H^+ ions are spent on the measuring process.

The reference electrode, even when the pH sensor is submerged, is never in direct contact with the solution to be measured: its glass casing holds a buffer solution (usually potassium chloride) and, this way, the sensor hardly gets decalibrated.

The other electrode, on the other hand, gets in contact with both the buffer solution and the solution to be measured: this last contact is maintained not directly but rather through a H^+-permeable glass (doped with Li^+).

The acquired values vary according to the solution's temperature and so pH sensors are usually equipped with a temperature calibration circuit (knowing the pH fluctuations on the buffer solution, an offset can be applied and the values can be corrected). A preamplifier can also be found on the conditioning circuit: this element takes the electrodes' signal and processes it (lowering its impedance and increasing its strength and stability) so it no longer is easily affected by electrical noise (Kuphaldt, n.d.).

Dissolved Oxygen Sensoring

There are several types of dissolved oxygen (DO) sensors. They can be made to operate based on optical sensing methods or electrochemical sensing methods, quite similar to the pH sensor. Optical sensors measure luminescence either regarding its lifetime or its intensity. Clark electrochemical sensors or membrane-covered sensors are oxygen sensitive and their behavior is similar to that of a pH sensor. Atlas Scientific™ sells a galvanic DO sensor: in this electrochemical configuration, an anode (Zinc) and a cathode (Silver) are confined in an electrolyte solution delimited by an oxygen permeable membrane. When the probe containing these elements contacts with oxygen molecules, they diffuse through the membrane and get reduced at the cathode. The reduction rate creates an electric potential which is proportional to the oxygen diffusing through the membrane. The cycle keeps repeating because oxygen keeps being reduced at the cathode and new oxygen molecules are used for repeating measurements (YSI, 2009).

Oxidation-Reduction Potential Sensoring

The oxidation-reduction potential sensor has the same working principle has the other two sensors. The main difference is that the sensor's permeability is not as ion-selective. The sensor answers to the balance of oxidation-reduction reactions of where it is placed. The two electrodes on the system have two separate functions: the reference electrode provides a constant stable output value for comparison; the measuring electrode (made out of platinum) acts as an electron donor (reducing agent) or acceptor (oxidizing agent) (Myron L Company, 2012; Walker & Bier, 2009).

The sensor will output values between -1023.99 [mV] and +1023.99 [mV] and the lower the outputted value, the more reductive character (incites the reduction of solution species, the gain of electrodes) is the solution where the sensor is used (Atlas Scientific, n.d.). The ORP sensor at the cathode chamber of a MFC is expected to have negative values.

Load Varying Assembly

The load varying assembly consists of a digitally controlled potentiometer and a switch. In order to accurately establish a relationship between the internal resistance of the cell and the load at its terminals, the open circuit voltage measurement is essential. Therefore, the use of a switch is of the uttermost importance as it allows for the recovering of this measurement. The digital potentiometer used was a Maxim 5450 and the switch a Maxim 4644. The digital potentiometer has 255 resistive discrete values and at power on is at midscale (position 127). The wiper makes a make-before-break transition and guarantees, therefore, that the cell's load is never left floating.

After receiving the configuration data from a PC, instrumentation platform decodes the data and proceeds with the setup of the polarization trials. The trials begin by setting the resistance value to the smallest load value input by the user with the interface. The load increment step is calculated according to the maximum and minimum load values and the number of points of the trial. Before measuring the load voltage, the switch opens and the microprocessor measures the open circuit voltage after a configurable time interval. After this, the switch closes and applies the load to the cell. The voltage measurement is retrieved after another settable time interval. The cycle is repeated until all the load values are retrieved.

Fast Prototyping Platform: mbed

To manage all the peripherals and to allow the configuration of the platform and the storage and display of the retrieved data, a microprocessor was chosen.

The mbed technology is a platform based on an ARM processor (Cortex M0 or Cortex M3) with a free remote compiler accessible with a web browser and internet connection (also being possible to have an offline toolchain like Code Red, GCC or IAR). Alongside this tool there are discussion forums, blogs and online documentation to use freely; all of which are neatly organized and appropriately detailed. There is also a vast library set (both for hardware and software) allowing for each user to quickly and efficiently implement some basic operations.

The mbed hardware available is updated very frequently. From all the available processors, two are highlighted: the NXP LPC1768 and the NXP LPC11U24. Both have a built-in USB Flash programming interface, making its use identical to using a common USB Flash Drive. They also support a virtual serial port, enabling communication with any programming language capable of communicating with a COM port (Labview or Matlab for instance).

The NXP LPC1768 and the LPC11U24 are of the same size and quite similar aspect: the first has Ethernet and a M3 processor and the second has a low power M0 processor without CAN or PWM capabilities (Mbed, n.d.).

The NXP LPC1768 was the chosen platform as it had already been used in other related works. Its intuitive use was the most determining aspect for choosing this hardware as the microprocessor used for data acquisition and management.

Adapting and Integrating Peripherals

Considering the characteristics of the platform, a printed circuit board (PCB) was developed.

According to the Atlas Scientific™ sensors' manufacturer, the embedded circuits responsible for the biochemical data acquisition should be connected to a BNC plug: these plugs allow for better noise immunity, a safer and sturdier connection and impedance matching between peripherals.

The PCB had to be able to integrate the embedded sensoring circuits, the BNC plugs and associated integrated circuits, the load varying assembly and the microprocessor in the smallest possible area (higher PCB area imply higher production costs). To accommodate all these devices, the PCB was idealized to have two layers: a top layer for the microprocessor and load varying assembly, and a bottom layer for the BNC plugs and the embedded sensoring circuits.

The MFC monitoring and acquisition system is shown in Figure 18. Careful study of the circuit's positioning and their connections was needed: the sensoring circuits needed to be connected to the mbed plataform serial ports (TX port of the sensor IC to the RX port of the NXP LPC1768 and so on). Though they are not represented (for an improved clarity of the connections), each top and bottom PCB layer was filled with a copper filling (which assured the ground connections) to improve connectivity and decrease PCB production costs.

Figure 18. Experimental setup platform. At left, the block diagram and, at right, a picture of the assembly with the biochemical sensors

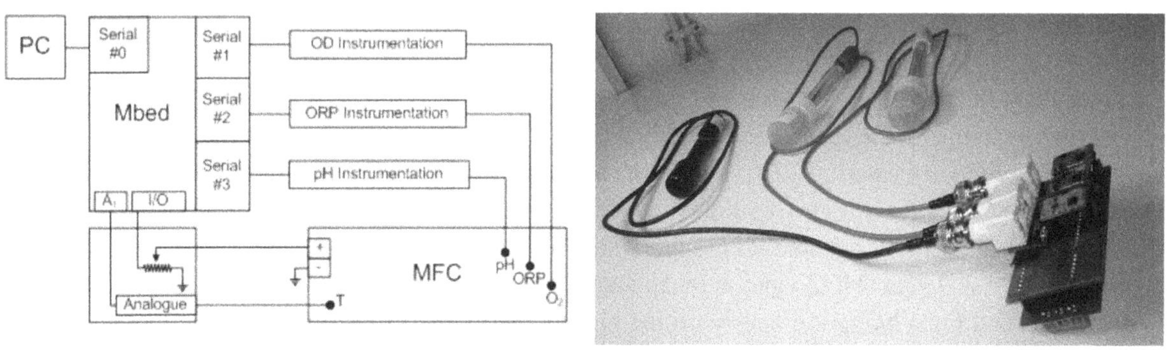

FUTURE RESEARCH DIRECTIONS

Work With the Instrumentation Platform

The set-up of a MFC should account for the reactor acclimation, considering that the microbial population adapts better to the substrate and the cell's electron transfer characteristics (metabolic conditions) if it is subjected to such conditions for adequate time. That time can vary between days, weeks or months and compromises have to be made in order to limit the extent of that timeframe. Usually, the cell's maximum voltage determines when the acclimation is adequate: a load, dimensioned according to previous studies of cell with similar characteristics as the one to be studied, is connected between the cell's terminals and its voltage is monitored at constant time intervals. When the current through it has smoother fluctuations, then the cell can be considered acclimated. Typical acclimation times can be of 10 to 20 days but a choice of electro active engineered bacteria can decrease this period to 15 to 30 minutes.

The developed tool can be used as early as the bacteria are introduced at the anode. The biochemical data recovered during the acclimation time can, not only help to determine if the cell is acclimated, but also track changes to the anolyte during that period. Biochemists can use this data to bioengineer improved versions of the bacteria. On top of that, the polarization curves recovered for this timeframe are expected to help to dimension the adequate load for the acclimation, also decreasing the time needed for such procedure. The acquisition period for both the sensors and the polarization curves can be equal to the one chosen for the cell's voltage measurement. This time is, however, dependent on previously described parameters (relationship between bacteria, substrate and metabolic conditions) and should be adapted after experimental data is retrieved. If no previous trials are available, typical periods are of about 30 minutes. The flow of the substrate to the cell, in continuous or batch mode, also plays a part in the acquisition period configuration: in continuous mode, smaller periods should be applied; for batch operation, a period smaller than that of the batch exchange will be enough. The load maximum and minimum limits and the number of points for the polarization curve can be set according to previous studies, according to the polarization curves during the acclimation period or by overall estimation. This last choice will consider load values with no or small influence on the cell's behaviour.

Research Trends on Microbial Fuel Cells

According to the ISI Web Of Knowledge data (Thomson Reuters, n.d.), the first article published on microbial fuel cells dates back to 1962 and belongs to Davis and Yarbrough (J. B. Davis & Yarbrough, 1962). Their investigation studied if ethane and glucose could feed bacteria and produce an electrical output.

Since then, the number of publications on the subject has increased dramatically: from all the 7894 results for a search with the topic "microbial fuel cells", 5352 (about 68%) of them have been produced since 2010. For this same period (2010-2014), the most preeminent authors are Bruce E. Logan, from the Pennsylvania State University (USA), Xia Wang (China) currently at Nankai University though originally from Logan's team, and Bert Hamelers (Netherlands) from the center of excellence for sustainable water technology (wetsus). The MFC topic is under intensive investigation: countries like the USA, China, South Korea and India being the ones with the highest number of contributions. Figure 19 shows the publications affected to each main researcher on the field and Figure 20 the country-wise distribution (for up to 19 publications since 2010) of publications.

Figure 19. Publications, from 2010 onwards, for the main researchers on the MFC area

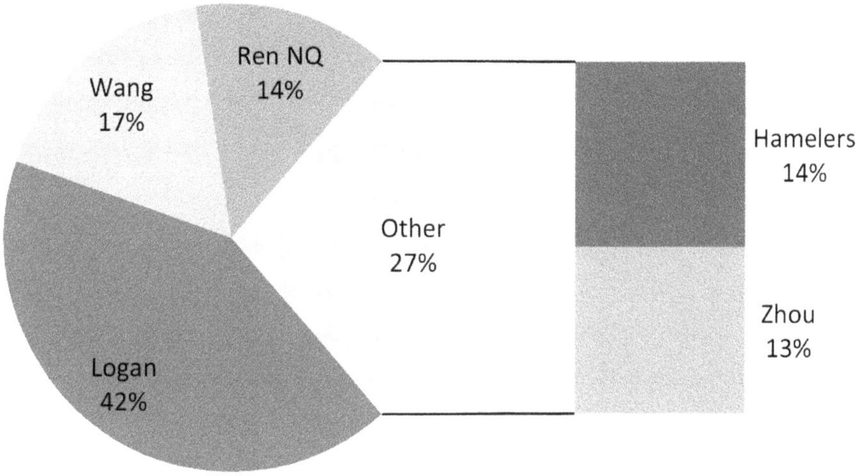

The research being developed is covering a wide range of topics. This new technology still has a number of key aspects not thoroughly understood: microbial communities and substrate choices, electrodes' geometry and composition, chamber characteristics or even which energy harvesting techniques should be applied.

In 2013, Logan was connected to several other researchers and research initiatives. Pairing up with Valerie J. Watson and Cesar Nieto Delgado, the team studied the use of activated carbon compounds for cathode applications (Watson, Nieto Delgado, & Logan, 2013a, 2013b). With Xia Wuang, Fang Zhang and other researchers, other studies of using activated carbon compounds for air-cathode MFCs were also developed (Chen, Zhang, Logan, & Hickner, 2013; Xia et al., 2013). Anode geometry (Lanas & Logan, 2013), anode potential and power overshoot (Peng, Yu, Yu, & Wang, 2013; Zhu, Tokash, Hong, & Logan, 2013), and electrode separation (Ahn & Logan, 2013a, 2013b) studies were also conducted by team's partnering with Logan, mostly to understand which arrangements are the most successful. Even reviewing the publications back in 2012 (Chen, Wei, Logan, & Hickner, 2012; Liping Huang, Chai, Quan, Logan, & Chen, 2012; Liping Huang, Gan, et al., 2012; Ishii, Logan, & Sekiguchi, 2012; Wei, Tokash, Chen, Hickner, & Logan, 2012; Yang, Feng, & Logan, 2012; F. Zhang, Chen, Hickner, & Logan, 2012), Logan, and the teams he belonged to, address electrodes studies, particularly air-cathode investigations, more regularly than any other. Wang, who worked alongside Logan, seems to be following this tendency as well since 4 of his publications were related to anode modifications. These works reported on the manipulation of chemical and physical properties of compounds (nickel-coated sponges (X. Liu et al., 2013), mesoporous anodes (Nguyen, Yu, Wang, Wang, & Song, 2013) and polyaniline networks on grapheme nanoribbons-coated carbon paper anodes (Zhao et al., 2013)) in order to increase anode efficiency.

Efforts are also being made to prove MFC technology can also be applied to more unconventional substrates like urine(Ieropoulos et al., 2013; C. Santoro et al., 2013; Carlo Santoro et al., 2013) or terracotta and earthenware (Winfield, Greenman, Huson, & Ieropoulos, 2013). The validation of MFC uses with wastewaters continues throughout 2014 (Ma, Wang, Li, Wang, & Wu, 2014; Muthukumar &

Figure 20. Publications on MFC, from 2010 onwards, distributed country-wise

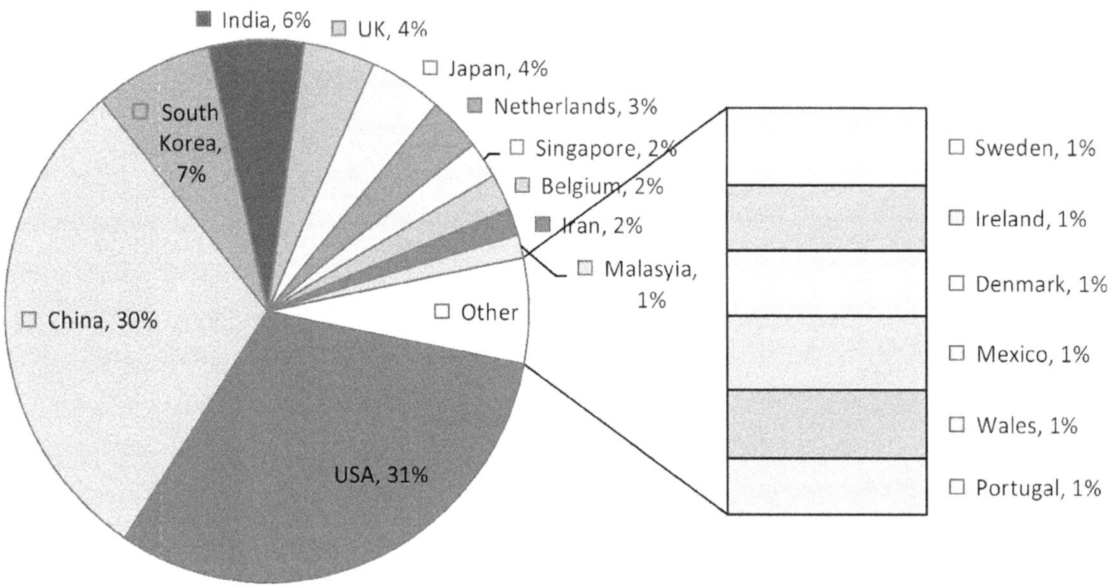

Sangeetha, 2014), just as it happened before (Du et al., 2007; Yuan Liu, Liu, Wang, Hou, & Yang, 2013; Pant et al., 2010; Rachinski et al., 2010; Rozendal et al., 2008).

Reports of work being developed in Portugal started in 2010 (Inoue et al., 2010; Martins et al., 2010). According to the available information at the ISI Web of Knowledge website, research seems to be more focused on bacterial characterization and electron transfer mechanics (Alves, Paquete, Fonseca, & Louro, 2011; da Silva, Roseiro, & Reis, 2012; Fonseca et al., 2013; Louro & Paquete, 2012; Matos & Lopes da Silva, 2013; Morgado, Dantas, Bruix, Londer, & Salgueiro, 2012) than on any other field.

Patents have been filed (and approved) for MFC related devices. Logan has had a significant role in branding this knowledge to his name and has presented patents for desalination devices and methods, materials and configuration for scalable microbial fuel cells, electrodes and methods for microbial fuel cells and other elements (Logan, n.d.).

These initiatives, the publications' yearly numbers and the several meetings and conferences on the subject, are key indicators that this technology is quickly growing and will soon be presented in a commercial form, either for private use (domestic) or public groups (wastewater management services and other suitable partners).

CONCLUSION

The operation of waste water treatment plants can be highly improved with the introduction of a monitoring system. Typically, the area of operation to cover is wide and, at the same time, data acquisition spots are inaccessible. These two reasons justify by themselves the introduction of wireless energy networks.

Energy availability can restrict the usefulness of widely disperse monitoring systems. Devices powered by batteries require the periodical replacement of this component. This task adds cost to the operation of the system. At the same time, a monitoring system's service time is usually mainly limited by its power source's operating life.

Microbial Fuel cells technology adapts perfectly to the operation of waste water plants. The working principle here described is able to provide the energy required by the monitoring system, in order to operate both the data acquisition system and the wireless communication network.

ACKNOWLEDGMENT

Pedro Serra and António Espírito-Santo are partially supported by the Portuguese Foundation for Science and Technology (FCT) and Instituto de Telecomunicações strategic funds project PEst-OE/EEI/LA0008/2013.

REFERENCES

Aelterman, P., Rabaey, K., Clauwaert, P., & Verstraete, W. (2006). Microbial fuel cells for wastewater treatment. *Water Science and Technology: A Journal of the International Association on Water Pollution Research, 54*(8), 9–15. Retrieved from http://www.ncbi.nlm.nih.gov/pubmed/17163008

Ahn, Y., & Logan, B. E. (2013a). Altering Anode Thickness To Improve Power Production in Microbial Fuel Cells with Different Electrode Distances. *Energy & Fuels, 27*(1), 271–276. doi:10.1021/ef3015553

Ahn, Y., & Logan, B. E. (2013b). Domestic wastewater treatment using multi-electrode continuous flow MFCs with a separator electrode assembly design. *Applied Microbiology and Biotechnology, 97*(1), 409–416. doi:10.1007/s00253-012-4455-8 PMID:23053104

Alajmi, B. N., Ahmed, K. H., Finney, S. J., & Williams, B. W. (2011). Fuzzy-Logic-Control Approach of a Modified Hill-Climbing Method for Maximum Power Point in Microgrid Standalone Photovoltaic System. *IEEE Transactions on Power Electronics, 26*(4), 1022–1030. doi:10.1109/TPEL.2010.2090903

Alves, A. S., Paquete, C. M., Fonseca, B. M., & Louro, R. O. (2011). Exploration of the "cytochromome" of Desulfuromonas acetoxidans, a marine bacterium capable of powering microbial fuel cells. *Metallomics : Integrated Biometal Science, 3*(4), 349–353. doi:10.1039/c0mt00084a PMID:21298162

Angenent, L. T., Karim, K., Al-Dahhan, M. H., Wrenn, B. A., & Domíguez-Espinosa, R. (2004). Production of bioenergy and biochemicals from industrial and agricultural wastewater. *Trends in Biotechnology, 22*(9), 477–485. doi:10.1016/j.tibtech.2004.07.001 PMID:15331229

Arends, J. B. A., & Verstraete, W. (2012). 100 years of microbial electricity production: Three concepts for the future. *Microbial Biotechnology, 5*(3), 333–346. doi:10.1111/j.1751-7915.2011.00302.x PMID:21958308

Atlas Scientific. (n.d.). *Atlas Scientific ORP Kit*. Retrieved January 15, 2014, from https://www.atlas-scientific.com/product_pages/kits/orp-kit.html

Balato, M., & Vitelli, M. (2014). A new control strategy for the optimization of Distributed MPPT in PV applications. *International Journal of Electrical Power & Energy Systems*, *62*, 763–773. doi:10.1016/j.ijepes.2014.05.032

Bennetto, H. (1990). Electricity generation by microorganisms. *Biotechnology Education*. Retrieved from http://www.ncbe.reading.ac.uk/NCBE/MATERIALS/METABOLISM/PDF/bennetto.pdf

Boghani, H. C., Kim, J. R., Dinsdale, R. M., Guwy, A. J., & Premier, G. C. (2013). Control of power sourced from a microbial fuel cell reduces its start-up time and increases bioelectrochemical activity. *Bioresource Technology*, *140*, 277–285. doi:10.1016/j.biortech.2013.04.087 PMID:23708786

Boghani, H. C., Papaharalabos, G., Michie, I., Fradler, K. R., Dinsdale, R. M., Guwy, A. J., & Premier, G. C. et al. (2014). Controlling for peak power extraction from microbial fuel cells can increase stack voltage and avoid cell reversal. *Journal of Power Sources*, *269*, 363–369. doi:10.1016/j.jpowsour.2014.06.059

Bullen, R., Arnot, T., Lakeman, J., & Walsh, F. (2006). Biofuel cells and their development. *Biosensors and …*. Retrieved from http://www.sciencedirect.com/science/article/pii/S0956566306000406

Carannante, G., Fraddanno, C., Pagano, M., & Piegari, L. (2009). Experimental Performance of MPPT Algorithm for Photovoltaic Sources Subject to Inhomogeneous Insolation. *IEEE Transactions on Industrial Electronics*, *56*(11), 4374–4380. doi:10.1109/TIE.2009.2019570

Chan, D. S. H., & Phang, J. C. H. (1987). Analytical methods for the extraction of solar-cell single- and double-diode model parameters from I-V characteristics. *IEEE Transactions on Electron Devices*, *34*(2), 286–293. doi:10.1109/T-ED.1987.22920

Chao, K.-H. (2014). An extension theory-based maximum power tracker using a particle swarm optimization algorithm. *Energy Conversion and Management*, *86*, 435–442. doi:10.1016/j.enconman.2014.05.018

Chen, G., Wei, B., Logan, B. E., & Hickner, M. A. (2012). Cationic fluorinated polymer binders for microbial fuel cell cathodes. *RSC Advances*, *2*(13), 5856. doi:10.1039/c2ra20705b

Chen, G., Zhang, F., Logan, B. E., & Hickner, M. A. (2013). Poly(vinyl alcohol) separators improve the coulombic efficiency of activated carbon cathodes in microbial fuel cells. *Electrochemistry Communications*, *34*, 150–152. doi:10.1016/j.elecom.2013.05.026

Da Silva, T. L., Roseiro, J. C., & Reis, A. (2012). Applications and perspectives of multi-parameter flow cytometry to microbial biofuels production processes. *Trends in Biotechnology*, *30*(4), 225–232. doi:10.1016/j.tibtech.2011.11.005 PMID:22257766

Davis, F., & Higson, S. P. J. (2007). Biofuel cells--recent advances and applications. *Biosensors & Bioelectronics*, *22*(7), 1224–1235. doi:10.1016/j.bios.2006.04.029 PMID:16781864

Davis, J. B., & Yarbrough, H. F. (1962). Preliminary Experiments on a Microbial Fuel Cell. *Science*, *137*(3530), 615–616. doi:10.1126/science.137.3530.615 PMID:17836545

Debabov, V. G. (2008). Electricity from microorganisms. *Microbiology*, *77*(2), 123–131. doi:10.1134/S002626170802001X PMID:18522314

Degrenne, N., Allard, B., & Buret, F. (2011). Comparison of 3 self-starting step-up DC: DC converter topologies for harvesting energy from low-voltage and low-power microbial fuel cells. *Proceedings of the ...*. Retrieved from http://ieeexplore.ieee.org/xpls/abs_all.jsp?arnumber=6020291

Degrenne, N., Buret, F., Allard, B., & Bevilacqua, P. (2012). Electrical energy generation from a large number of microbial fuel cells operating at maximum power point electrical load. *Journal of Power Sources*, *205*, 188–193. doi:10.1016/j.jpowsour.2012.01.082

Dewan, A., Ay, S. U., Karim, M. N., & Beyenal, H. (2014). Alternative power sources for remote sensors: A review. *Journal of Power Sources*, *245*, 129–143. doi:10.1016/j.jpowsour.2013.06.081

Dewan, A., Beyenal, H., & Lewandowski, Z. (2009). Intermittent Energy Harvesting Improves the Performance of Microbial Fuel Cells. *Environmental Science & Technology*, *43*(12), 4600–4605. doi:10.1021/es8037092 PMID:19603683

Du, Z., Li, H., & Gu, T. (2007). A state of the art review on microbial fuel cells: A promising technology for wastewater treatment and bioenergy. *Biotechnology Advances*, *25*(5), 464–482. doi:10.1016/j.biotechadv.2007.05.004 PMID:17582720

Enrique, J. M., Durán, E., Sidrach-de-Cardona, M., & Andújar, J. M. (2007). Theoretical assessment of the maximum power point tracking efficiency of photovoltaic facilities with different converter topologies. *Solar Energy*, *81*(1), 31–38. doi:10.1016/j.solener.2006.06.006

Esram, T., & Chapman, P. L. (2007). Comparison of Photovoltaic Array Maximum Power Point Tracking Techniques. *IEEE Transactions on Energy Conversion*, *22*(2), 439–449. doi:10.1109/TEC.2006.874230

Fonseca, B. M., Paquete, C. M., Neto, S. E., Pacheco, I., Soares, C. M., & Louro, R. O. (2013). Mind the gap: Cytochrome interactions reveal electron pathways across the periplasm of Shewanella oneidensis MR-1. *The Biochemical Journal*, *449*(1), 101–108. doi:10.1042/BJ20121467 PMID:23067389

He, Z., Minteer, S., & Angenent, L. (2005). Electricity generation from artificial wastewater using an upflow microbial fuel cell. *Environmental Science & ...*. Retrieved from http://pubs.acs.org/doi/abs/10.1021/es0502876

Huang, L., Chai, X., Quan, X., Logan, B. E., & Chen, G. (2012). Reductive dechlorination and mineralization of pentachlorophenol in biocathode microbial fuel cells. *Bioresource Technology*, *111*, 167–174. doi:10.1016/j.biortech.2012.01.171 PMID:22357291

Huang, L., Gan, L., Wang, N., Quan, X., Logan, B. E., & Chen, G. (2012). Mineralization of pentachlorophenol with enhanced degradation and power generation from air cathode microbial fuel cells. *Biotechnology and Bioengineering*, *109*(9), 2211–2221. doi:10.1002/bit.24489 PMID:22392229

Huang, L., Pop, V., de Francisco, R., Vullers, R., Dolmans, G., de Groot, H., & Imamura, K. (2010). Ultra low power wireless and energy harvesting technologies — An ideal combination. In *2010 IEEE International Conference on Communication Systems* (pp. 295–300). IEEE. doi:10.1109/ICCS.2010.5686436

Ieropoulos, I. A., Greenman, J., Melhuish, C., & Hart, J. (2005). Comparative study of three types of microbial fuel cell. *Enzyme and Microbial Technology*, *37*(2), 238–245. doi:10.1016/j.enzmictec.2005.03.006

Ieropoulos, I. A., Ledezma, P., Stinchcombe, A., Papaharalabos, G., Melhuish, C., & Greenman, J. (2013). Waste to real energy: The first MFC powered mobile phone. *Physical Chemistry Chemical Physics : PCCP*, *15*(37), 15312–15316. doi:10.1039/c3cp52889h PMID:23939246

Inoue, K., Qian, X., Morgado, L., Kim, B.-C., Mester, T., Izallalen, M., & Lovley, D. R. et al. (2010). Purification and characterization of OmcZ, an outer-surface, octaheme c-type cytochrome essential for optimal current production by Geobacter sulfurreducens. *Applied and Environmental Microbiology*, *76*(12), 3999–4007. doi:10.1128/AEM.00027-10 PMID:20400562

Ishaque, K., Salam, Z., & Taheri, H. (2011). Simple, fast and accurate two-diode model for photovoltaic modules. *Solar Energy Materials and Solar Cells*, *95*(2), 586–594. doi:10.1016/j.solmat.2010.09.023

Ishii, S., Logan, B. E., & Sekiguchi, Y. (2012). Enhanced electrode-reducing rate during the enrichment process in an air-cathode microbial fuel cell. *Applied Microbiology and Biotechnology*, *94*(4), 1087–1094. doi:10.1007/s00253-011-3844-8 PMID:22223104

Jain, S., & Agarwal, V. (2007). Comparison of the performance of maximum power point tracking schemes applied to single-stage grid-connected photovoltaic systems. *IET Electric Power Applications*, 753–763. doi:10.1049/iet-epa

Jiang, Y. (2013). Adaptive step size with adaptive-perturbation-frequency digital MPPT controller for a single-sensor photovoltaic solar system. *Power Electronics, IEEE, 28*(7), 3195–3205. Retrieved from http://ieeexplore.ieee.org/xpls/abs_all.jsp?arnumber=6316180

Kamarzaman, N. A., & Tan, C. W. (2014). A comprehensive review of maximum power point tracking algorithms for photovoltaic systems. *Renewable & Sustainable Energy Reviews*, *37*, 585–598. doi:10.1016/j.rser.2014.05.045

Karmakar, S., Kundu, K., & Kundu, S. (2010). Design and Development of Microbial Fuel cells. *Technology and Education Topics in …*. Retrieved from http://www.formatex.info/microbiology2/1029-1034.pdf

Karube, I., Matsunaga, T., Tsuru, S., & Suzuki, S. (1977). Biochemical fuel cell utilizing immobilized cells of clostridium butyricum. *Biotechnology and Bioengineering*, *19*(11), 1727–1733. doi:10.1002/bit.260191112

Kim, B., Chang, I., & Gadd, G. (2007). Challenges in microbial fuel cell development and operation. *Applied Microbiology and Biotechnology*. Retrieved from http://link.springer.com/article/10.1007/s00253-007-1027-4

Kim, J. R., Cheng, S., Oh, S.-E., & Logan, B. E. (2007). Power Generation Using Different Cation, Anion, and Ultrafiltration Membranes in Microbial Fuel Cells. *Environmental Science & Technology*, *41*(3), 1004–1009. doi:10.1021/es062202m PMID:17328216

Kim, S.-T., Bang, T.-H., Lee, S.-C., & Park, J.-W. (2014). Real-Time Maximum Power Point Tracking Method Based on Three Points Approximation by Digital Controller for PV System. *Journal of Electrical Engineering and Technology*, *9*(5), 1447–1453. doi:10.5370/JEET.2014.9.5.1447

Kim, Y., Hatzell, M. C., Hutchinson, A. J., & Logan, B. E. (2011). Capturing power at higher voltages from arrays of microbial fuel cells without voltage reversal. *Energy & Environmental Science*, *4*(11), 4662. doi:10.1039/c1ee02451e

Kollimalla, S. K., & Mishra, M. K. (2014). A Novel Adaptive P&O MPPT Algorithm Considering Sudden Changes in the Irradiance. *IEEE Transactions on Energy Conversion*, *29*(3), 602–610. doi:10.1109/TEC.2014.2320930

Kuphaldt, T. R. (n.d.). *pH measurement: Electrical Instrumentation Signals - Electronics Textbook.* Retrieved January 14, 2014, from http://www.allaboutcircuits.com/vol_1/chpt_9/6.html

Lanas, V., & Logan, B. E. (2013). Evaluation of multi-brush anode systems in microbial fuel cells. *Bioresource Technology*, *148*, 379–385. doi:10.1016/j.biortech.2013.08.154 PMID:24063821

Liang, P., Wu, W., Wei, J., Yuan, L., Xia, X., & Huang, X. (2011). Alternate charging and discharging of capacitor to enhance the electron production of bioelectrochemical systems. *Environmental Science & Technology*, *45*(15), 6647–6653. doi:10.1021/es200759v PMID:21710972

Liu, H., & Logan, B. E. (2004). Electricity Generation Using an Air-Cathode Single Chamber Microbial Fuel Cell in the Presence and Absence of a Proton Exchange Membrane. *Environmental Science & Technology*, *38*(14), 4040–4046. doi:10.1021/es0499344 PMID:15298217

Liu, H., Ramnarayanan, R., & Logan, B. E. (2004). Production of Electricity during Wastewater Treatment Using a Single Chamber Microbial Fuel Cell. *Environmental Science & Technology*, *38*(7), 2281–2285. doi:10.1021/es034923g PMID:15112835

Liu, X., Du, X., Wang, X., Li, N., Xu, P., & Ding, Y. (2013). Improved microbial fuel cell performance by encapsulating microbial cells with a nickel-coated sponge. *Biosensors & Bioelectronics*, *41*, 848–851. doi:10.1016/j.bios.2012.08.014 PMID:22939511

Liu, Y., Li, M., Ji, X., Luo, X., Wang, M., & Zhang, Y. (2014). A comparative study of the maximum power point tracking methods for PV systems. *Energy Conversion and Management*, *85*, 809–816. doi:10.1016/j.enconman.2014.01.049

Liu, Y., Liu, H., Wang, C., Hou, S.-X., & Yang, N. (2013). Sustainable energy recovery in wastewater treatment by microbial fuel cells: Stable power generation with nitrogen-doped graphene cathode. *Environmental Science & Technology*, *47*(23), 13889–13895. doi:10.1021/es4032216 PMID:24219223

Logan, B. E. (2008). *Microbial Fuel Cells.* Retrieved from http://books.google.pt/books/about/Microbial_Fuel_Cells.html?id=cXnc2wmrE9gC&pgis=1

Logan, B. E. (2009). Exoelectrogenic bacteria that power microbial fuel cells. *Nature Reviews. Microbiology*, *7*(5), 375–381. doi:10.1038/nrmicro2113 PMID:19330018

Logan, B. E. (n.d.). *Patents Related to Microbial Fuel Cells and Microbial Electrolysis Cells.* Retrieved February 11, 2014, from http://www.engr.psu.edu/ce/enve/logan/bioenergy/mfc_patents.htm

Logan, B. E., Hamelers, B., Rozendal, R., Schröder, U., Keller, J., Freguia, S., & Rabaey, K. et al. (2006). Microbial Fuel Cells: Methodology and Technology †. *Environmental Science & Technology*, *40*(17), 5181–5192. doi:10.1021/es0605016 PMID:16999087

Logan, B. E., & Regan, J. M. (2006). Microbial fuel cells-challenges and applications. *Environmental Science & Technology*, *40*(17), 5172–5180. doi:10.1021/es0627592 PMID:16999086

Lopez-Lapena, O., Penella, M. T., & Gasulla, M. (2010). A New MPPT Method for Low-Power Solar Energy Harvesting. *IEEE Transactions on Industrial Electronics*, *57*(9), 3129–3138. doi:10.1109/TIE.2009.2037653

Louro, R. O., & Paquete, C. M. (2012). The quest to achieve the detailed structural and functional characterization of CymA. *Biochemical Society Transactions*, *40*(6), 1291–1294. doi:10.1042/BST20120114 PMID:23176470

Lovley, D. R. (2006). Bug juice: Harvesting electricity with microorganisms. *Nature Reviews. Microbiology*, *4*(7), 497–508. doi:10.1038/nrmicro1442 PMID:16778836

Ma, J., Wang, Z., Li, X., Wang, Y., & Wu, Z. (2014). Bioelectricity generation through microbial fuel cell using organic matters recovered from municipal wastewater. *Environmental Progress & Sustainable Energy*, *33*(1), 290–297. doi:10.1002/ep.11745

Martins, G., Peixoto, L., Ribeiro, D. C., Parpot, P., Brito, A. G., & Nogueira, R. (2010). Towards implementation of a benthic microbial fuel cell in lake Furnas (Azores): Phylogenetic affiliation and electrochemical activity of sediment bacteria. *Bioelectrochemistry (Amsterdam, Netherlands)*, *78*(1), 67–71. doi:10.1016/j.bioelechem.2009.07.003 PMID:19716775

Matos, C. T., & Lopes da Silva, T. (2013). Using multi-parameter flow cytometry as a novel approach for physiological characterization of bacteria in microbial fuel cells. *Process Biochemistry*, *48*(1), 49–57. doi:10.1016/j.procbio.2012.11.003

Mbed. (n.d.). Platforms. Retrieved November 02, 2013, from http://developer.mbed.org/platforms/

Meehan, A., Gao, H., & Lewandowski, Z. (2011). Energy Harvesting With Microbial Fuel Cell and Power Management System. *IEEE Transactions on Power Electronics*, *26*(1), 176–181. doi:10.1109/TPEL.2010.2054114

Mei, Q., Shan, M., Liu, L., & Guerrero, J. M. (2011). A Novel Improved Variable Step-Size Incremental-Resistance MPPT Method for PV Systems. *IEEE Transactions on Industrial Electronics*, *58*(6), 2427–2434. doi:10.1109/TIE.2010.2064275

Membranes International Inc. (n.d.). *CMI-7000 Cation Exchange Membranes Technical Specifications*. Retrieved June 16, 2014, from http://www.membranesinternational.com/tech-cmi.htm

Molognoni, D., Puig, S., Balaguer, M. D., Liberale, A., Capodaglio, A. G., Callegari, A., & Colprim, J. (2014). Reducing start-up time and minimizing energy losses of Microbial Fuel Cells using Maximum Power Point Tracking strategy. *Journal of Power Sources*, *269*, 403–411. doi:10.1016/j.jpowsour.2014.07.033

Morgado, L., Dantas, J. M., Bruix, M., Londer, Y. Y., & Salgueiro, C. A. (2012). Fine Tuning of Redox Networks on Multiheme Cytochromes from Geobacter sulfurreducens Drives Physiological Electron/Proton Energy Transduction. *Bioinorganic Chemistry and Applications*, *298739*. doi:10.1155/2012/298739 PMID:22899897

Muthukumar, M., & Sangeetha, T. (2014). The Harnessing Of Bioenergy From A Dual Chambered Microbial Fuel Cell (Mfc) Employing Sago-Processing Wastewater As Catholyte. *International Journal of Green Energy*, *11*(2), 161–172. doi:10.1080/15435075.2013.771581

Muthuramalingam, M., & Manoharan, P. S. (2014). Comparative analysis of distributed MPPT controllers for partially shaded stand alone photovoltaic systems. *Energy Conversion and Management*, *86*, 286–299. doi:10.1016/j.enconman.2014.05.044

Myron, L. Company. (2012). *Application Bulletin: Oxidation Reduction Potential (ORP)/Redox and Free Chlorine*. Retrieved from http://www.myronl.com/PDF/application_bulletins/orp_ab.pdf

Nguyen, T.-H., Yu, Y.-Y., Wang, X., Wang, J.-Y., & Song, H. (2013). A 3D mesoporous polysulfone-carbon nanotube anode for enhanced bioelectricity output in microbial fuel cells. *Chemical Communications (Cambridge, England)*, *49*(91), 10754–10756. doi:10.1039/c3cc45775c PMID:24108240

Nilsson, J. W., & Riedel, S. A. (2001). *Electric Circuits*. Retrieved from http://books.google.pt/books/about/Electric_Circuits.html?id=sehSAAAAMAAJ&pgis=1

Pant, D., Van Bogaert, G., Diels, L., & Vanbroekhoven, K. (2010). A review of the substrates used in microbial fuel cells (MFCs) for sustainable energy production. *Bioresource Technology*, *101*(6), 1533–1543. doi:10.1016/j.biortech.2009.10.017 PMID:19892549

Park, J.-D., & Ren, Z. (2011). Efficient energy harvester for microbial fuel cells using DC/DC converters. In 2011 IEEE Energy Conversion Congress and Exposition (pp. 3852–3858). IEEE. doi:10.1109/ECCE.2011.6064292

Park, J.-D., & Ren, Z. (2012a). High efficiency energy harvesting from microbial fuel cells using a synchronous boost converter. *Journal of Power Sources*, *208*, 322–327. doi:10.1016/j.jpowsour.2012.02.035

Park, J.-D., & Ren, Z. (2012b). Hysteresis controller based maximum power point tracking energy harvesting system for microbial fuel cells. *Journal of Power Sources*, *205*, 151–156. doi:10.1016/j.jpowsour.2012.01.053

Peng, X., Yu, H., Yu, H., & Wang, X. (2013). Lack of anodic capacitance causes power overshoot in microbial fuel cells. *Bioresource Technology*, *138*, 353–358. doi:10.1016/j.biortech.2013.03.187 PMID:23624054

Petrone, G., Spagnuolo, G., & Vitelli, M. (2012). An analog technique for distributed MPPT PV applications. *Industrial Electronics, IEEE*, *59*(12), 4713–4722. Retrieved from http://ieeexplore.ieee.org/xpls/abs_all.jsp?arnumber=6092483

Potter, M. (1911). Electrical effects accompanying the decomposition of organic compounds. *Proceedings of the Royal Society of London. Series B*. Retrieved from http://www.jstor.org/stable/80609

Rabaey, K., Rodríguez, J., Blackall, L. L., Keller, J., Gross, P., Batstone, D., & Nealson, K. H. et al. (2007). Microbial ecology meets electrochemistry: Electricity-driven and driving communities. *The ISME Journal*, *1*(1), 9–18. doi:10.1038/ismej.2007.4 PMID:18043609

Rabaey, K., & Verstraete, W. (2005). Microbial fuel cells: novel biotechnology for energy generation. *TRENDS in Biotechnology*. Retrieved from http://www.sciencedirect.com/science/article/pii/S0167779905000922

Rachinski, S., Carubelli, A., Mangoni, A. P., & Mangrich, A. S. (2010). Pilhas de combustíveis microbianas utilizadas na produção de eletricidade a partir de rejeitos orgânicos: Uma perspectiva de futuro. *Quimica Nova*, *33*(8), 1773–1778. doi:10.1590/S0100-40422010000800026

Rahmani, R., & Seyedmahmoudian, M. (2013). Implementation of Fuzzy Logic Maximum Power Point Tracking Controller for Photovoltaic System. *American Journal of Applied Sciences*, *10*(3), 209–218. doi:10.3844/ajassp.2013.209.218

Raju, M., & Grazier, M. (2008). *Energy harvesting. ULP meets energy harvesting: a game-changing combination for design engineers*. Retrieved from http://scholar.google.pt/scholar?hl=pt-PT&q=ULP+Meets+Energy+Harvesting&btnG=&lr=#0

Rawat, R., & Chandel, S. S. (2013). Hill climbing techniques for tracking maximum power point in solar photovoltaic systems - a review. *Special Issue of International Journal of Sustainable Development and Green Economics, 2*(1-1), 90–95. Retrieved from http://irnet.sg/irnet_journal/IJSDGE/IJSDGE_doc/IJSDGE_V2I1,2_papers/17.pdf

Ren, S., Xia, X., Yuan, L., Liang, P., & Huang, X. (2013). Enhancing charge harvest from microbial fuel cells by controlling the charging and discharging frequency of capacitors. *Bioresource Technology*, *146*, 812–815. doi:10.1016/j.biortech.2013.08.055 PMID:23993288

Rodrigo, M. A., Cañizares, P., García, H., Linares, J. J., & Lobato, J. (2009). Study of the acclimation stage and of the effect of the biodegradability on the performance of a microbial fuel cell. *Bioresource Technology*, *100*(20), 4704–4710. doi:10.1016/j.biortech.2009.04.073 PMID:19487121

Rodrigues, E., & Melício, R. (2011). Simulation of a solar cell considering single-diode equivalent circuit model. *... on Renewable Energies ...*. Retrieved from http://icrepq.com/icrepq'11/339-rodrigues.pdf

Rozendal, R. A., Hamelers, H. V. M., Rabaey, K., Keller, J., & Buisman, C. J. N. (2008). Towards practical implementation of bioelectrochemical wastewater treatment. *Trends in Biotechnology*, *26*(8), 450–459. doi:10.1016/j.tibtech.2008.04.008 PMID:18585807

Salas, V., Olías, E., Barrado, A., & Lázaro, A. (2006). Review of the maximum power point tracking algorithms for stand-alone photovoltaic systems. *Solar Energy Materials and Solar Cells*, *90*(11), 1555–1578. doi:10.1016/j.solmat.2005.10.023

Santoro, C., Ieropoulos, I., Greenman, J., Cristiani, P., Vadas, T., Mackay, A., & Li, B. (2013). Current generation in membraneless single chamber microbial fuel cells (MFCs) treating urine. *Journal of Power Sources*, *238*, 190–196. doi:10.1016/j.jpowsour.2013.03.095

Santoro, C., Ieropoulos, I., Greenman, J., Cristiani, P., Vadas, T., Mackay, A., & Li, B. (2013). Power generation and contaminant removal in single chamber microbial fuel cells (SCMFCs) treating human urine. *International Journal of Hydrogen Energy*, *38*(26), 11543–11551. doi:10.1016/j.ijhydene.2013.02.070

Sell, D., Kramer, P., & Kreysa, G. (1989). Use of an oxygen gas diffusion cathode and a three-dimensional packed bed anode in a bioelectrochemical fuel cell. *Applied Microbiology and Biotechnology*, *31*(2), 211–213. doi:10.1007/BF00262465

Sera, D., Mathe, L., & Kerekes, T. (2013). On the perturb-and-observe and incremental conductance MPPT methods for PV systems. *IEEE Journal of, 3*(3), 1070–1078. Retrieved from http://ieeexplore.ieee.org/xpls/abs_all.jsp?arnumber=6517453

Shmilovitz, D. (2005). On the control of photovoltaic maximum power point tracker via output parameters. *IEE Proceedings. Electric Power Applications, 152*(2), 239. doi:10.1049/ip-epa:20040978

Smith, P. (2011). *Comparing Low-Power Wireless Technologies | DigiKey*. Retrieved December 03, 2014, from http://www.digikey.com/en/articles/techzone/2011/aug/comparing-low-power-wireless-technologies

Thomson Reuters. (n.d.). *ISI Web of Knowledge*. Retrieved January 01, 2014, from http://wokinfo.com/

Todar, K. (2008). *Online Textbook of Bacteriology*. Retrieved from http://textbookofbacteriology.net/index.html

Tsai, H., Tu, C., & Su, Y. (2008). Development of generalized photovoltaic model using MATLAB/SIMULINK. *Proceedings of the World Congress on ...*. Retrieved from http://s2i.bordeaux.free.fr/Espace Terminale/Ressources/Projet/Projet 2 2013-2014/2.DevelopmentofGeneralizedPhotovoltaicModelusingMatlab.pdf

Ultra low power wireless connectivity technology backgrounder. (n.d.). Retrieved October 18, 2014, from http://www.nordicsemi.com/eng/News/Press-Center/Press-Backgrounders/ULP-wireless-connectivity-backgrounder

Walker, D., & Bier, A. W. (2009). *Oxidation Reduction Potential: Understanding a Challenging Measurement*. Retrieved from http://www.hach.com/asset-get.download.jsa?id=7639984589

Wang, H., Park, J.-D., & Ren, Z. (2012). Active energy harvesting from microbial fuel cells at the maximum power point without using resistors. *Environmental Science & Technology, 46*(9), 5247–5252. doi:10.1021/es300313d PMID:22486712

Wang, H., & Ren, Z. J. (2013). A comprehensive review of microbial electrochemical systems as a platform technology. *Biotechnology Advances, 31*(8), 1796–1807. doi:10.1016/j.biotechadv.2013.10.001 PMID:24113213

Watson, V. J., Nieto Delgado, C., & Logan, B. E. (2013a). Improvement of activated carbons as oxygen reduction catalysts in neutral solutions by ammonia gas treatment and their performance in microbial fuel cells. *Journal of Power Sources, 242*, 756–761. doi:10.1016/j.jpowsour.2013.05.135

Watson, V. J., Nieto Delgado, C., & Logan, B. E. (2013b). Influence of chemical and physical properties of activated carbon powders on oxygen reduction and microbial fuel cell performance. *Environmental Science & Technology, 47*(12), 6704–6710. doi:10.1021/es401722j PMID:23692057

Wei, B., Tokash, J. C., Chen, G., Hickner, M. A., & Logan, B. E. (2012). Development and evaluation of carbon and binder loading in low-cost activated carbon cathodes for air-cathode microbial fuel cells. *RSC Advances, 2*(33), 12751–12758. doi:10.1039/C2RA21572A

Winfield, J., Greenman, J., Huson, D., & Ieropoulos, I. (2013). Comparing terracotta and earthenware for multiple functionalities in microbial fuel cells. *Bioprocess and Biosystems Engineering, 36*(12), 1913–1921. doi:10.1007/s00449-013-0967-6 PMID:23728836

Wired, U. K. (2012). *NASA Wants to Power Robots With Microbes*. Retrieved from http://www.wired.com/2012/01/navy-microbial-fuel-cells/

Woodward, L., Perrier, M., Srinivasan, B., Pinto, R. P., & Tartakovsky, B. (2010). Comparison of real-time methods for maximizing power output in microbial fuel cells. *AIChE Journal. American Institute of Chemical Engineers*, *56*(10), 2742–2750. doi:10.1002/aic.12157

Xia, X., Zhang, F., Zhang, X., Liang, P., Huang, X., & Logan, B. E. (2013). Use of pyrolyzed iron ethylenediaminetetraacetic acid modified activated carbon as air-cathode catalyst in microbial fuel cells. *ACS Applied Materials & Interfaces*, *5*(16), 7862–7866. doi:10.1021/am4018225 PMID:23902951

Xiao, W., & Elnosh, A. (2011). Overview of maximum power point tracking technologies for photovoltaic power systems. *IECON 2011-37th*. Retrieved from http://ieeexplore.ieee.org/xpls/abs_all.jsp?arnumber=6119946

Xu, W., Mu, C., & Jin, J. (2014). Novel Linear Iteration Maximum Power Point Tracking Algorithm for Photovoltaic Power Generation. *IEEE Transactions on Applied Superconductivity*, *24*(5), 1–6. doi:10.1109/TASC.2014.2333534

Yang, Q., Feng, Y., & Logan, B. E. (2012). Using cathode spacers to minimize reactor size in air cathode microbial fuel cells. *Bioresource Technology*, *110*, 273–277. doi:10.1016/j.biortech.2012.01.121 PMID:22342583

YSI. (2009). *The Dissolved Oxygen Handbook*. Retrieved from https://www.fondriest.com/pdf/ysi_do_handbook.pdf

Zhang, F., Chen, G., Hickner, M., & Logan, B. (2012). Novel anti-flooding poly (dimethylsiloxane) (PDMS) catalyst binder for microbial fuel cell cathodes. *Journal of Power Sources*. Retrieved from http://www.sciencedirect.com/science/article/pii/S0378775312010981

Zhang, L., Hurley, W., & W[UNKNOWN ENTITY &odie;]lfle, W. (2011). A new approach to achieve maximum power point tracking for PV system with a variable inductor. *Power Electronics, IEEE, 1*, 948–952. Retrieved from http://ieeexplore.ieee.org/xpls/abs_all.jsp?arnumber=5609212

Zhao, C., Gai, P., Liu, C., Wang, X., Xu, H., Zhang, J., & Zhu, J.-J. (2013). Polyaniline networks grown on graphene nanoribbons-coated carbon paper with a synergistic effect for high-performance microbial fuel cells. *Journal of Materials Chemistry A*, *1*(40), 12587. doi:10.1039/c3ta12947k

Zhou, Z., Holland, P. M., & Igic, P. (2014). MPPT algorithm test on a photovoltaic emulating system constructed by a DC power supply and an indoor solar panel. *Energy Conversion and Management*, *85*, 460–469. doi:10.1016/j.enconman.2014.06.007

Zhu, X., Tokash, J. C., Hong, Y., & Logan, B. E. (2013). Controlling the occurrence of power overshoot by adapting microbial fuel cells to high anode potentials. *Bioelectrochemistry (Amsterdam, Netherlands)*, *90*, 30–35. doi:10.1016/j.bioelechem.2012.10.004 PMID:23178374

Zhuang, L., Zheng, Y., Zhou, S., Yuan, Y., Yuan, H., & Chen, Y. (2012). Scalable microbial fuel cell (MFC) stack for continuous real wastewater treatment. *Bioresource Technology*, *106*, 82–88. doi:10.1016/j.biortech.2011.11.019 PMID:22197329

KEY TERMS AND DEFINITIONS

Acclimation Time: The time for an individual organism to adjust its metabolism to new environmental conditions and maintain or regain its performance.

Anaerobic Digester: A system developed for microorganisms to decompose biodegradable matter in oxygen free environments. Yogurts are produced in anaerobic digestion processes and wastewater residues can also be partially cleaned using this same process.

Bioelectrochemical Systems: Devices designed to work with electricity and biological phenomena.

Charge Pump: Similar to DC/DC converters, they are simpler and usually used to start-up the converters. They, however, do not have the same regulation freedom and ability than DC/DC converters.

DC/DC Converter: Electronic circuit that converts a DC signal to one with the same nature and different voltage level. A current conversion is also possible but not as frequent.

Maximum Power Point: An IV curve point that results in the highest product between a device's current and voltage in a given instant. This point isn't related with the highest current or highest voltage, rather the highest result from the product between these two variables.

Microbial Fuel Cell: An energy harvesting device that has a similar working principle to those of batteries. On Microbial Fuel Cells, bacteria are the oxidizing agents, working with an organic anolyte, and electrons from that oxidation flow to a cathode where they react to produce water.

Wireless Sensor Networks: A network of sensor nodes capable of exchanging data wirelessly. Additionally, the same network can also be powered by local energy sources, completely removing the needs for wires.

Chapter 8
Modelling Formalisms for Green Transportation Systems

Calin Ciufudean
Stefan cel Mare University, Romania

ABSTRACT

Protection of environment implies the coordination of the local and regional authorities to implement actions for conservation of all natural resources in the actual context of globalisation. One of the most important direction for applying the above mentioned desiderates is a better planning of transportation, as we all know that globalisation implies an accelerated process of externalisation of ideas, know-how, products and labours. In this chapter, we consider discrete event systems (DES) modeled via a state space representation. The model objective is the avoidance of a given set of states, or equivalently the fact that certain predicates, specified in terms of states are always false. We address the state space controlled Petri Nets (PN) and a technique to reduce the complexity of these nets, by taking into account the fact that the complexity of the considered nets depends mainly on the representation of the control design, respectively on the forbidden sets of places. Our goal is to describe in an efficient way the forbidden sets of the modeled system, e.g. to describe in an efficient way the DES modeled with controlled Petri nets. An example from the railway systems illustrates this approach.

INTRODUCTION

In this chapter, we consider discrete event systems (DES) modeled via a state space representation. The model objective is the avoidance of a given set of states, or equivalently the fact that certain predicates, specified in terms of states are always false. We address the state space controlled Petri Nets (PN) and a technique to reduce the complexity of these nets, by taking into account the fact that the complexity of the considered nets depends mainly on the representation of the control design, respectively on the forbidden sets of places (Laftit, 1992). Our goal is to describe in an efficient way the forbidden sets of the modeled system, e.g. to describe in an efficient way the DES modeled with controlled Petri nets (Yee & Ventura, 2000). An example from the railway systems illustrates this approach: the Petri net model of railway station incompatible tracts. We mention that every railway station has such a table,

DOI: 10.4018/978-1-4666-9792-8.ch008

given to the general traffic rules allocated to a certain station. A railway transport system consists of several concurrent units such as railway engines, wagons, automated guided vehicles, logic controllers, which function asynchronously to meet the dynamically changing needs of the transport plan. Methods aiming for modelling, analysis, control and simulation of such systems are important. The typical functions of the control are to sequence the operations, monitor the system functioning, and determine the states of different elements in a transport system in respect to real time. Petri nets graphically describe the system behavior in terms of the successive states realized after the occurrence of events. Events correspond to a change from one state to the next state. The number of tokens in each place represents the state of the Petri net. When a transition fires, tokens are moved from one place to another place. The firing of a transition corresponds to an event. In a controlled Petri net, the transitions fire only when all their input control places are marked. The control objective is to prevent a set of forbidden states from being reached, while at the same time enabling a maximal set of achievable state sequences. The PNs property to allow a graphical and analytical representation, closely related to the technological process being modeled, makes them a very useful tool for enabling the application of the theory. We address our work with a comparison between two most used formalisms for modeling manufacturing systems, including transportation ones: Timed Petri Nets (TPN) versus Ladder Diagrams (LLD) (Falcione & Krogh, 1993). We notice that as the specification changes, the TPN requires fewer changes compared to the LLD. Using TPNs the control logic can be qualitatively analyzed to check properties such as absence of deadlocks and presence of reversibility in the system. Using LLDs qualitative analysis is not possible until it is stimulated or implemented. Using TPNs, the initial state of the system can be directly represented by its initial marking. The procedure for controlling a system using TPNs is straightforward, simple, and can be applied to control any discrete event system that has digital in/out interface and a computer. Adding more attributes to places and transitions can extend TPNs in order to control complex transport systems. Designing a discrete event controller with choices to perform a control task can extend the present approach.

2. CONTROLLED PETRI NETS

2.1 Basic Properties

A controlled Petri net is defined as a five-tuple $G=(P, T, F, C, B)$, where P and T are finite sets of state places and transitions, respectively; $F \subset (PxT) \cup (TxP)$ is a set of directed arcs connecting state places and transitions; C is a finite set of control places; $B \subset (CxT)$ is a set of directed arcs associating control places with transitions. A state place $p \in P$ is said to be an input to (respectively output from) transition $t \in T$ if $(p, t) \in F$ (respectively $(t, p) \in F$). Denote by ${}^{(p)}t$, respectively $t^{(p)}$ the sets of these input, respectively output places. The sets ${}^{(t)}p$, $p^{(t)}$, $c^{(t)}$, and ${}^{(c)}t$ are defincd similarly. Each controllable transition is influenced by one control place and this control place influences only this one single transition. A control $u: C \rightarrow \{0, 1\}$ indicates whether control place c is marked or not ($u(c) = 1$ or 0). Two special controls are $u_1(c) = 1, \forall c$, enabling all transitions, and $u_0(c) = 0, \forall c$, disabling all controllable transitions. A transition, in a controlled Petri net, is enabled if it is both state and control enabled, that means that under marking m must have: $m(p) \geq 1, \forall p \in {}^{(p)} t$, if we assume that there is at most one arc in B and F between any place and any transition. When an enabled transition t_i fires, a token can be removed from each state place, which has an arc to t_i. Then, the token will be added to any place, which has an arc from t_i. We assume that several transitions can fire simultaneously provided they are always enabled, and the new state, after the occurrence of these events is:

$$m'(p) = m(p) - \sum_{t \in T} 1(p \text{`} \in {}^{(p)}t) + \sum_{t \in T} 1(p \in t^{(p)}) \qquad (1)$$

One denotes by a cycle (Gaffari, 2003) a sequence of places ($p_1, p_2, ..., p_n$) which does not contain the same place twice and where there exists a transition connecting the first place to the last place. Cycles are uncontrollable if, for each pair of consecutive places in T, there is at least one uncontrollable transition connecting them.

2.2 Sets of Forbidden States

As in most papers on forbidden state problem (Holloway & Krogh, 1990), (Ciufudean, s.a., 2003) we represent the set of forbidden states via unions and intersections of simpler sets. In the language of predicates this means that complicated predicates are decomposed into simple predicates via conjunction and disjunction. This leads to a simple, modular control design. The simplest possible forbidden set $M_{p,k}$ corresponds to a particular place containing k or more tokens:

$$M_{p,k} = \{m \in M \mid m(p) \geq k\} \qquad (2)$$

If we consider simple forbidden sets, i.e. one-dimensional specification in (Ciufudean, 2003) where the total token load from a set of places is limited: for $r : P' \subset P \rightarrow R_+$, we have:

$$M_{p,k,r} = \{m \in M \mid \sum_{p \in P'} r(p).m(p) \geq k\} \qquad (3)$$

Equation (2) represents a buffer where at most *k-1* pieces can be stored. Equation (3) represents a case where a finite buffer contains different types of work-pieces, where different work-pieces require a different amount of space, so that the buffer overflows if the weighted sum $\sum_{p \in P'} r(p).m(p)$ became k or more. In a railway line, for example, if $k = 1$, $r(p) = 1$ and P' represents neighbouring sections, then $M_{p,k,r}$ represents the fact that two trains being simultaneously in neighbouring section is a dangerous situation. In intersections of simple forbidden sets, a state $m \in M$ is forbidden when it belongs to each of the simple forbidden sets $M_{pi,ki,ri}$ at the same time. This corresponds to a conjunction in the predicate formulation. Besides conjunctions, one can also take disjunction of predicates. That means that a state is forbidden if it belongs to at least one of a group of simple forbidden sets or of intersections of simple forbidden sets:

$$M = \bigcup_{j=1}^{v} \bigcap_{i=1}^{w} M_{P'_{ij}, k_{ij}, r_{ij}} \qquad (4)$$

Where v, w, k_{ij} are positive integers.

One may say (Gaffari, s.a., 2003) that a marking m can be reached uncontrollably from an initial marking m_0, if, even if u_{zero} is used and thus all controllable transitions are disabled, the initial marking m_0 can be transformed via enabled transitions into the final marking m. An influence path inf_p of a transition t is a sequence of places $inf_p = (p_1, p_2, ..., p_m) \subset P$ such that:

- $t^{(p)} = (p_1)$;
- For $j = 2, 3,..., n$, there exists an uncontrollable transition $t_j \in T$ leading from p_{j-1} to p_j;
- All transition following p_n are either controllable or lead to a place p_j already encountered earlier in the influence path $inf._p$.

Once a token has entered in an influencing path $inf._p$, it can uncontrollably reach p. The complexity of the control synthesis algorithms depends strongly on the number of places in the influencing zones for the different simple forbidden sets. It is thus useful to try to reduce the size of the network, before starting the control design. We can replace several parallel transitions between a pair of places, by one single transition. This transition should be uncontrollable since one of the parallel transitions is uncontrollable. If in a controlled Petri net in certain condition exists influence paths, where a sequence of uncontrollable transitions moves the tokens from any upstream place in $inf._p$ to p_i. This implies that places of an uncontrollable path are indistinguishable from a control viewpoint. Thus, from a control point of view, it is allowed to replace all the places in $inf._p$, and the transitions connecting these places, by one single place. The reduced net is characterized by a new finite marking set M'. We call this equivalence between functional properties the functional abstraction equivalence (F.A.E.). An example will facilitate the understanding of this concept.

3 AN EXAMPLE OF CONTROLLED PRODUCTION FLOWS OPTIMIZATION

Two trains, two railway engines (REs), are placed in the railway station given in Figure 1, composed of five lines (sections) $l_1, l_2, \ldots, l_5$ and eight switches $sw_1, sw_2, ..., sw_8$. In the given railway (as in the most railway stations) it is forbidden to shunt wagons by pushing them, which means that a train enters a section on an end (odd switches, for example) and leaves it on the other end (even switches). The discrete events are generated by line sensors indicating when a line belongs to a train route generate the discrete events. Controlled Petri net models the system made up of union of two independent controlled Petri nets: one net represents the train number 1 (with wagons, for example) path, and the other one represents the train number 2 (with carriages) path.

Our Petri net can, therefore, be used to represent concurrent systems with no shared events. The different places $P_w = (P_{wi}, i = 1, ..., 5)$ for the train number 1 and $P_{ca} = (P_{cai}, i = 1, ..., 5)$ for the train number 2 identify possible locations of those two trains (see Figure 2 and 3).

Analogous we have different transitions $T_w = (t_{wi}, i = 1, ..., 8)$, and $T_{ca} = (t_{cai}, i = 1,..., 6)$ and different control places $C_w = (c_{wi}, i = 1, ..., 6)$, and $C_{ca} = (c_{cai}, i = 1, ..., 6)$ allocated to the above given transitions. The controlled Petri nets are given in Fig. 2, and in Fig. 3. The reduced net without uncontrollable cycles is given in Figure 4. For railway system we require that train n_{01} and train n_{02} never be in the same section at the same time (Bacelli & Lin, 1992):

Figure 1. Schematic of railway station with five lines and eight switches

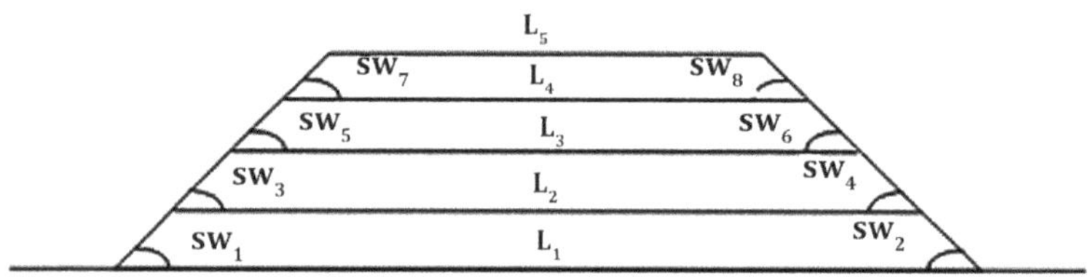

Figure 2. Controlled Petri nets for movements of wagons

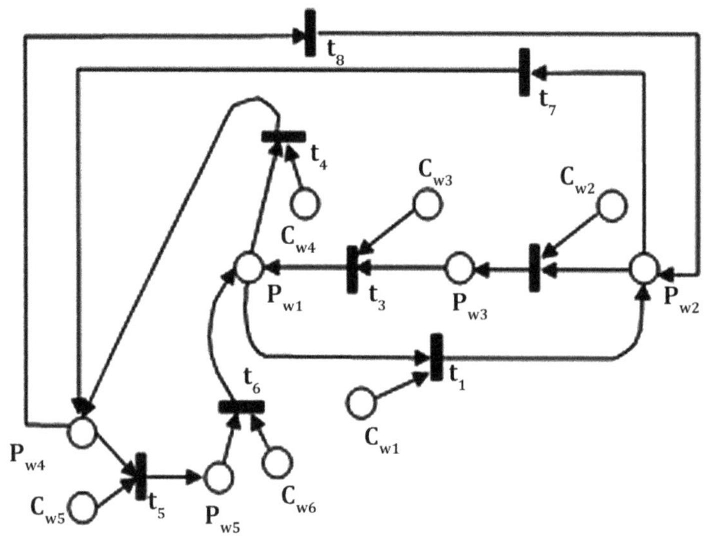

Figure 3. Controlled Petri nets for movements of carriages

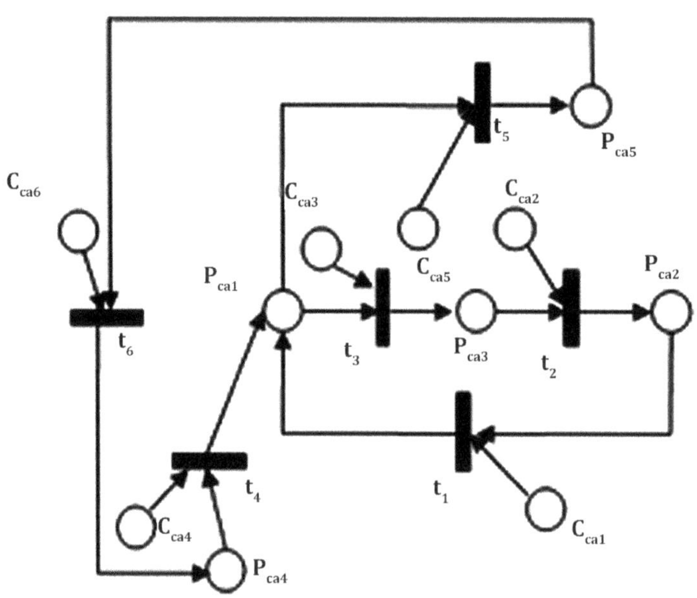

Figure 4. Reduced net for wagons movement

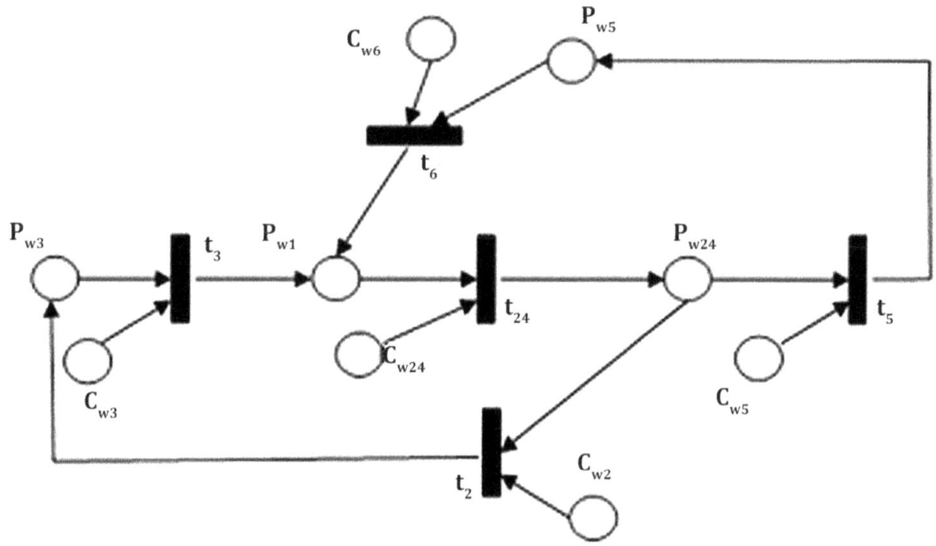

$$M = \bigcup_{j=1}^{5} \{m \in M \mid m(p_{wj}) + m(p_{caj}) \geq 2\} \quad (5)$$

The Petri net model of the table of the incompatible tracts of the railway station given in Fig. 1 contains the states representing the forbidden tracts of two different trains. For example, suppose we prevent two trains (wagon and carriage) to go simultaneously to line 1, the set of forbidden markings M_{Pl} is:

$$\mathrm{M}_{\mathrm{pl}} = \{m \in M \mid m(P_{w1}) \geq 1\} \cap \{m \in M \mid m(P_{ca1}) \geq 1\} \quad (6)$$

The influencing path $inf._{pl}$ the sets of neighbouring sections of section 1; $P^{`}_{l}$ and the control places $C_{P^{`}l}$ are given in Table 1.

Following the example given in Table 1 we can build the table of all the incompatible tracts of the railway station in Figure 1. We proposed a model for the forbidden state problem of DES modeled by controlled Petri nets. Our goal is to present, by an adequate control, a given set of forbidden states from being reached. Forbidden markings characterize constrain sets expressing that some places of the Petri net cannot contain more than a certain amount of tokens. This concept allows us to obtain a maximally

Table 1. Incompatible tracts to line 1 of the railway station in Figure 1

i	P_1	Inf. p_1	$Cp^{`}_1$
1	(P_{w1}, P_{ca1})	(P_{w3}, P_{w5}) (P_{ca2}, P_{ca4})	(C_{w4}, C_{w7}) (C_{ca4}, C_{ca7})

permissive control law. An example, dealing with a railway system, respectively the railway station incompatible tract of the Petri net model illustrates the theoretical concepts. Future sections will extend the results by comparing controlled Petri nets with another well-known formalism for modelling discrete event controllers, e.g. ladder logic diagrams (LLDs).

4 LLDS AND PNS COMPARISON CRITERIA

Traditionally, ladder logic diagrams (LLDs) are used to capture the sequence of operations executed by the system's controller. Realizing the potential of PNs for control, in France, GRAFCET – a PN – like representation tool is proposed as a standard specification of sequence controllers (Holloway & Krogh, 1990; Li & Wonham, 1993; Van Breusegem & Boel, 1992; Gaffari, 2003). The international standard versions are called sequential function charts (Ciufudean, 1993; Bacelli & Lin, 1992; Laftit, 1992). More details and advantages of PNs for controlling manufacturing system can be found in (Proth & Xie, 1994; Campos et al., 1991; Zurawski & Zhon, 1994). However, the comparison between LLDs and PNs for the design of sequence controllers is not encountered in the earlier studies. The sequence controller here means a class of discrete event controllers without choices in executing the operations. Two of the important factors for comparison of PN and LLD for discrete event control are identified as design complexity and response time as described below. Design complexity is defined as the complexity associated in designing the control logic for a given specification. The number of nodes and links for a given graphical control logic design mainly determines graphical complexity. The important factor that influences graphical complexity and adaptability is the physical appearance (size) of the model, whereas not only the physical appearance but also the method of implementation influences the response time. The method of implementation constitutes the software and hardware used to control the system using either PNs or LLDs. Graphical complexity and adaptability cannot be quantified, whereas response time can be measured accurately, given a logic design and implementation. The number of nodes and links used in a control logic model gives an idea about the graphical complexity, adaptability, and response time. For PNs the nodes are places and transitions and links are areas; whereas in LLDs, nodes are normally operations, timers, relays, and the links are connections. If more nodes and links are used in a design, it is graphically more complex and thus, it may need more response time. In a similar manner, control logic is more adaptable if it needs fewer changes in the number of nodes and links compared with logic to meet a change in the specification. Hence, this study uses the number of nodes and links in LLD and PN as a measure to compare their design complexity and response time.

5 PETRI NET BASED CONTROLLERS

Earlier studies use a variety of places to model timers and counters (Zurawski & Zhon, 1994; Ciufudean & Filote, 2005). This might make the model difficult to understand. In our paper, neither new places nor transitions are introduced for modeling them. Timers are modeled by assigning attributes to transitions and counters by places with initial markings and weights on certain arcs. In (Ciufudean, 2005) new sets of places are introduced for modelling I/O signals, which increases the number of places in the PN model. In our paper, I/O signals are modelled as attributes for places and transitions respectively. Hence, due to the use of attributes, our TPNs have fewer nodes and links compared to PNs (Falcione & Krogh, 1993) thereby reducing the graphical complexity. Specifically, the input signals are put in a vector (*X*) that reads the state of the input signals from digital input interface X, associates attributes to every place. $X_i = X(p_i)$ and is an attribute associated with place p_i and represents the input channel number associated with p_i. For example, if p_i models a limit switch, the TPN reads the status of that switch from the digital input interface through the channel number represented by X_i. The initial marking, $m_0(p_i)$ is considered as the first attribute of p_i and X_i is the second one. The contents of any input channel X_i are either "0" or "1". Output signal vector (*Y*) is intended to send output signals through digital output interface. For example, t_i may model the activity "activate a permissive color to signal *A*". Writing a specific number on to the digital output interface activates each color of a signal. During the program execution when a transition fires TPN writes the decimal number corresponding to the output channel to digital output interface. The contents of any output channel are either "0" or "1". There are two events for a transition firing, start firing and end firing. Between these the firing is in progress. The removal of tokens from a transition's input place(s) occurs at start firing. The deposition of tokens to a transition's output place(s) occurs at end firing. While transition firing is in progress, the time to end firing, called the remaining firing time, decreases from firing duration to zero at which its firing is completed. The execution rules of a TPN include enabling and firing rules:

1. A transition $t \in T$ (*T* is a finite set of transitions) is enabled if $\forall p \in P$ and $I(p, t)..0$, $m(p) \geq I(p, t)$ and $X(p)$ has content "1", where *P* is a finite set of places, $I: PXT \rightarrow N$ is an input function that defines the set of directed arcs from *P* to *T*, where $N = \{0, 1, 2,..., n\}$, $m: P \rightarrow N$, is a marking whose i^{th} component represents the number of tokens in the ith place. An initial marking is denoted by m_0.
2. Enabled in a marking *m*, *t* fires and results in a new marking *m'* following the rule: $m'(p)=m(p) + O(p, t) - I(p, t)$, $\forall p \in P$, where, $O: PXT \rightarrow N$, is an output function that defines the set of directed arcs from *T* to *P* (Ciufudean, 2006).

The design procedure for formulating a TPN based controller is briefed in the following five steps:

1. Model the control sequence using PNs to obtain the PN model of the sequence controller.
2. Assign input channels to inputs of the system such as limit switches, sensors, etc. to formulate an input-mapping table.
3. Assign output channels to outputs of the system such as relays, switches, etc. Also, identify timing information for activities to obtain an output-mapping table.
4. In the input-mapping table, assign an input channel number to each place in the PN based controller. The initial state of the system decides the initial marking of TPN. In the PN model some places do not represent the inputs of the system as they represent the intermediate state of the system or logical places to model counters in the sequence. No channel has to be assigned to these places, represented by "-".
5. Using the output mapping table and the action(s) that are modelled by a transition, assign a number to each transition in a PN based controller. The operations and the time delays given in the sequence to be controlled decides firing time function of TPN.

By following the above procedure, a TPN based controller can be formulated, as it is illustrated in Figure 5, for a given sequence.

Figure 5. A TPN based controller

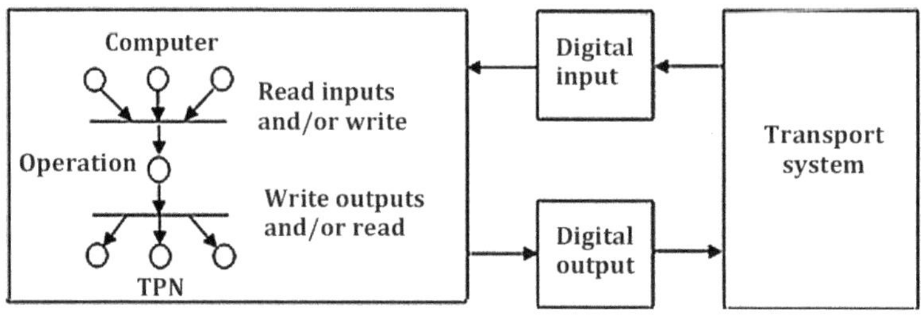

Table 2. The input mapping table

Operation	a_1	b_1	c_1	a_0	b_0	c_0	sw1	ES
X_i	0	1	2	4	5	6	8	10

Table 3. The output mapping table

Operation	Output channel number	Y_i	
		NA	ND
A	0	1	-1
B	1	2	-2
C	2	4	-4

Where, NA (ND) is the number to be written on digital output interface for a forward (or a return) stroke of an RE. Sequence 1: A+, B+, {C+, A-}, {B-, C-}.

Figure 6. LLD for sequence ST, A+, B+, {C+, A-}, {B-, C-}

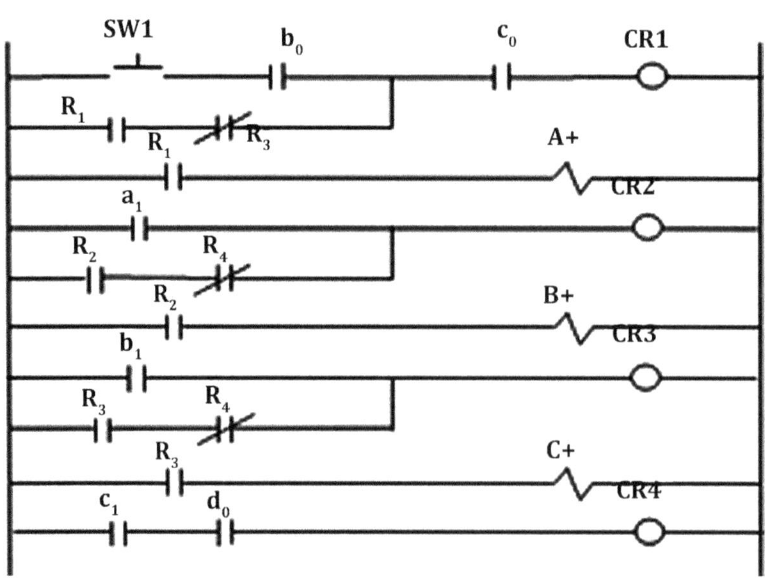

Figure 7. TPN for sequence ST, A+, B+, {C+, A-}, {B-, C-}

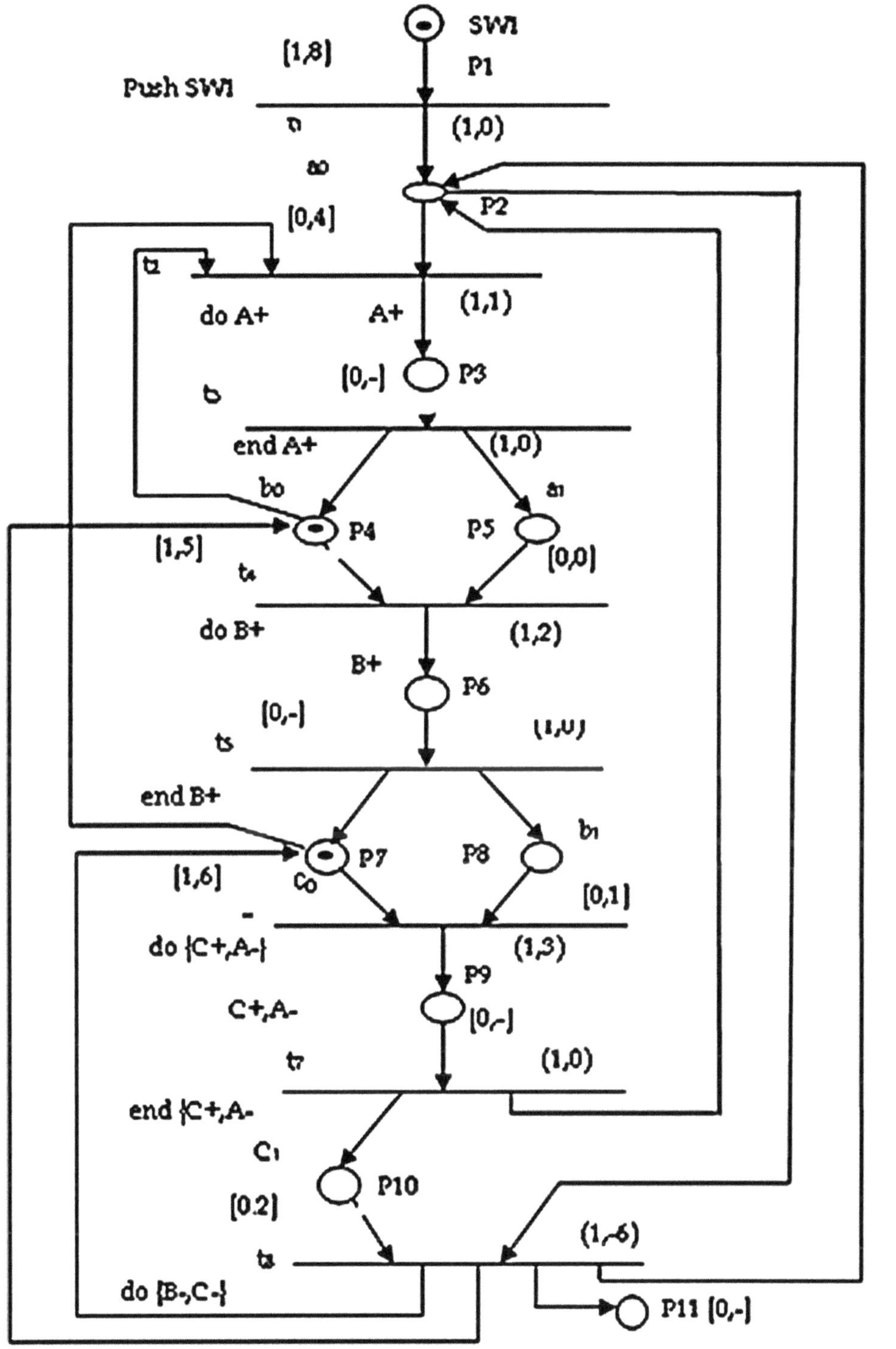

Figure 8. LLD for sequence 2

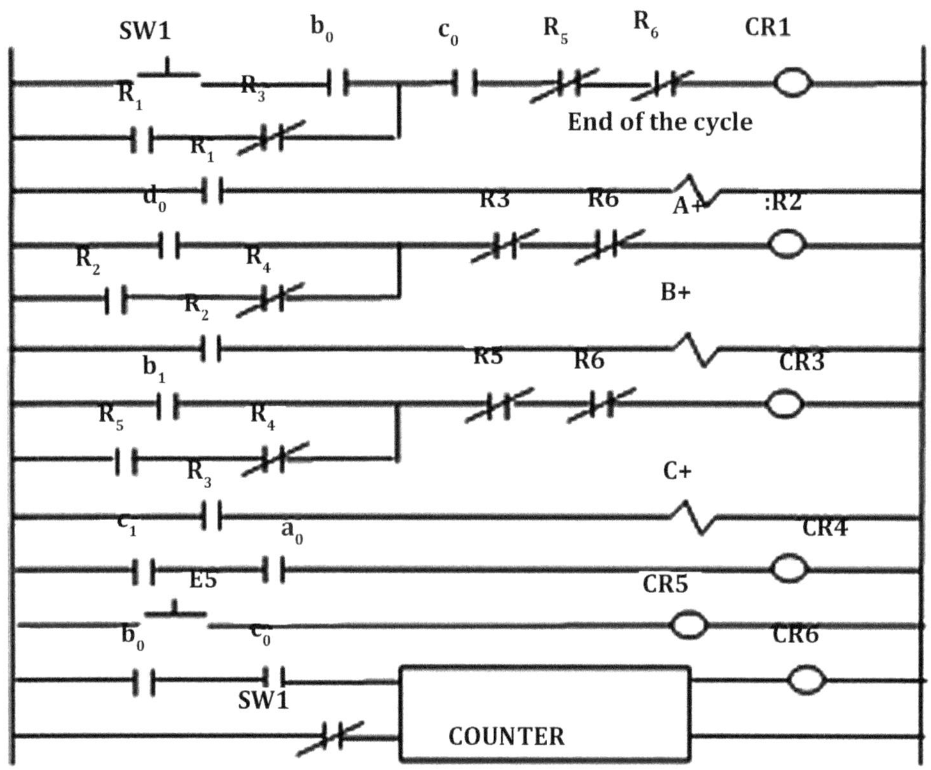

6 COMPARING LLDS AND PNS MODELS OF A RAILWAY SYSTEM

One effective way to perform the comparison between LLDs and TPNs is through a railway-automated system. The system considered in this paper consists of four railway engines (e.g. *A, B, C* and *D*) used to form a wagon convoy. Each RE has two normally open limit switches (physically these limit switches are isolated railway sections). For example, when RE *A* contacts limit switch $a_0(a_1)$, it indicates that the RE *A* is at the and of its return stroke (forward stroke). Three push buttons are provided to start the system (switch SW_1), to stop the system normally (switch SW_2) and to stop the system immediately in emergency (switch SW_3, ES). The system has 11 inputs corresponding to 8 limit switches (two for each RE) and 3 push buttons. The system has 6 outputs corresponding to 4 relays that give the senses of the RE's strokes and two lights that indicate the status of the system (a permissive and a restrictive color). Tables 2 and Table 3 show the I/O mapping of the system.

Consider that the system has to be controlled to execute the above sequence where A+ means that the RE has to do forward stroke and A- return one. {C+, A-} represents two concurrent actions taking place simultaneously: RE *C* to do a forward stroke and RE A to do a return one. Figure 6 shows the LLD and Figure 7 shows the TPN corresponding to this sequence.

Note that in TPN a place has attributes $[n_1, n_2]$ where n_1 is the first attribute representing an n number of tokens and n_2 is the second one mapping an input channel number. A transition has attributes $[n_1', n_2']$ where n_1' is the firing duration and n_2' is to be written on the digital output interface. The LLD shown in Figure 6 has 56 basic elements (23 nodes and 33 links), whereas the TPN shown in Figure 7 has

Figure 9. TPN for sequence 2

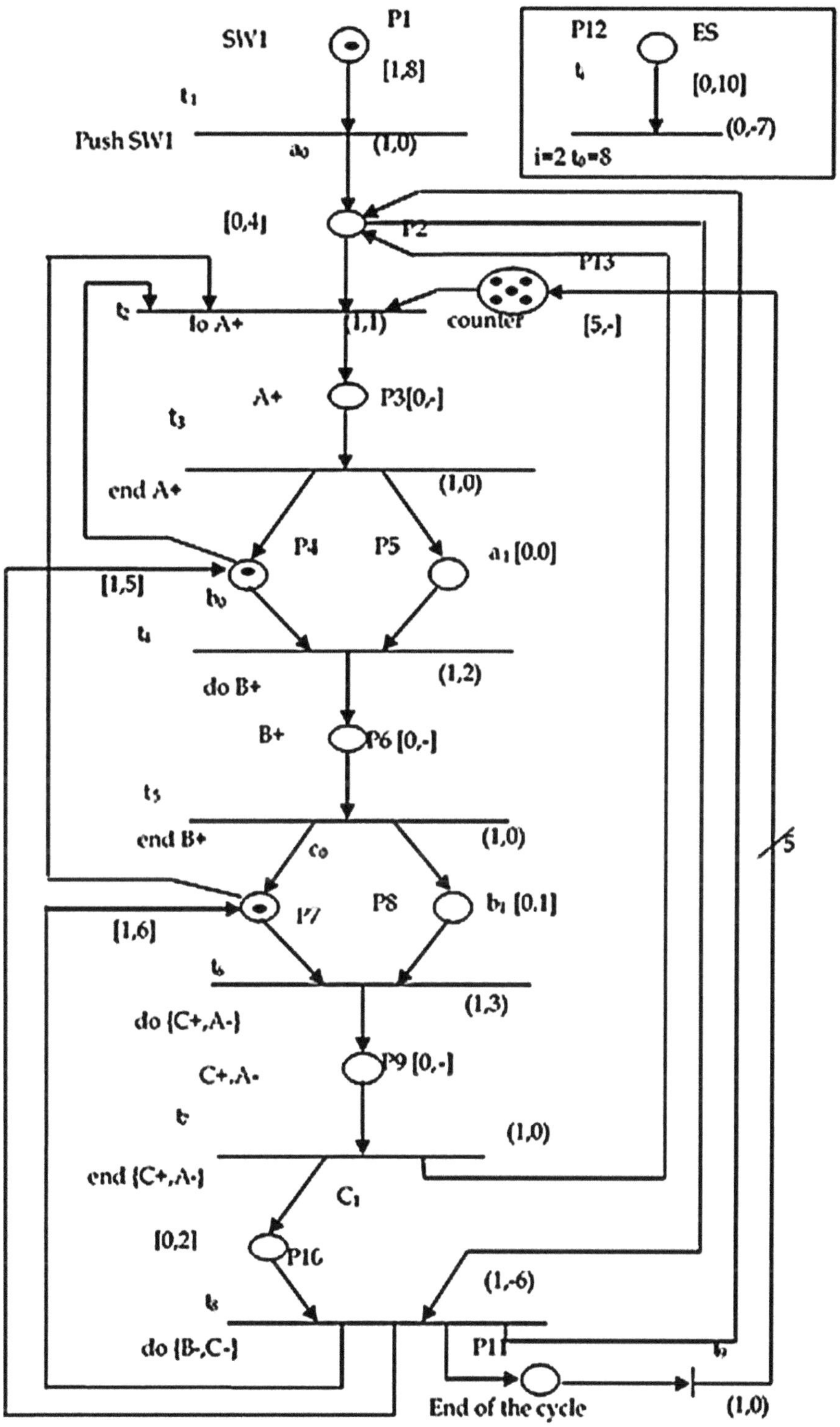

46 basic elements (19 nodes and 27 links). In order to compare the LLDs and TPNs another sequence with increasing complexity is considered sequence 2: START, 5[A+, B+, {C+, A-}, {B+, C-}] with Emergency Stop and Counter. In this sequence, there is a need to provide emergency stop and a counter.

In this system, both the LLD and TPN are implemented, so that when the emergency stop switch, ES is pressed, the whole system, including the active elements, are immediately stopped. Figure 8 shows the LLD and Figure 9 shows the TPN corresponding to this sequence.

The LLD shown in Figure 8 has 86 basic elements (36 nodes and 50 links), whereas the TPN shown in Figure 9 has 59 basic elements (22 nodes and 37 links). The LLD needs more additional basic elements compared to TPN.

7 CONCLUSION

We notice that as the specification changes, the TPN requires fewer changes compared to the LLD. Using TPNs the control logic can be qualitatively analyzed to check properties such as absence of deadlocks and presence of reversibility in the system. Using LLDs qualitative analysis is not possible until it is stimulated or implemented. Using TPNs, the initial state of the system can be directly represented by its initial marking. The procedure for controlling a system using TPNs is straightforward, simple, and can be applied to control any discrete event system that has digital in/out interface and a computer. Adding more attributes to places and transitions can extend TPNs in order to control complex transport systems (Ciufudean, 2007). Designing a discrete event controller with choices to perform a control task can extend the present approach.

REFERENCES

Baccelli, F., & Liu, Z. (1992). Comparison properties of stochastic decision free Petri nets. *IEEE Transactions on Automatic Control*, *37*(12), 1905–1920. doi:10.1109/9.182477

Campos, J., Chiola, G., & Silva, M. (1991). Properties and performance bounds for closed free choice synchronized monoclass quering networks. *IEEE Transactions on Automatic Control*, *36*(12), 1368–1382. doi:10.1109/9.106153

Ciufudean, C. (2000). Modeling the reliability of the interaction man-machine in railway transport. The Annals of the "Stefan cel Mare" University of Suceava, (13), 80-84.

Ciufudean, C. (2002). *Discrete Event Systems for Modelling the Railway Traffic*. Bucharest: Matrix Rom Publishing House. (in Romanian)

Ciufudean, C. (2003). Discrete event control of railway systems. In *Proc of International Conference AQTR Cluj Napoca.*

Ciufudean, C., Filote, C., & Amarandei, D. (2008). Control Charts of Workflows. *Advances in Data Mining*, *8th Industrial Conference ICDM 2008*, (pp. 330-344). Leipzig, Germany: Springer Verlag Berlin-Heidelberg.

Ciufudean, C., & Larionescu, A. (2002). Safety criteria for production lines modeled with Petri nets. Advances in Electrical and Computer Engineering, 2(2), 15-20.

Falcione, A., & Krogh, B. H. (1993). Design recovery for relay ladder logic, *IEEE Trans. on Cont. Syst. Mag.*, *24*(4), 90–98.

Filote, C., & Ciufudean, C. (2010). Discrete Event Models for Flexible Manufacturing Cells. In Future Manufacturing Systems. Sciyo.

Ghaffari, A., Rezg, N., & Xie, X. (2003). Feedback control logic for forbiden state problems of marked graphs: Application to a real manufacturing system. *IEEE Transactions on Automatic Control*, *48*(1), 18–29. doi:10.1109/TAC.2002.806651

Holloway, L.E., & Krogh, B., (1990). Synthesis of feedback control logic for a class of controlled Petri nets. *IEEE Trans. Autom. Contr.*, (5), 514-523.

Laftit, S., Proth, J. M., & Xie, X. L. (1992). Optimization of invariant criteria for event graphs. *IEEE Transactions on Automatic Control*, *37*(12), 547–555. doi:10.1109/9.135488

Li, J., & Wonham, W. M. (1993). Control of vector discrete event systems i – The base model. *IEEE Transactions on Automatic Control*, *38*(8), 1214–1227. doi:10.1109/9.233154

Proth, J. M., & Xie, X. L. (1994). Cycle time of stochastic event graphs: Evaluation and marking optimization. *IEEE Transactions on Automatic Control*, *39*(7), 1482–1486. doi:10.1109/9.299640

van Breusegem, V., & Boel, R. K. (1992). Maximally permissive feedback controls for controlled state machine. In *Proc. WODES'92*.

Xie, X. (1998). Perturbation analysis of stochastic Petri nets. *IEEE Transactions on Automatic Control*, *43*(1), 76–80. doi:10.1109/9.654889

Yee, S. T., & Ventura, J. A. (2000). Phase-type approximation of Petri nets for analysis of manufacturing systems. *IEEE Trans. on Rob. and Autom.*, (3), 318-322.

Zurawski, R., & Zhon, M. C. (1994). Petri nets and industrial applications: A tutorial. *IEEE Transactions on Industrial Electronics*, *41*(6), 567–583. doi:10.1109/41.334574

KEY TERMS AND DEFINITIONS

Diagnosis: Diagnosis is the identification of the nature and cause of anything.

Discrete Event Systems: Discrete event system is a discrete-state which contains solely discrete state spaces and event-driven state transition mechanisms.

Green Transportation: Green transportation system is a sustainable transport system which refers to the broad subject of transport that is sustainable in the senses of social, environmental and climate impacts and the ability to, in the global scope, supply the source energy indefinitely.

Ladder Diagram: Ladder diagram is a programming language that represents a program by a graphical diagram based on the circuit diagrams of relay logic hardware. Ladder logic is used to develop software for programmable logic controllers (PLCs) used in industrial control applications.

Model Controller: Model controller is a software architectural pattern for implementing user interfaces. It divides a given software application into three interconnected parts, so as to separate internal representations of information from the ways that information is presented to or accepted from the user.

Petri Net: Petri net is a mathematical modeling language for the description of discrete event systems.

Qualitative Formalism: Qualitative formalism it aims at describing the common-sense background knowledge on which our human perspective on the physical reality is based.

Railway Engineering: Railway engineering is a multi-faceted engineering discipline dealing with the design, construction and operation of all types of railway systems. It encompasses a wide range of engineering disciplines, including civil engineering, computer engineering, electrical engineering, mechanical engineering, industrial engineering and production engineering.

Railway System: Railway system is the commercial organization responsible for operating a system of transportation for trains that pull passengers or freight.

Subsequence: Subsequence is a sequence (set of elements) that can be derived from another sequence by deleting some elements without changing the order of the remaining elements.

Chapter 9
On the Decision Criteria for "Greening" Information Systems

Tagelsir Mohamed Gasmelseid
International University of Africa, Sudan

ABSTRACT

The impacts of climate changes are significantly influencing the approaches of organizations and governments to use resources, develop appropriate environment-friendly strategic frameworks and adopt a holistic approach to understand their operating environment. These impacts can be seen in the search for energy effective solutions, outsourcing processes and engaging into partnerships and alliances. At the same time, governments have been stepping up policy and legislative initiatives, assessment frameworks, and engagement in international conventions to cut carbon emissions and promote sustainability.

1. INTRODUCTION

The impacts of climate changes are significantly influencing the approaches of organizations and governments to use resources, develop appropriate environment-friendly strategic frameworks and adopt a holistic approach to understand their operating environment. These impacts can be seen in the search for energy effective solutions, outsourcing processes and engaging into partnerships and alliances. At the same time, governments have been stepping up policy and legislative initiatives, assessment frameworks, and engagement in international conventions to cut carbon emissions and promote sustainability. Governments and businesses have a wide range of initiatives dealing with the impacts of information and communication technologies (ICTs) on the environment and climate change. Initiatives concentrate on greening ICTs by directly reducing the emissions of computers and servers. However, ICT applications through their capabilities to record analyze and report, also have an important role to play in reducing global warming and environmental degradation. However, only about one-fifth of green initiatives in business have measurable targets and their frequency is also higher in government lead initiatives rather than business associations. Even fewer governments and business associations focus on measuring the quality and impact of their policies and programmes (OECD, 2009A).

DOI: 10.4018/978-1-4666-9792-8.ch009

In addition to the direct effects representing environmental issues directly related to ICTs, their applications can greatly enable energy savings through the use of "smart" ICTs and sensor-based networks and the Internet. As enablers, ICT applications can contribute to more sustainable use of global resources, for instance by tracking and monitoring water use, biodiversity, land use, pollution. Advances in ICTs and other technologies facilitate behavioral and organizational changes towards sustainability (OECD, 2009B).

With reference to climate change and the importance of ICT greening, there has been a wide agreement on some issues including:

1. The performance of ICT has to improve because it constitutes a major part of the solution in tackling climate change and related environmental challenges; its performance has to improve. Smart applications in transport, buildings and urban environments, energy generation, distribution and production need to be increasingly, ICT-enabled (ITU, OECD and GeSI 2009).
2. There needs to be a better fit between environmental policies and ICT policy pathways to improve the contribution of ICT to the mitigation of climate change activities. While such fit is essential to ensure the orchestration of functions it also determines the extent of innovation to be undertaken and the drivers of its initiation and diffusion. Different types of innovation usually require significant changes in the behavior of employees, task systems, new knowledge to be embodied in policy formulation processes, status quo, and information, values, and incentives, among other things (Nystrom, Ramamurthy & Wilson, 2002).
3. Better information is crucial for greater efficiency, to reap the undoubted benefits of ICT applications across the economy. The lack of information and ignorance about environmental issues will engender concern to be translated into both personal and political behavior changes (Bartiaux, 2008).
4. Green growth policies need to address issues of equity and the digital divide through use of ICTs. Harnessing the capabilities of ICTs to empower consumers is essential to measure and manage individuals' environmental footprints. To serve this purpose, affordable and relevant ICT applications need to be diffused and used globally.

ICT is an obvious target with its relentless growth and high turnover of technology, but it is also a key tool for delivering green services and implementing a green policy across the organization. Because ICT is directly and inexorably related to sustainable development, the formulation of an organization-wide carbon reduction strategy is gaining paramount importance in improving organizational green profile and competitiveness. Such organization-wide strategies will invariably include the Management Information Systems (MIS) of the organization. This chapter investigates and discusses the impact of the green movement on the MIS of an organization. Therefore, this chapter is focused more on the information systems side of the green movement rather than the information technology aspect of it.

2. GREEN COMPUTING

Green computing is emerging as an innovative paradigm concerned with the use of information systems and technologies (such as servers, CPUs and peripherals, among others) in an environment-friendly and energy-efficient manners that contribute to global warming and the reduction of CO2 emission. It also dictates new axioms for the use of information technologies and the advocating the change of the behavior of users to achieve power saving and reduction of emissions. Using power management features to power-down computing machines and peripherals after use, using energy-efficient monitors, minimizing paper use and adopting recycling procedures and using alternative energy sources are examples of green-related tools to change the behavior of users. Within this context, "greening" processes are not limited to green infrastructure, data centers and communication networks it also covers "greening" of user behavior and resource-use metrics in their effort to use virtualized energy-efficient infrastructure.

3. CURRENT OPTIONS ON THE GREENING MENU

The question of greening ICT is looming very big within the context of the growing impacts of climate change. However, while "greening" processes are being regarded as adaptation measures, different "greening" options and initiatives are being considered as shown below. Each of these initiatives has to deal with corresponding information systems – supported by data bases, processes and interfaces.

1. Minimizing work-related journeys and commuting in order to reduce carbon emissions and make significant savings and productivity improvements (Terry, 2007). Emphasis tends to be made on introducing full time home based working and the use of electronic document and records management (EDRM) and workflow monitoring systems. The aim is to improve the productivity of staff, overall well being, and return on investment. Introduction of mobile work technologies including Tablet PCs and video conferencing to manage organizational interactions with customers has also been growing. Using such technologies to collect information (through meeting and interviews) also aims at reducing work-related journeys.
2. Focusing on the incorporation of "greening" dimensions in the processes of the entire supply chain through "green" procurement and outsourcing. The aim of the concerned Environmental Purchasing Policies (EPP) used by some organizations is to actively encourage suppliers to supply environmentally friendly products. They also promote the incorporation of environmental criteria in the selection of suppliers and award of contracts.
3. Conducting regular replacement of existing equipments with more environmentally friendly equipments.
4. Reducing the energy use of ICT equipment to reduce carbon emissions including the use of ultra low energy thin clients and servers to minimize power.

4. CHALLENGES OF THE GREENING APPROACH AND OPTIONS

Despite the potentials of the greening options to be used by organizations and governments, the real time outcomes of such greening processes should be thoroughly investigated. The main problem is that such interventions continued to be technology-oriented with a remarkable lack of focus on the development and use of relevant assessment and monitoring techniques. Greening information and communication technology should not be regarded as an end in itself but rather a catalyst for improving sustainable development and, most importantly, realizing organizational objectives. However, each of the greening options has its own impacts on the capacity of the organization to achieve its objectives and its corresponding processes.

Greening ICT can facilitate more collaborative and less carbon intensive ways of working. On the other hand, the emerging technologies used for mobile, remote, home and flexible working processes dictate new working styles that call for:

1. A dramatic cultural change and;
2. A transformation of the organization's information management strategy in order to align the "greened" technology with on-going business processes, working behavior and style.

On the other hand, "greened" alignment also demands the development of "green" procedures. The problem of such "greening" process is the challenge of restricting the benefits only to carbon reduction. Because of the "political overheating" of the term green ICT, organizations tend to under estimate the assessment benefits in terms of:

1. The connection of greening processes on organizational objectives and the capacity of the organization to realize them. Such connection sheds light on the entire (and potential) competitive matrix and resource use strategies.
2. The capacity of the organization to select and deploy (conventional and electronic) business models. However, the models to be adopted also reflect structure-based technology adoption decisions. The development of societal and organizational structures that enable well-informed choices of technologies which promote climate stability, adaptation to the effects of climate change and is essential (Charikleia, Doukas & Psarras, 2010).
3. The capacity of the organization to function as an open system and manage its connections with the change agents, partners and intermediaries interacting in its operating environment. Practically, "greening" ICT significantly affect both the sell side denoting the Customers Relationships Management (CRM) and buy-side denoting the Supply Chain Management (SCM) processes of organizational functions. The challenge therefore, is how to understand the contribution of carbon reduction towards the optimization (and sometimes the maximization) of the organizational relationships with its partners (suppliers, distributor, agents, and intermediaries) and customers. As a result, the impact on the organization extends to include its sourcing and outsourcing procedures and competitiveness.

The reduction of energy use of ICT equipment, as a "greening" mechanism deserves a comprehensive organization-wide (and sometimes industry wide) carbon reduction strategy. But the development and deployment of such a strategy is challenged by the following factors:

1. The scale of analysis dictated by the guidelines used for distributing and allocating carbon reduction loads. This brings a controversy regarding (i) "who" distributes loads: the government? Or the city council? Or the industry? Or the organization? Etc and (ii) the acceptability and universality of load allocation mechanisms and regulatory and/or obedience measures.
2. The perception of "impacts" differentials originating from the incorporation of carbon emission reductions in the entire organizational intervention policies. Such perception differentials are complicated further by the difficulty of defining "reduction" commitments, and whether such reduc-

tion should be implemented against "commitments" only other factors. It worth mentioning that perception differentials follow different style for manufacturing and non-manufacturing processes and profit and nonprofit organizational objectives. Especially for organizations managed using the concepts of profit, cost and responsibility centers, such differentials may lead to sub-optimization and organizational conflicts.

3. Because ICT is directly and inexorably related to the realization of development goals, reducing the number and type of information technologies in use as a measure to counter carbon emissions negatively affects different sectors. Especially in education, medicine, media, e-services, among others, are examples of such affected sectors. It also affects the organizational processing capacity of the entire organization. The use of energy saving software, on the other hand, is challenged by licensing and automatic upgrades that may occur at different time intervals. Moreover, especially for ERP-enabled and/or internet-based organizations the functionality of the entire applications is also affected by any change in the architecture and assembly of the entire ERP including prompt response and information exchange through messaging.

5. MANAGEMENT INFORMATION SYSTEMS ORIENTED GREENING DECISION CRITERIA

Approaching the greening process through emphasis on management information systems provides a road map for understanding the dynamism of greening and its impact on organizational functionality. This, as mentioned earlier in this chapter, is vital for the success of the green movement within the organization. While pure technology focus can help reduce the emissions from equipments and other operational gadgets, focus on MIS is crucial supplement to the overall energy reduction effort of the organization. However, despite the existence of a wide range of factors to be considered, the impacts of ICT greening (and the decision to adopt any greening procedure) can be viewed based on the following broad factors:

1. The impact of ICT greening procedures and techniques on the role of MIS in the achievement of organizational objectives and maintenance of competitive advantage edges:

ICT has been a driving course in the achievement of competitive advantages in organizations but such use has been guided by understanding the five forces described by Porter in the Five Forces Model: the power of the customer, the power of suppliers, the existence of close substitutes, rivalry among rivals and new market entrants. However, the challenge for organizations is how to align the impact of such factors in favor of its improved organizational competitive capacity. The conceptualization of ICT as "ends" rather than "means" and the lack of integrated implementation models challenges the role of MIS in enhancing competitiveness. Because MIS has outstanding focus on the internal functionality of the enterprise, the situation is complicated further by any technology-oriented "greening" procedures that underestimate the impact of organizational, informational and managerial concerns. For an organization to think about adopting "greening" its information technology, emphasis should be made on the extent to which such greening procedures are expected to influence the ability of its MIS to influence competitiveness factors in favor of the enterprise. The misconception that technology, in and of itself, can provide advantage; therefore, in greening MIS the focus in not on pure technology but on systems and processes surrounding the technologies. Within the context of the digital economy, the role of MIS is not limited to information provision but it extends to the conversion of the strategic grid and strategic opportunity matrix specifically to the environmentally-conscious effort of the organization. The failure to capture the change and dynamism of the competitive climate; requires the organization to think of green awareness as a true and competitive tool.

2. The improvement of the information processing capacity of the organization:

Information processing capacity refers to the volume of data that an organization can utilize in its efforts to devise actions that enable the organization to prosper (Christopher et al, 2009). It plays a key role in shaping the interpretations that managers make of events and trends and their attempts to understand how organizations develop knowledge (Egelhoff, 1982; Daft & Lengel, 1986; Grant, 1996).

In relation to the process of providing management information, information systems play six significant roles (Philip & Booth, 2001) mainly:

1. Survival which assumes that the technology is an essential part of doing business, without which the organization would be unable to function. It deals with such operations as accounting, payroll, automation of manufacturing functions, etc and focuses on improvement of operational efficiencies, cost reductions and internal integration. Survival can also incorporate the use of systems
2. Sources and resources Being guided with the Resource Dependency theory (Sabherwal & Tsoumpas, 1993), the organization is regarded as a system which interacts with its external environment and using inputs to process them and generate outcomes and feedback. Inter-organizational information systems play significant roles in the acquisition, processing, and marketing of products/services. Adopting a competence-driven view that organizations are basically collections of resources, information systems technology allows for improved access to resources, increased integration and enhanced competencies development strategies (Clemons & Row, 1991; Klein & Kromen, 1995). This is because of the improved organizational capacity to manage those relationships, ensuring a free internal and external flow of resources (Pfeffer & Salancik, 1978) and the satisfaction of the demand originating from its environment.
3. Strategy where information systems and technologies to allow organizations to create (not only improve) competitive advantages through improved skills, tools and mechanisms. As a result, the organization can exploit the ideas and strategic visions of talented individuals.
4. Service Value Analysis refers to the identification of new ways of doing business through the redesigning or complete rethinking of the way in which individual processes are performed. Information systems and technologies operate as enablers in the move to more flexible and cross-functional process arrangements.
5. S(c)yberspace which provides organizations with a platform to implement operation, and build network or computer-mediated relationships with their stakeholders and partners in a flexible, customizable, innovative way. As a result, organizations enjoy more enhanced responsiveness, adaptability and coordination.

6. Sustainability which is concerned with the management processes associated with the others because it determines the success or failure of the roles of information systems. The more sustainable the role of MIS in the organization, the more will the improved organizational capacity to enjoy benefits.

The impact of greening processes on the six roles to be played by information systems are described in accordance with the impacts of such processes on its architecture, information processing capacity and knowledge management as shown in Table 1.

Using the systems approach, the impacts shown in the above mentioned two dimensions can be described by viewing impacts on greening procedures of MIS based on the following factors:

5.1 MIS Architecture

Information system architecture is composed of the definition of the components of the IS, a description of their interconnections (both ''logical'' and ''physical'' within a network, for instance) and finally, their interaction in time (system dynamics) (Virginie et al, 2006). It is important for the identification of relevant technological acquisitions and other related resources. It is also important for locating technological and organizational modifications to be adopted during the development process (Brancheau & Wetherbe, 1986). However, the architecture of MIS is directly and inexorably related to the software architecture and organizational architecture. Software architecture is the structure of the components of a program or system (together with their interrelationships and dependencies) and the principles and guidelines that govern their design and evolution (Faheem & Capretz, 2008). It plays a key role in bridging the gap between requirements and implementation. It also provides a tractable guide to the overall system, permits designers to reason about the ability of a system to satisfy certain requirements, and suggests a blueprint for system construction and composition (Shifeng & Goddard, 2007; Garlan, 2001; Allen et al, 1998). According to Nadler and Gerstein (1992), organizational architecture can be defined as ''the art of shaping 'behavioral' space to meet the needs and aspirations of a business.'' It describes models and guides organizational change using the categories of purpose, structural materials, style, and collateral technologies. It is one aspect of a dominant design.

Table 1. Greening impacts on IS roles using IS indictors

	Survival	Sources and Resources	Strategy	Service Value Analysis	S(c)yberspace	Sustainability
MIS Architecture	Automation of processes & internal integration.	CRM & SCM integration & efficiency	Alternative conventional &electronic business models.	Process & product redesign	CRM & SCM coordination and organizational control	Functional optimization
Information processing capacity	Competitiveness and responsiveness	Rapid response and JIT	Policy and information analysis	High added value	Functional and process integration	Process and policy issues
Knowledge management	Added capacity	Modified ways	Knowledge production and use	Added values	New models	leadership and accountability

Figure 1. Greening dimension: MIS Architecture

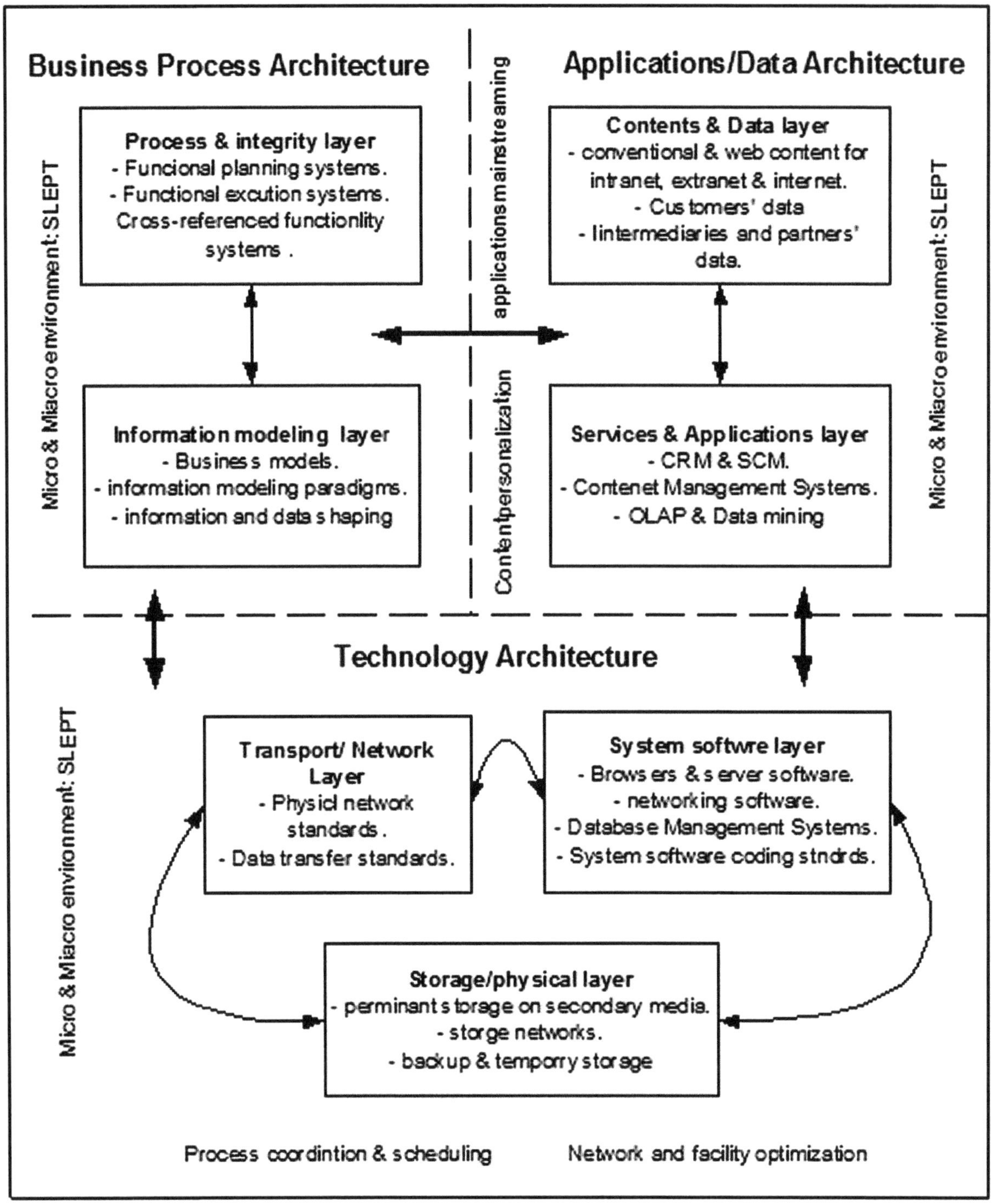

As shown in Figure 1, the entire architecture includes the following components:

1. **Business Process Architecture:** This component represents the connections among different organizational processes and functionalities. It incorporates two main layers:
   a. Process and integrity layer depicting planning and execution systems. Functional planning systems include supply chain planning, manufacturing planning, and demand planning. Functional execution systems include procurement, manufacturing, warehouse management, and order planning. Each aspect of these systems needs to be modeled, studied, and optimized for Green ICT.
   b. Information modeling layer describing the variety of business models in use, information modeling paradigms, and data shaping techniques in use.
2. **Application and Data Architecture:** This component describes the applications to be offered throughout the architectural model based on two types of layers: (i) Services application layer to be used for the operationalization of services such as customer relationship management; supply chain management; data mining and analysis; and content management systems. (ii) Content and data layer describing network-specific content and data for intranets, internet and extranet including customers' data.
3. **Technology Architecture:** The technology architecture represents a major component of the entire MIS because it describes its hardware, software, telecommunication and databases. It incorporates three main layers:
   a. System and applications software layer which describes operating systems, web browser, server software and standards, network software, and database management systems.
   b. Transport/network layer incorporating the dimensions of physical data transfer and management, physical network and transport standards (mainly TCP/IP), coding standards, physical interoperability and functional orchestration.
   c. Storage/physical layer representing the level of storage on permanent magnetic storage on web servers, optical backup, temporary storage in memory (RAM) and storage on network and Virtual networks.

2. Mainstreaming Knowledge Management Procedures and Techniques

Greening procedures have to be viewed in terms of their impacts on Knowledge management decisions involving the integration and transfer of knowledge. The importance of such decisions stems from the fact that they shape the overall process of formal and informal design, implementation and sharing of Knowledge. One of the main knowledge management related decisions is knowledge codification through which greening practices may have significant impacts on the information processing capacity of the enterprise (Fernando et al, 2009; Soo, Devinne, & Midgley, 2002). Knowledge codification is defined as the process of converting the codifiable tacit knowledge into messages (patents, databases, user manuals, etc.) that can then be processed as information (R. Cowan & Foray, 1997; Balconi, 2002; Cohendet & steinmueller, 2000; Albino, Garavelli, & Schiuma, 2001). Because the codification process modifies the proportions of tacit and explicit knowledge present in the firm and their location by transferring some of the knowledge from the minds of the workers to the organization's data warehouses (Subramanian & Youndt, 2005), impacts of greening on the entire MIS architecture may be negative.

3. Organizational Innovation, Change Management, and Modeling

According to Dainty et al, (2007), innovation is defined as the process of bringing new and improved products and processes to market; developing, adopting and adapting manufacturing processes to enhance productivity and product quality; developing, adopting and adapting business practices to enhance the performance of the firm. Therefore, it affects the capacity of the organization to model its internal and external interactions. At the level of organizational innovation, the challenges include conflicts of vested interests, project based working patterns, lack of technology and other resources, lack of top level management support, functional sub optimization, incompatibility of innovation with organization structure, inappropriate organizational culture and distorted social structure (Barnes, et al. 2001; Terziovski, et al. 2003; Loewe & Dominiquini 2006; Leifer, et al. 2000; Belland, 2009)

As shown in Figure 1, architecture oriented impacts of greening processes should be understood in accordance with internal and external change agents. Internal change agents refer to impacts and consequences of change of organizational objectives, resource base and resource use matrix. As shown in Figure 2, external change agents are social, legal, environmental, political and technological factors. The impacts of such factors may originate at local (organization-specific or country based) and/or international landscapes. However, both sets of change agents tend to significantly influence the role of MIS to support competitiveness as well as the information processing capacity of the entire organization. This is because they affect the entire organization's corporate and support strategies and the perceived business models in use (both conventional and electronic). The impact of greening procedures should be understood in terms of the nature and magnitude of organizational change to take place, organizational

Figure 2. Modeling Organizational innovation

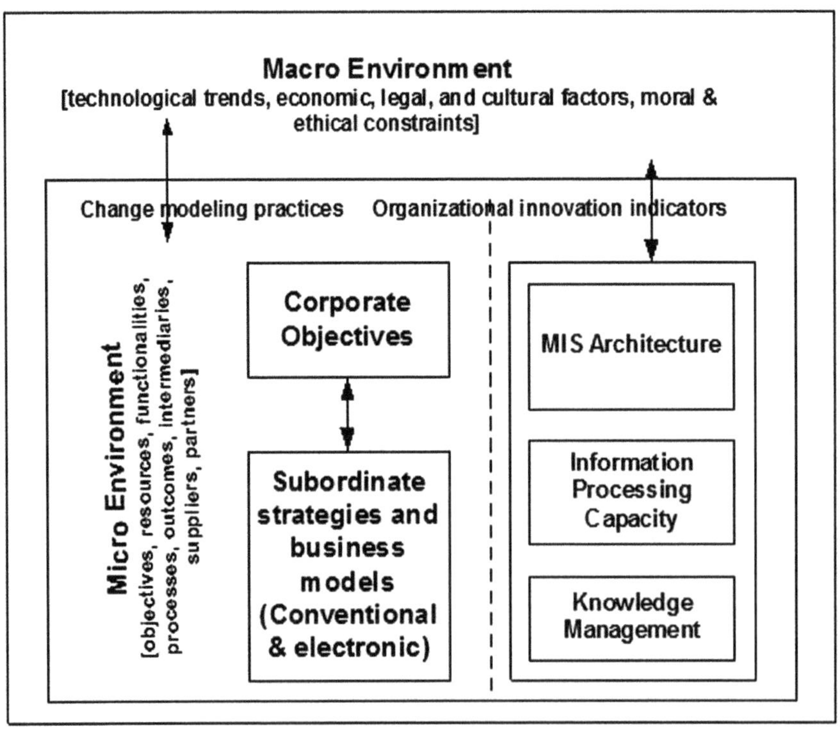

change modeling and management capacity and the degree of organizational innovation to be incorporated. On the other hand, both internal and external factors influence the entire MIS architecture and its functionality. Generally speaking, the dramatic organizational shift necessitates the incorporation of environment-friendly factors.

The lack of an integrated greening approach complicates enterprise-wide organizational change management and innovation. The impacts of internal and external factors dictate new axioms for technology infusion and diffusion and generate comprehensive information systems management issues. Whether organizations are prepared to adapt to the resulting changes consequences and requirements originating from greening processes remains questionable. Within this context, organizations must approach change in a holistic manner and avoid partial solutions. Therefore, simply using higher powered technological gadgets to increase profit margin is unlikely to serve the organization well in the carbon-focused future.

FUTURE TRENDS AND DIRECTIONS

The emphasis on green information and communication technology is growing due to the dramatic impact of climate change natural systems, economic activities, food systems and the lives of business organizations. The impact follows an integrated "hyper" fashion where consequences taking place at the top of the hierarchy of analysis (e.g., natural systems) dramatically affect populated units, structures and systems at the bottom. Within this context, climate-related, IT-based and business organizations tend to emphasize on the adoption of technological mechanisms and technical interventions. Such emphasis is being motivated by the technological advancements that took place over the last century and the lack of integrated frameworks that move beyond their boundaries. However, the complexity of climate change patterns, the intensity of consequences at different landscapes and the wide range of limitations, technological and technical interventions they are not enough to understand the potentials and consequences of climate change and, accordingly, decide on ICT greening processes. The limited view resulting from the adoption of technological and technical approaches reduces the ability of organizations to crystallize and understand non-technical mechanisms and their consequences. It is technical, technological, social, cultural dimensions and mechanisms which lead appropriate decision making regarding ICT greening. Understanding the impacts of such greening procedures on the organization's MIS provides some guidance in this regard.

Climatic changes are expected to continue at varying rates of development in terms of magnitude, scale and impacts. In the light of the dominance of "scaling" rather than "analysis" orientations of organizations involved in climate change assessment, lack of a comprehensive approach will continue to threaten effective ICT greening processes. While emphasis on ICT greening using technological intervention mechanisms is a candidate for escalation, additional emphasis will be made on the development of integrated frameworks. The near future will exhibit a growing emphasis on understanding the process of ICT greening through multiple dimensions including in addition to technology, organizational, institutional, resource-based and socio-cultural dimensions. Emphasis on understanding the contextual impacts of ICT greening procedures on organizational management information systems assumes a paramount important priority for business organizations especially under the existing highly competitive, networked digital economy.

REFERENCES

Albino, V., Garavelli, A., & Schiuma, G. (2001). A metric for measuring knowledge codification in organizational learning. *Technovation*, *1*(2), 413–422. doi:10.1016/S0166-4972(00)00058-4

Allen, R., Douence, R., & Garlan, D. (1998). *Specifying and analyzing dynamic software architectures*. Proc. of Fundamental Approaches to Software Engineering.

Balconi, M. (2002). Tacitness, codification of technological knowledge, and the organization of industry. *Research Policy*, *31*(3), 357–379. doi:10.1016/S0048-7333(01)00113-5

Barnes, J., Bessant, J., Dunne, N., & Morris, M. (2001). Developing manufacturing competitiveness within South African industry: The role of middle management. *Technovation*, *21*(5), 293–309. doi:10.1016/S0166-4972(00)00047-X

Brancheau, J., & Wetherbe, J. (1986). Information architectures: Methods and practice. *Information Processing & Management*, *22*(6), 453–463. doi:10.1016/0306-4573(86)90096-8

Brian, B. (2009). Using the theory of habitus to move beyond the study of barriers to technology integration. *Computers & Education*, (52): 353–364.

Charikleia, K., Haris, D., & John, P. (2010). Technology transfer through climate change: Setting a sustainable energy pattern. *Renewable & Sustainable Energy Reviews*, *14*(6), 1546–1557. doi:10.1016/j.rser.2010.02.001

Christopher, W., Craighead, G., Tomas, M., & David, J. (2009). The effects of innovation–cost strategy, knowledge, and action in the supply chain on firm performance. *Journal of Operations Management*, (27): 405–421.

Clemons, E., & Row, M. (1991). Sustaining IT advantage: The role of structural differences. *Management Information Systems Quarterly*, *15*(3), 275–292. doi:10.2307/249639

Cohendet, P., & Steinmueller, W. (2000). The codification of knowledge: A conceptual and empirical exploration. *Industrial and Corporate Change*, *9*(9), 195–209. doi:10.1093/icc/9.2.195

Cowan, R., & Foray, D. (2002). The economics of codification and the diffusion of knowledge. *Industrial and Corporate Change*, (6): 595–622.

Daft, L., & Lengel, H. (1986). Organizational information requirements, media richness, and structural design. *Management Science*, *32*(5), 554–571. doi:10.1287/mnsc.32.5.554

Egelhoff, G. (1982). Strategy and structure in multinational corporations: An information-processing approach. *Administrative Science Quarterly*, *27*(2), 435–458. doi:10.2307/2392321

Faheem, A., & Luiz, F. (2008). The software product line architecture: An empirical investigation of key process activities. *Information and Software Technology*, (50): 1098–1113.

Fernando, E., García, M., Eva, P., José, E., & Navas, L. (2009). Knowledge codification and technological innovation success: Empirical evidence from Spanish biotech companies. *Technological Forecasting and Social Change*, *76*(1), 141–153. doi:10.1016/j.techfore.2008.03.016

Françoise, B. (2008). Does environmental information overcome practice compartmentalization and change consumers' behaviors? *Journal of Cleaner Production*, *16*(11), 1170–1180. doi:10.1016/j.jclepro.2007.08.013

Garlan, D. (2001). Software architecture. In J. Marciniak (Ed.), *Wiley Encyclopedia of Software Engineering*. John Wiley & Sons.

Grant, R. (1996). Toward a knowledge-based theory of the firm. *Strategic Management Journal*, *17*(S2), 109–122. doi:10.1002/smj.4250171110

ITU, OECD, & GeSI. (2009). *ICTs and climate change: Official side-event by ITU, OECD, and GeSI.* United Nations Climate Change Talks. Retrieved on February 2, 2010 form: OECD: www.oecd.org/sti/ict/green-ict

Klein, S., & Kromen, J. (1995). IT-enabled co-operations: a resource-based approach. In *Proceedings of the 3rd European Conference on Information Systems.*

Leifer, R., McDermott, C., O'Connor, G., Peters, L., & Rice, et al. (2000). *Radical Innovation How Mature Companies Can Outsmart Upstarts.* Boston: Harvard Business School Press.

Loewe, P., & Dominiquini, J. (2006). Overcoming the barriers to effective innovation. *Journal of Strategy and Leadership*, *34*(1), 24–31. doi:10.1108/10878570610637858

Morton, S., Burns, D., & Dainty, A. (2007). Beyond Lean: Overcoming *Resistance* to Innovation to Improve Productivity. *POMS 18th Annual Conference.*

Nadler, D., & Gerstein, M. (1992). *What is organizational architecture?* Harvard Business Review September–October.

OECD. (2009A). *Towards Green ICT Strategies. Assessing Policies and Programmes on ICT and the Environment.* DSTI/ICCP/IE (2008)3/FINAL, May 2009. Retrieved February 5, 2010 from www.oecd.org/sti/ict/green-ict

Paul, C., Ramamurthy, K., & Alla, L. (2002). Organizational context, climate and innovativeness: Adoption of imaging technology. *Journal of Engineering and Technology Management*, *19*(3-4), 221–247. doi:10.1016/S0923-4748(02)00019-X

Pfeffer, J., & Salancik, G. (1978). *The External Control of Organizations: A Resource Dependence Perspective*. New York: Harper and Row Publishers.

Philip, G., & Booth, M. E. (2001). A new six 'S' framework on the relationship between the role of information systems (IS) and competencies in 'IS' management. *Journal of Business Research*, *51*(3), 233–247. doi:10.1016/S0148-2963(99)00051-X

Sabherwal, R., & Tsoumpas, P. (1993). The development of strategic information systems: Some case studies and research proposals. *European Journal of Information Systems*, *2*(24), 240–259. doi:10.1057/ejis.1993.36

Shifeng, Z., & Steve, G. (2007). A software architecture and framework for Web-based distributed Decision Support Systems. *Decision Support Systems*, (43): 1133–1150.

Soo, C., Devinne, Y., Midgley, D., & Deering, A. (2002). Knowledge management: Philosophy, processes, and pitfalls. *California Management Review*, *44*(4), 129–150. doi:10.2307/41166146

Subramanian, M. (2005). The influence of intellectual capital on the types of innovative capabilities. *Academy of Management Journal*, *48*(3), 450–463. doi:10.5465/AMJ.2005.17407911

Terry Street. (2007). Green ICT –taking the strategic approach. *Socitm Consulting*. Retrieved on February 2, 2010 from socitm.gov.uk/consulting

Terziovski, M., Fitzpatrick, P., & O'Neill, P. (2003). Successful predictors of business process reengineering (BPR) in financial services. *International Journal of Production Economics*, *84*(1), 35–50. doi:10.1016/S0925-5273(02)00378-X

Virginie, G., & Francois, G. (2006). Design of information system architectures using a key-problem framework. *Computers in Industry*, *57*(2), 189–200. doi:10.1016/j.compind.2005.09.001

Chapter 10
A Biologically-Inspired Computational Solution for Protein Coding Regions Identification in Noisy DNA Sequences

Muneer Ahmad
King Faisal University, Saudi Arabia

ABSTRACT

Biologically inspired computational solutions for protein coding regions identification are termed as optimized solutions that could enhance regions of interest in noisy DNA signals contrary to contemporary identification. Exponentially growing genomic data needs better protein translation. The solutions proposed so far rely on statistical, digital signal processing and Fourier transforms approaches lacking the reflection for optimal biologically inspired identification of coding regions. This paper presents a peculiar biologically inspired solution for coding regions identification based on wavelet transforms with notion of a peculiar indicator sequence. DNA signal noise has been reduced considerably and exon peaks can be discriminated from introns significantly. A comparative analysis performed over datasets commonly used for protein coding identification revealed the outperformance of proposed solution in power spectral density estimation graphs and numerical discrimination measure's calculations. The significant results achieved depict 75% reduction in computational complexity than Binary indicator sequence method and 32% to 266% improvement than other methods in literature (as a comparison with standard NCBI range). The significance in results has been achieved by efficiently denosing the target DNA signal employing wavelets and peculiar indicator sequence.

DOI: 10.4018/978-1-4666-9792-8.ch010

INTRODUCTION

Biologically inspired computational solutions have been adopted in variety of computational problems. Contrary to contemporary solutions, biologically inspired solutions provide better enhanced identification while applied to protein regions identification in noisy sequences. It is well known that DNA sequence contains genes and gene comprises genic and intergenic regions. RNA translation from DNA is an important and critical task because exact identification of protein helps in knowing information regarding protein structure and cell functions. Exons are the regions in gene that are translated to protein and exons boundaries are diffused in intron-exon noise (Mahmood Akhtar et al. 2007). Optimal identification of exons from 1/f noise needs careful attention and adoption of suitable methodology.

Protein is composed of small scale units called amino acids. There are 20 types of amino acids and the sequence of these units determines the type and function of individual protein molecule.

There are 64 possible codon (tri-nucleotide structure of bases) values (Ahmad and Mathkour, 2009) that transcribe the DNA chains to protein chains at regions known as exon in several clusters of non genic regions introns. Exons are the regions responsible for carrying nucleotide bases for protein translation. A codon "ATG" identifies the start of the sequence that contains the protein coding regions and codons "TAA", "TGA" and "TAG" are stop codons of this sequence where T is normally replaced with U (called Uracil). It is worth mentioning that mere start codon may not help in protein sequence identification, perhaps some other factors are also required in certain species. Learning the exact location of coding regions leads to the provision of optimal solution of the underlying problem.

Table 1 describes the possible combinations of codons for transcription of protein. These combinations help in exonic distinguishing from introns.

Following are the several different solutions proposed by researchers in literature that rely on Fourier Transforms (Tina and Tessamma, 2010, Shuo and Ye, 2008, Akhter et al., 2008, Hota and Srivastava, 2008, Akhtar, 2008, Datta, 2004, Vaidyanathan, 2002), Wavelet Transforms (Tina and Tessamma, 2010, Mena, 2008 and Hang, 2005), digital filters (Kakumani, 2008 and Hang, 2005), indicator sequences (Akhtar, 2008 and Hota, 2008), support vector machine (Guo, 2009), fuzzy models (Shuo, 2008) and hidden markov models (Hazrina, 2008). These solutions don't reflect the noise suppression in DNA signals so a clear picture demonstrating the optimization of exonic peaks is unclear. This phenomenon has been described in results and discussions section.

Fourier Transforms (Sho, 2008 and Akhtar, 2008) provides the frequency contents of signal without any time domain information and lacks the information for occurrences of frequency components at specified intervals. The Heisenberg uncertainty principal is solved by the concepts of wavelets (Mena, 2008). A more sophisticated option for a global scaled window is selected using this phenomenon. The designed window can be shifted to entire signal piecewise (by sliding a few bases over the signal) for

Table 1. Tri-nucleotide composition for codon

TTT TTC TTA TTG	TCT TCC TCA TCG	TAT TAC TAA TAG	TGT TGC TGA TGG
CTT CTC CTA CTG	CCT CCC CCA CCG	CAT CAC CAA CAG	CGT CGC CGA CGG
ATT ATC ATA ATG	ACT ACC ACA ACG	AAT AAC AAA AAG	AGT AGC AGA AGG
GTT GTC GTA GTG	GCT GCC GCA GCG	GAT GAC GAA GAG	GGT GGC GGA GGG

one cycle and the same process is repeated by changing the window parameters. This makes able to gain the frequency component resolution at different scale levels.

(Tina and Tessamma, 2010) have performed that discrete wavelet transform can be used for better minimization of noise and maximization of prediction accuracy. (Roy et al., 2009) proposed a generic algorithm for frequency distribution of various spectral values relevant to individual nucleotide bases. (Guo and Zhu, 2009) presented a support vector machine method for identification of protein regions.

(Hazrina and Siti, 2008) present gene prediction system based on Hidden Markov Model (HMM) using Perl and PHP. (Shuo and Zhu, 2009) aimed an integrative method using Takagi-Sugeno fuzzy model for identification of exonic regions.

(Kakumani et al., 2008) proposed a method using statistically optimal null filter to maximize the SNR (signal to noise ratio). The presence of exonic region in DNA strands has been detected by employment of least square optimization criteria. (Akhtar et al., 2008) demonstrated an optimization for DFT based methods relying on effect of window lengths for signal processing based exonic identification. Hota and Srivastava, 2008) proposed a complex indicator sequence methodology for exon prediction in DNA. (Akhtar et al., 2008) have described a digital signal processing methodology for exonic and intronic region prediction with comparisons to the existing techniques. (Grandhi and Vijay, 2008) proposed a 2-simplex mapping method for identification of exon regions in DNA. (Mena-Chalco et al., 2008) used Modified Gabor-Wavelet Transform for exonic analysis. (Gupta et al., 2007) proposed a time series approach for exon and intron prediction. Changchuan and Stephen, 2007) predicted the exonic regions based on period three property of exons. DFT has been used for extraction of Fourier coefficients from four indicator sequences made from the DNA stretch. (Vaidyanathan et al., 2002) suggested an approach based on antinoch IIR filter and compared results against traditional approaches applying windowed Discrete Fourier Transforms. (Datta and Asif, 2004) formulated a fast DFT based methodology for genetic region search in DNA. (Hang and Feng, 2005) predicted protein secondary structure using continuous wavelet transforms and Chou-Fasman method. (Suprakash et al., 2004) employed a DFT based algorithm for detection of exonic regions in DNA strand. (Suprakash et al., 2004) presented an empirical observation of DFT approach for protein regions. (Vaidyanathan and Byang, 2002) described digital filters for gene prediction in DNA. The designed filter has been used for the prediction of period 3 components and elimination of background noise 1/f spectrum shown by all DNA sequences.

(Meng-Hsiun Tsai et al., 2009) described a simple method for protein spots detection in 2D-GE images. They find the protein edges in image contrast and the spots are detected by the edges. (Zhang et al., 2007) proposed a denoising method that combines Adaptive Line Enhancer and takes advantages of Complex Discrete Wavelet Transform. The method removes the noise from the datasets. (Imran et al., 2009) proposed a method for reversible water marking using integer wavelet transform to accomplish the retrieval of hidden information along with recovering the original image contents.

The filters that carry a finite response to impulse signals are called FIR filters. The FIR filter of length k can be described (Akhtar et al., 2008) as

$$y[n] = \sum_{k=0}^{K-1} a_k x_{[n-k]}$$

Y is the transformed data and x is the input data. The filter takes a summation over input vector multiplied by a constant factor.

$$y[n] = -\sum_{k=1}^{N-1} a_k y[n-k] + \sum_{k=0}^{M-1} b_k x[n-k]$$

y represents a vector of length n that contains the transformed values for IIR filter (Kakumani et al., 2006). The filter used two kinds of coefficients, feed forward and feed backward represented by a_k and b_k.

$$H(z) = \frac{Y(z)}{X(z)} = \frac{B(z)}{A(z)} = \frac{\sum_{k=0}^{M-1} b_k z^{-k}}{1 + \sum_{k=1}^{N-1} a_k z^{-k}}$$

H is the transform function over z-transform when output vector is divided by input vector. The main difference between the two filters is stability, band width and order of filter. IIR filter with its extension is widely used in DSP techniques for DNA signal analysis.

Binary indicator sequence method prices 1 and 0 for the existence or non existence of a specific nucleotide in strand (Datta and Asif, 2004).

Sum of all binary indicator sequences is 1,

uA[n] + uG[n] + uC[n] + uT[n] = 1 for n=0, 1, 2,....N-1.

Let UA[k], UG[k], UC[k] and UT[k] be DFT's of the binary sequences, then

$$U_x[k] = \sum_{n=0}^{N-1} u_x[n] e^{-j2\pi kn/N} \qquad k = 1,\ 2, \ldots, N$$

Where Ux may be one of indicator sequences.

Electron-Ion Interaction Potential (EIIP) is proposed as against four binary indicator sequences which computationally reduce the overhead by 75%.

$$Y_{EIIP} = W_A X_A + W_T X_T + W_C X_C + W_G X_G$$

Where numerical values are A = 0.1260, T = 0. 1335, G= 0. 0806 and C = 0.1340 [19]

The Complex indicator sequence is also a replacement of binary indicator sequence. Complex indicator sequence [8] uses sequence of values namely X (A) = +1, X (T) = +j, X (G) = -1 and X (C) = -j

Proposed Approach

Our proposed biologically inspired approach takes advantages of biological genetic code context of nucleotides. We have assumed each nucleotide as a biological competitor that biologically competes to suppress those nucleotides that produce noisy channels in sequence. This inspired approach denoise the target DNA signal by employing the discrete wavelet transforms and mapping the nucleotide sequence with a peculiar indicator sequence.

Figure 1. Biologically inspired component based architecture

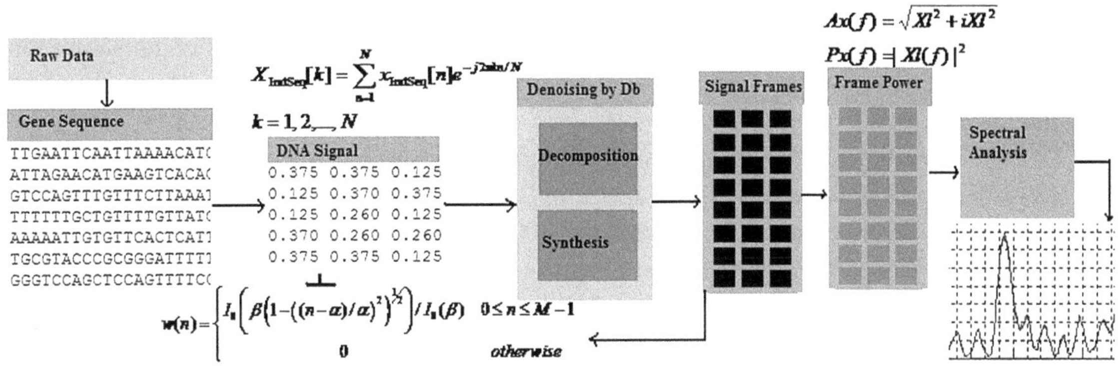

Figure 1 shows the biologically inspired architecture of proposed framework for protein coding region identification. In this architecture, the genetic sequence is the first component presented in the form of raw sequence of nucleotide bases (either processed or un-processed). This component is mapped into a signal with the help of a suitable indicator sequence for a DNA signal. The DNA signal is operated through low scale (high frequency) and high scale (low frequency) filters to be analyzed by wavelet transforms for minimization of spectral leakage. The decomposed signal is again passed through the low pass and high pass filters for perfect reconstruction. The denoised signal is then sectioned into frames. Each individual frame contains the fragment of signal windowed with Kaiser Window of length 351 base pairs. The frames are passed through calculations for STFT, magnitude and power. Frames are buffered in a resultant vector that is displayed in the form of exonic peaks. The discrimination measure for this spectral analysis shows the strength of peak values distinguished from exon-intron 1/f background noise. The power spectral estimate is done by plotting the target signal. In this plot, high peaks describe the presence of exonic regions. The PSD of peaks are calculated and corresponding intron peaks are identified. The discrimination measure D is the ratio of lowest peak of coding region to the highest peak of non coding region. The magnitude of discrimination measure shows the strength of applied methodology and degree of prediction accuracy. Greater the value of discrimination measure, greater is the distinguishing between exons and introns.

The peculiar indicator sequence is defined as by setting the values for base pairs as follows,

Adenine (A) = X (A) = 0.260

Thymine (T) = X (T) = 0.375

Guanine (G) = X (G) = 0.125

Cytosine (C) = X (C) = 0.370

With these measures the corresponding transform can be stated as,

$$X_{UTP}[k] = \sum_{n=0}^{N-1} x_{UTP}[n] e^{-j2\pi kn/N}$$
$$k = 1,\ 2, \ldots, N$$

The defined peculiar indicator sequence is a result of minute examination for the formation of clusters containing the tri-nucleotide codon composition in exonic regions.

We employed discrete wavelet transforms for denosing our DNA signal. There are thresholds for up and down frequency components being used as filters to decompose the DNA signal into two vectors containing low and high frequency components.

Figure 2 describes that DNA signal is passed through high scale (low frequency) and low scale (high frequency) filters each down sampled by a factor of 2. The target vectors for approximate and detail coefficients at first level contain half of the signal information. Signal components operated by low pass filter are then passed through the same process to obtain the vectors for approximate and detail coefficients at level 2. This phenomenon is repeated to further down sample the signal as per chosen threshold frequency levels. The signal has been decomposed in a series of high scale and low scale frequency components. These frequency components can be plotted and we can observe the approximate and detail components at granular level.

Figure 3 shows the actions for upsampling of approximate coefficients in a vector "AC" and detail coefficients in another vector "DC". The up-sampling passes the coefficients from low pass and high pass filters and generate the portion of signal that was decomposed in original transform.

Mathematically the up-sampling (convolution) can be written as

Figure 2. Daubechies DWT of order 3

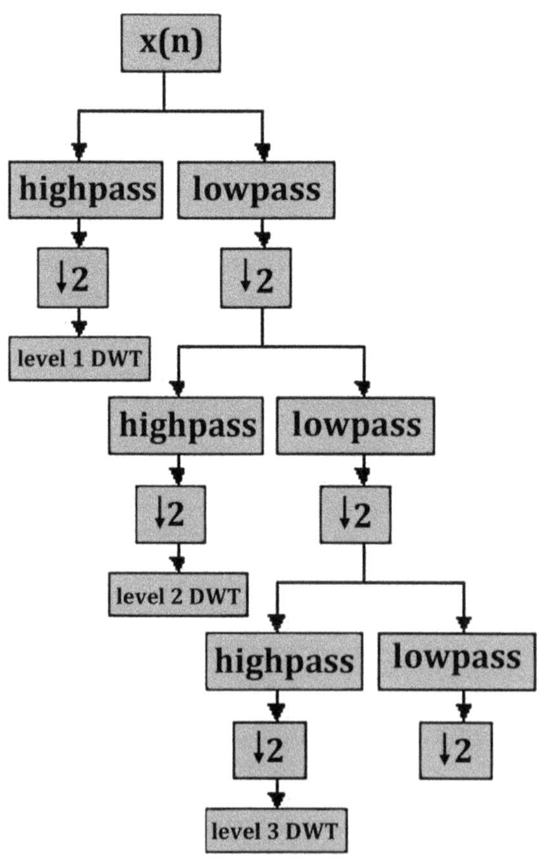

Figure 3. Upsampling by low and high pass filters

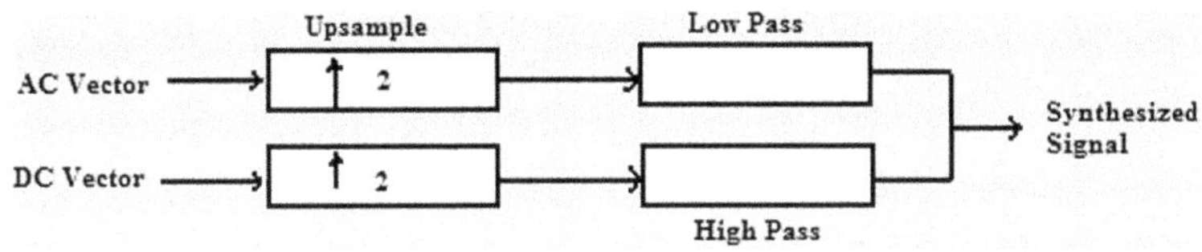

$$a(n) = \sum_k cA_1(k)h_1(n-2k)$$

$$b(n) = \sum_k cD_1(k)h_0(n-2k)$$

Repeating the same process as above in inverse and we reach at level 1 of the original transform as

$$A_1(n) = \sum_k cA_1(k)h_0(n-2k)$$

$$D_1(n) = \sum_k cD_1(k)h_1(n-2k)$$

The denosing of DNA signal helps in better estimation of discrimination measure and suppresses 1/f noise.

We tested different combinations of window functions for selection of an appropriate window for this analysis and found Kaiser Window of size 351 bp for better minimization of spectral leakage of DNA signal frequency components.

$$w(n) = \begin{cases} I_0\left(\beta\left(1-((n-\alpha)/\alpha)^2\right)^{1/2}\right)/I_0(\beta) & 0 \le n \le M-1 \\ 0 & otherwise \end{cases}$$

Kaiser window contains two important parameters α and β, normally α is chosen over three sample spaces ranging from 0 to 21, 22 to 50 and 51 to entire length of signal. The default value for β is 0.5 and it determines the leakage factor and side lobes attenuations. We observed the optimal value of β to be 3.5 for high impacts of peak heights and low noise. The architecture calculates the frequency components of signal for magnitude and power of individual frames.

Absolute value of Frame = |Frame| = $Ax(f) = | Xl(f) |$

Where Xl (f) calculates the absolute value $Ax(f) = \sqrt{Xl^2 + iXl^2}$

Power of Frame = Absolute value of frame to the power 2 = $|Frame|^2 = Px(f) = | Xl(f) |^2$

The frequencies are normalized by $Px(f) = | Xl(f) |^2 \dfrac{1}{f_s L}$, where fs is the sampling frequency and L is the length of original signal. Increasing the normalization factor beyond fs*L doesn't produce the healthy affects over results rather it makes to rescale the frequency vector in spectral analysis.

The discrimination measure "D" is defined as a ratio of the lowest peak value of exon in the exon set to the highest peak value of intron in intron set. This estimation determines the strength of distinguishing factors between genic and intergenic regions.

RESULTS AND DISCUSSIONS

We have chosen very common datasets used by other researchers to measure the percentage improvement in coding regions identification in power spectral density analysis. The first dataset is Sus Scrofa mitochondrion (Accession: NC_012095, 7500 base pairs).

Figure 4 describes the PSD's in the form of exon, intron peaks and genic region boundaries for DNA sequence.

Figure 4. PSD's of Different Approaches

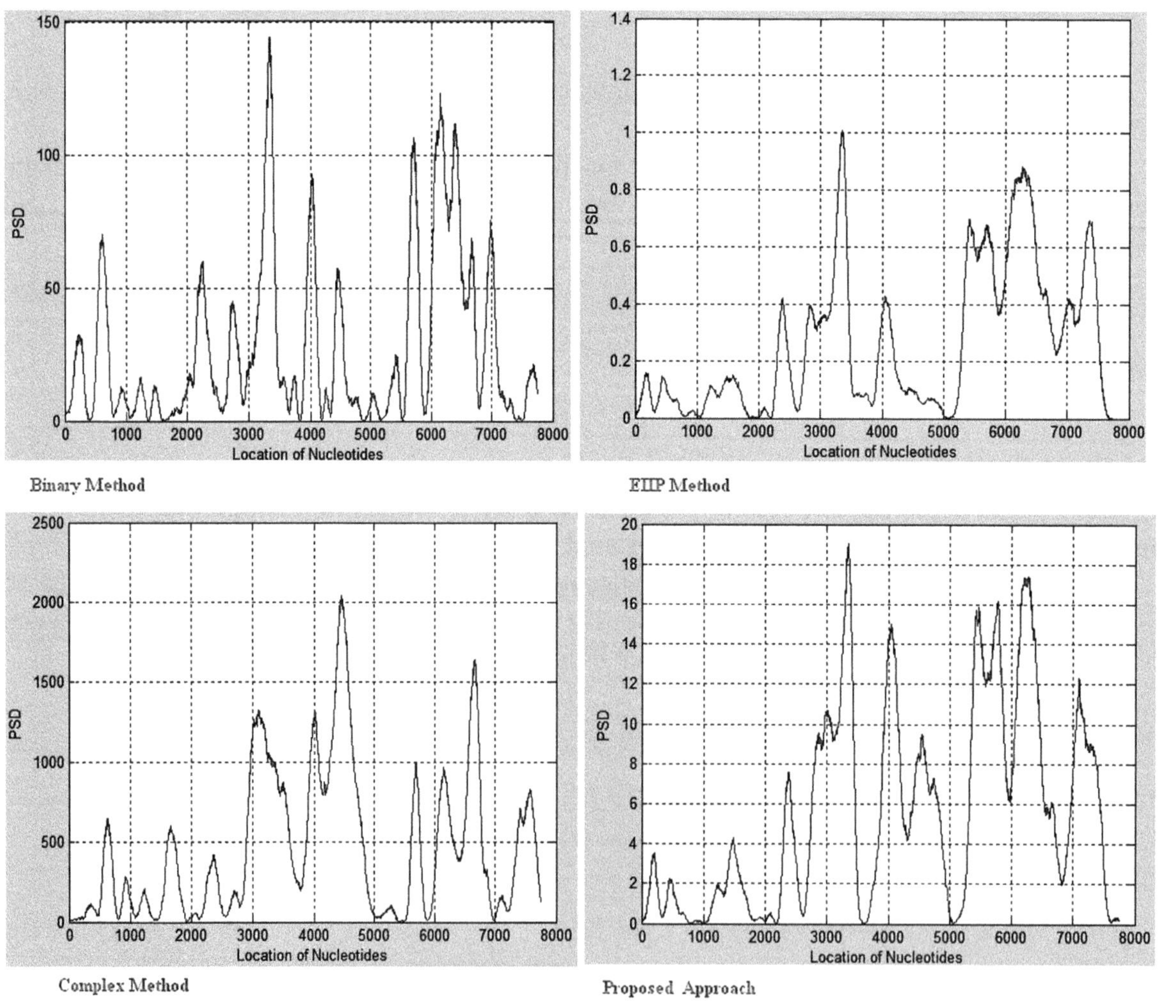

Binary indicator sequence method in Figure 4 shows the pitfalls concerning diffusion of exons in 1/f noise. It can be seen that an intron is more visible in regions 400-800 bp. Such an intron can't be seen in other PSD plots. Exon E_2 is not prominent in its entire range; rather it shows its peak from 3800-4200 bp and then a sudden discontinuity occurs with a rejoin between 4200-4900 bp. This phenomenon prevents from correct and transparent calculation of discrimination measure and range of coding regions bounds.

There is more comprehensive plot for EIIP method than Binary method. The intron peaks are reduced in the range 0 to 1000 bp. The last exon E_4 is having same range as Binary method (6800-7600) bp. Exonic peaks are not larger as compared to first method but the coding regions are more prominent for a comparable range to NCBI.

Complex method contains sharp intron peak in the same regions as Binary method but the coding regions boundaries are more visible that helps for a suitable analysis of discrimination measure and calculations of standard range of nucleotide pairs.

Proposed method demonstrates the reduction of noise (as well as spectral leakage) and improves the coding regions peaks and boundaries measures for nucleotides. There are no discontinuities found in exonic range and more comparable results are obtained (as set by the standard NCBI).

Table 2 describes the exons range calculated in PSD's of different approaches.

E_2 for Binary method contains a sharp discontinuity and rejoin. Nearly all the existing approaches encapsulate the range of genic regions compared with NCBI range. Fist exon is 100 bp ahead and 100-200 bp diffused, second exon is 100-300 bp ahead and 50-100 bp diffused, third exon is 100-3035 bp ahead and 80-100 bp diffused while fourth exon is 200 bp ahead and 80-120 bp diffused in all existing approaches than NCBI range. This shows some pre-start and post-end of nucleotide boundaries for exons. The proposed method falls closer with the standard range. The more clear comprehension for this comparison can be made in the form of discrimination measure for all the methods.

Table 3 shows the discrimination measure values calculated for different methods. There are four exons (as well as genes) in the sequence of Sus Scrofa with 7500 base pairs. It can be seen that a significant factor for discrimination measure is achieved with proposed approach. There is 32% gain in protein coding regions identification as compared to the highest factor of prevailing solution. In general, a gain of 40% was estimated than Filter 2 method, 58% than Filter 1, 64% than EIIP method and 56% than Binary method.

Figure 5 describes the discrimination measure calculated for different methods. It is clearly visible that proposed method contains the most significant factor for distinguishing genic peaks as against the

Table 2. Genic region boundaries for different approaches with NCBI range

Method	E_1	E_2	E_3	E_4
Binary Method	2600-3800	3800-4200 4200-4900	5500-6800	6800-7600
EIIP Method	2600-3500	3600-5000	5100-6800	6800-7600
Complex Method	2800-3800	3800-4980	6000-7000	7200-7500
Filter 1	2600-3800	3800-4800	5500-6750	6750-7600
Filter 2	2600-3950	3600-4900	5000-6750	6800-7550
UTP Method	2650-3650	3800-4990	5050-6800	6800-7600
NCBI Range	2746-3702	3911-4954	5335-6879	7027-7722

Table 3. Comparative analysis of various methods for discrimination measure

Method Employed	Exons and Intron Boundaries	Discrimination Measure	Percentage Improvement in Prediction over previous methods
Binary Indicator Sequence (STFT with Kaiser window of length 351)	E1 = 145 E2 = 90 E3 = 125 E4 = 75 Intron = 70	1.07	56%
EIIP Indicator Sequence (STFT with Kaiser window of length 351)	E1 = 1 E2 = 0.43 E3 = 0.88 E4 = 0.69 Intron = 0.42	1.02	64%
Complex Indicator Sequence (STFT with Kaiser window of length 351)	E1 = 1325 E2 = 2050 E3 = 1650 E4 = 825 Intron = 650	1.27	32%
IIR antinoch Filter (Filter 1)	E1 = 23.6 E2 = 63 E3 = 53.2 E4 = 47.8 Intron = 22.4	1.06	58%
Filter 2	E1 = 34.60 E2 = 113 E3 = 88.20 E4 = 77.8 Intron = 28.25	1.19	40%
Proposed Approach	E1 = 19 E2 = 15 E3 = 17.4 E4 = 12.35 Intron = 7.5	1.67	There is 32% to 64% gain in prediction accuracy over existing solutions for Sus Scrofa with proposed method

Figure 5. Discrimination measure versus methods

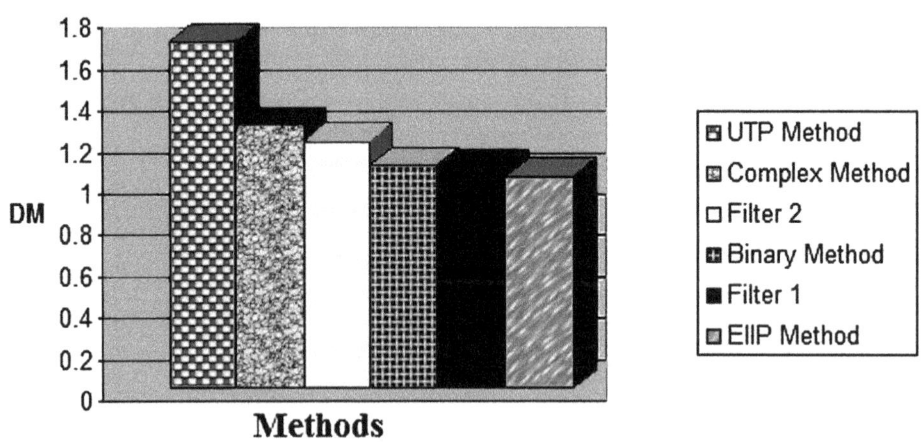

existing approaches. There is a gain of almost 32% than the second highest measure calculated in this prediction. A more comprehensive elaboration between the proposed and second highest measure can be made as follows,

Figure 6 demonstrates the power spectral density comparisons between Complex method and proposed approach. It is visible that Complex method has more prominent peaks for introns shown in red curve. Two intergenic regions in the range 0-1000 bp and 1000-2000 bp contain higher values compared with green curve. All genic region peak heights in green curves are higher than red curves. This phenomenon is the obvious demonstration of distinguishing factors between the two methods.

We selected another very common dataset Canis Lupus mitochondrion gray wolf (Accession: NC_008092, tested for first 7800 bp with four genes / exons). The estimation of power spectral density and discrimination measures over proposed and existing approaches revealed significant improvement in prediction.

Figure 7 (Binary Method) describes the power spectral density analysis for Binary indicator sequence method. The pitfall of this method starts with a larger intergenic peak between 0 to 1000 bp. This peak is not so prominent in PSD's of other approaches. There is also a disconnection of genic region from 2600-2900 and 2900-3600. This phenomenon hinders the correct calculation of exon boundaries and reduces the prediction in calculation of discrimination measure.

EIIP method in Figure 7 presents smaller intergenic peaks with distinguishable exonic peaks. The 1/f noise is minimized as compared with Binary method but there is also a dominant intron peak in the range 2400-2600 bp which reduced the discrimination measure factor.

Figure 6. PSD comparison of Complex and proposed methods

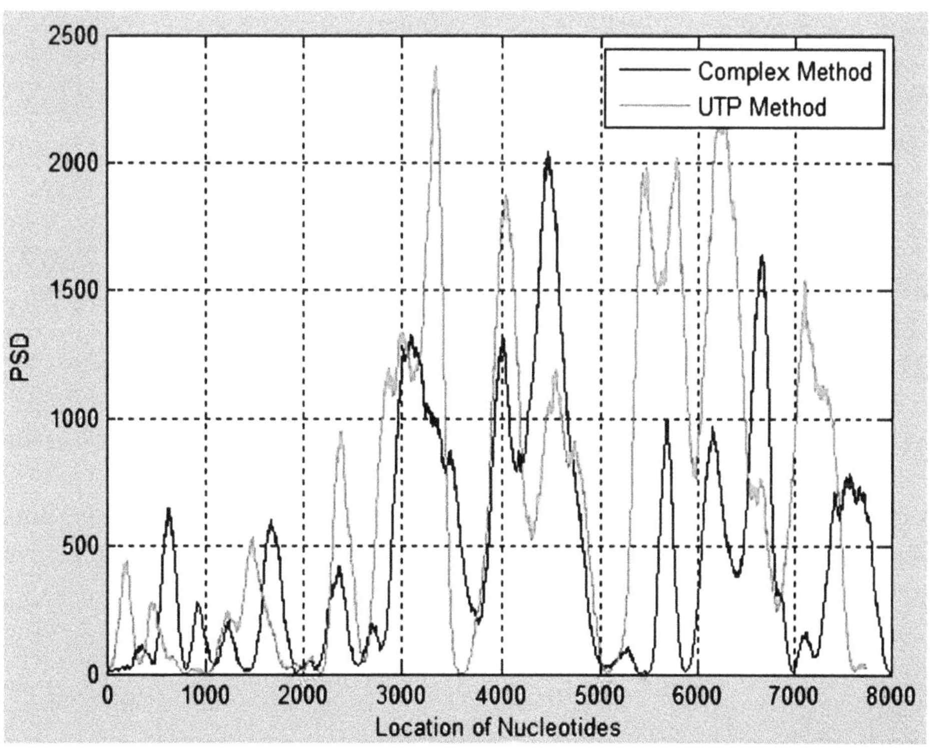

Figure 7. PSD's of Different Approaches

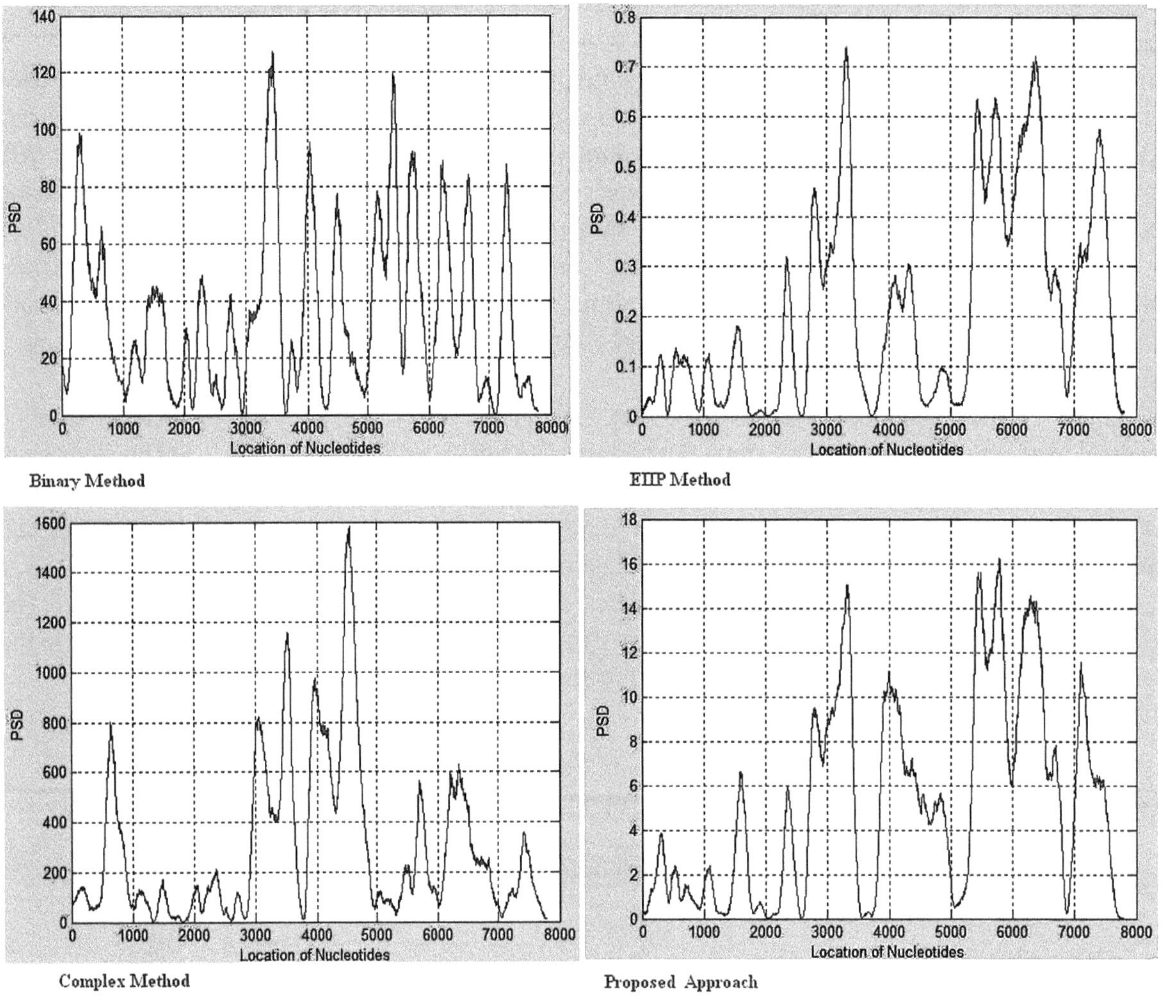

Figure 7 (Complex Method) plots a larger peak for intron in the same region (0-1000 bp) as Binary method and considerably reduced the distinguishing measure. Exons E3 and E4 are suppressed as compared with EIIP plot.

Figure 7 (Proposed Approach) describes the power spectral density estimate of proposed approach. We can see a considerable reduction in 1/f noise and enhancement in peaks for all exons. There is a large intron in range 1400-1600 but doesn't affect the significance of discrimination measure factor due to larger peak values for all genic regions.

Table 4 describes the exonic boundaries calculated against different approaches. Binary method contains a disconnection in first exon and exon peaks are far from the standard range. The proposed approach bestows the closer range comparable with NCBI range.

Table 5 presents the comparative analysis of proposed and existing approaches for discrimination measure. We can see a larger value for this factor for proposed approach. There is 75% gain in identification of coding regions as compared with EIIP method (with larger value of discrimination in all existing methods). In general, there is 75% to 266% gain in exonic identification compared with other approaches.

Table 4. Exon Boundaries in different approaches

Method	E1	E2	E3	E4
Binary Method	2600-2900 2900-3600	3600-4950	5000-7100	7100-7800
EIIP Method	2600-3650	3800-5000	5200-6900	6900-7800
Complex Method	2800-3950	3780-4900	5000-7000	7100-7800
Filter 1	2450-3600	3600-5000	5000-7100	7100-7750
Filter 2	2450-3700	3750-4950	5200-6950	6900-7700
UTP Method	2620-3550	3820-5000	5300-6850	6900-7710
NCBI Range	2748-3704	3915-4958	5350-6894	7035-7718

Table 5. Comparative analysis of various methods

Method Employed	Exons and Intron Peaks in PSD analysis	Discrimination Measure	Percentage Improvement in Prediction over previous methods
Binary Indicator Sequence (STFT with Kaiser window of length 351)	E1 = 128 E2 = 96 E3 = 120 E4 = 88 Intron = 98	0.9	83%
EIIP Indicator Sequence (STFT with Kaiser window of length 351)	E1 = 0.74 E2 = 0.3 E3 = 0.73 E4 = 0.57 Intron = 0.32	0.94	75%
Complex Indicator Sequence (STFT with Kaiser window of length 351)	E1 = 1150 E2 = 1590 E3 = 610 E4 = 360 Intron = 800	0.45	266%
Filter 1(IIR antinoch)	E1 = 25.6 E2 = 19.2 E3 = 24.5 E4 = 17.6 Intron = 19.4	0.91	81%
Filter 2 (Multistage)	E1 = 33.6 E2 = 27.2 E3 = 34.5 E4 = 25.6 Intron = 27.5	0.93	77%
Proposed Approach	E1 = 15 E2 = 11 E3 = 16.4 E4 = 11.5 Intron = 6.5	1.65	75% to 266% gain in prediction accuracy over existing solutions

Figure 8. PSD of EIIP versus proposed approach

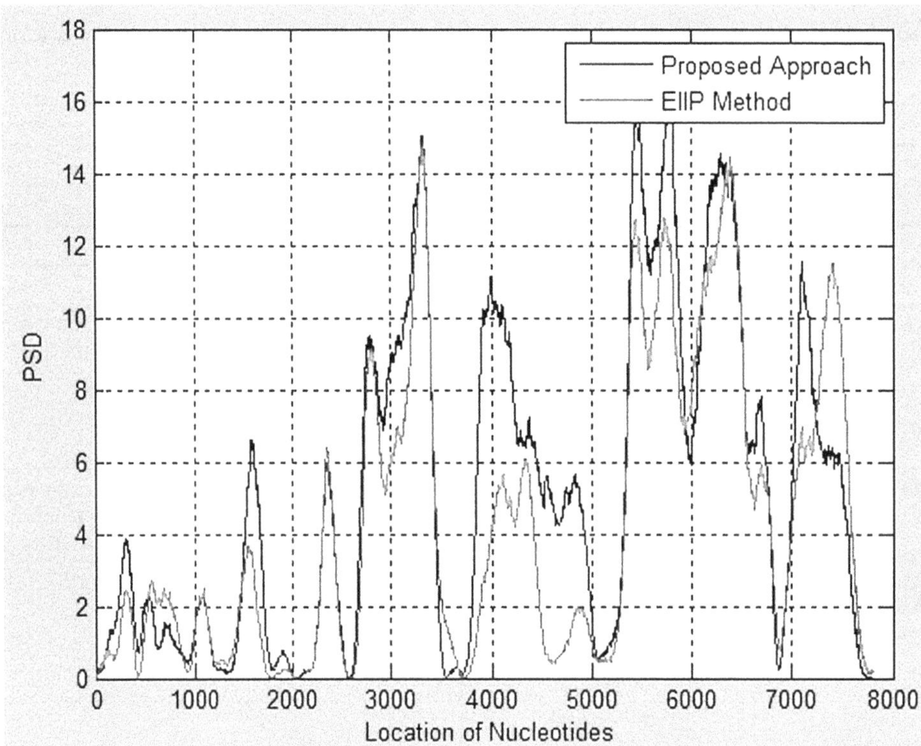

Figure 9. Discrimination measure for all approaches

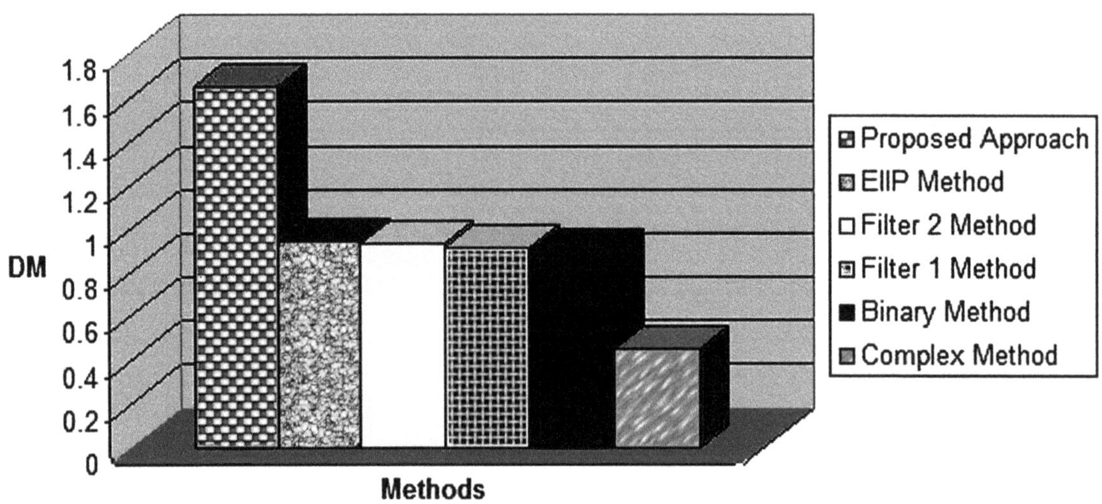

Figure 8 shows the PSD of EIIP method against the proposed approach. We can monitor larger exon peaks E2 and E3 which creates the difference in discrimination measure values in both plots.

Figure 9 depicts the discrimination measure factors for all approaches. It is worth mentioning that proposed method contains the most significant factor for distinguishing genic peaks as against the existing approaches. There is a gain of almost 75% than the second highest measure calculated in this prediction.

CONCLUSION

This paper presents a peculiar biologically inspired approach for protein coding regions identification using discrete wavelet transforms with notion of a peculiar indicator sequence for DNA signal mapping. It was observed that biologically inspired solution aided with discrete wavelets transforms greatly suppress the 1/f noise and enhances the discrimination ratio for exon peak values versus introns. A comparative analysis of results tested over commonly used datasets Sus Scrofa domesticus mitochondrion (Accession: NC_012095) and Canis Lupus mitochondrion gray wolf (Accession: NC_008092) revealed the significant improvement in prediction measures for proposed architecture as against the existing solutions. Better significance is expected to achieve employing lifting dyadic wavelet transforms as a future work.

REFERENCES

Akhtar, M., Ambikairajah, E., & Epps, J. (2008). Optimizing period-3 methods for eukaryotic gene prediction. *IEEE International Conference on Acoustics, Speech and Signal Processing*. doi:10.1109/ICASSP.2008.4517686

Akhtar, M., Epps, J. & Ambikairajah, E. (2008). Advances in Eukaryotic Gene Prediction. *Selected Topics in Signal Processing*, *2*(3), 310 – 321.

Chen, H., Gu, F., & Liu, F. (2005). Predicting protein secondary structure using continuous wavelet transform and Chou-Fasman method. *27th Annual International Conference of the Engineering in Medicine and Biology Society*.

Datta, S., Asif, A., & Wang, H. (2004). Prediction of protein coding regions in DNA sequences using Fourier spectral characteristics. *IEEE Sixth International Symposium on Multimedia Software Engineering Proceedings*. doi:10.1109/MMSE.2004.63

Datta, S., & Asif, A. (2005). A fast DFT based gene prediction algorithm for identification of protein coding regions. *IEEE International Conference on Acoustics, Speech, and Signal Processing*. doi:10.1109/ICASSP.2005.1416388

Datta & Asif. (2004). DFT based DNA splicing algorithms for prediction of protein coding regions. *Conference Record of the Thirty-Eighth Asilomar Conference on Signals, Systems and Computers*.

George, T. P., & Thomas, T. (2010). Discrete wavelet transform de-noising in eukaryotic gene splicing. *BMC Bioinformatics*, *11*(Suppl 1), S50. doi:10.1186/1471-2105-11-S1-S50 PMID:20122225

Grandhi, D. G., & Vijay Kumar, C. (2008). *2-Simplex mapping for identifying the protein coding regions in DNA. In IEEE region conference* (pp. 1–3). TENCON.

Guo & Zhu. (2009). Prediction of Protein Coding Regions by Support Vector Machine. *International Symposium on Intelligent Ubiquitous Computing and Education*.

Gupta, R., Mittal, A., Singh, K., Bajpai, P., & Prakash, S. (2007). A Time Series Approach for Identification of Exons and Introns. *10th International Conference on Information Technology*. doi:10.1109/ICIT.2007.54

Hazrina & Siti. (2008). Gene prediction system. *International Symposium on Information Technology.*

Hota, M.K. & Srivastava, V.K. (2008). DSP technique for gene and exon prediction taking complex indicator sequence. *IEEE Region 10 Conference.*

Imran, Asifullah, Asad, & Tae-Sun. (2009). Reversible Watermarking Based on Intelligent Coefficient Selection and Integer Wavelet Transform. *International Journal of Innovative Computing, Information and Control*, *5*(12), 4675-4682.

Kakumani, R., Devabhaktuni, V., & Ahmad, M. O. (2008). Prediction of protein-coding regions in DNA sequences using a model-based approach. *IEEE International Symposium on Circuits and Systems*. doi:10.1109/ISCAS.2008.4541818

Mahmood, E., & Ambikairajah. (2007), On DNA numerical representations for period-3 based exon prediction. *IEEE International Workshop on Genomic Signal Processing and Statistics.*

Mathkour, H., Ahmad, M., & Khan, H. M. (2009). An integrated statistical comparative analysis between variant genetic datasets of Mus musculus. *International Journal of Computational Intelligence in Bioinformatics and Systems Biology*, *1*(2), 163–176. doi:10.1504/IJCIBSB.2009.030647

Mena-Chalco, J.P., Carrer, H., Zana, Y. & Cesar, R.M. (2008). Identification of Protein Coding Regions Using the Modified Gabor-Wavelet Transform. *IEEE/ACM Transactions on Computational Biology and Bioinformatics, 5*(2), 198 – 207.

Ohara, Zhang,, & Toda, Miyake, & Imamura. (2007). De-noising Method by Combining Adaptive Line Enhancer and Complex Discrete Wavelet Transform. *ICIC Express Letters*, *1*(2), 145–151.

Roy, M., Biswas, S., & Barman, S. (2009). Identification and Analysis of Coding and Noncoding Regions of a DNA Sequence by Positional Frequency Distribution of Nucleotides (PFDN) Algorithm. *4th International Conference on Computers and Devices for Communication.*

Shuo & Yi-Sheng. (2008). An integrative algorithm for predicting protein coding regions. *IEEE Asia Pacific Conference on Circuits and Systems.*

Vaidyanathan & Byung-jun. (2002). Gene and exon prediction using allpass-based filters. *IEEE International Workshop on Genomic Signal Processing and Statistics GENSiPS.*

Vaidyanathan & Byung-jun. (2002). *Digital filters for gene prediction applications*. IEEE Asilomar on Signals, Systems, and Computers.

Vaidyanathan & Byung-jun. (2002). Digital filters for gene prediction applications. *Thirty-Sixth Asilomar Conference on Signals, Systems and Computers*. doi:10.1109/ACSSC.2002.1197196

Yin, C., & Yue, S. S.-T. (2007). Prediction of protein coding regions by the 3-base periodicity analysis of a DNA sequence. *Journal of Theoretical Biology*, *247*(4), 687–694. doi:10.1016/j.jtbi.2007.03.038 PMID:17509616

Chapter 11
Application of Self-Healing in Wireless Sensor Network: A Survey

Noor Zaman
King Faisal University, Saudi Arabia

Low Tang Jung
Universiti Teknologi PETRONAS, Malaysia

Vasaki Ponnusamy
Universiti Tunku Abdul Rahman, Malaysia

ABSTRACT

Wireless Sensor Networks (WSNs) are gaining popularity in many monitoring and event detection areas. But the limitation of low power sensors needs to be further researched to combat the self-healing aspect of sensor networks. Therefore this paper aims to present a survey on application of self-healing routing protocols in wireless sensor networks that have been designed by analyzing their operation, advantages and disadvantages. This presents an outline of each application of the protocol and an idea of how these features could be adapted into future research area in self-healing wireless sensor networks in terms of energy aspect. The self-healing characteristics are mainly derived from biologically-inspired mechanisms for autonomic behavior. Autonomic behavior systems are characterized into four functional systems as self-configuration, self-healing, self-optimization and self-protection. This survey mainly focuses on self-healing characteristics from the biologically-inspired perspectives as well as non-bio-inspired perspectives.

INTRODUCTION

Wireless sensor networks are created for monitoring situation such as temperature, sound, movement, location, to create "smart environments". (Youssef et al. 2002). Sensor network has its distinct features from traditional network by having limited energy resources. Most of the routing in wireless sensor is

DOI: 10.4018/978-1-4666-9792-8.ch011

done through lowest energy route this eventually exhausts energy at these nodes impacts the lifetime of the entire network. (Sha et al. 2005). Energy efficient routing becomes the prime importance when considering extending the lifetime of sensor network. (Gui & Mohapatra 2003). When sensor nodes exhaust their energy indefinitely, the nodes will eventually create a routing hole. Routing hole refers to the areas where data cannot be routed since there are no active nodes available. Therefore a self-healing mechanism is needed to detect this issue. So the technique of fighting routing holes in sensor network is denoted as self-healing methods. Therefore the two prominent features in wireless sensor network that need attention in research are energy-efficiency and self-healing. This paper targets to carry out an evaluation of several energy efficient and self-healing routing protocols in wireless sensor networks. This paper differs from other studies, as it performs a study on bio-inspired as well as non-bio inspired mechanisms for energy efficient routing protocols that have been submitted by others given the advantages and disadvantages of each proposed protocol. The paper is organized as follows: section 1 will look at some of the literature reviews based on the terms such as self-healing and energy efficiency. The literature starts by looking at what is meant by autonomic computing in which self-healing is one of the aspects of it. Further literature looks at what is meant by energy efficient routing in wireless sensor network. Section 2 consists of surveys on various energy efficient self-healing routing protocols discussing their protocol approach and results. Section 3 further gives a summary of each protocol giving a brief summary on its advantages and disadvantages. A complete comparison using various metrics is given at section 4. Section 5 concludes the paper with recommendation and future research issues.

Autonomic Behavior in Wireless Sensor Networks

Autonomic system is considered as a system that is capable of self-management services. Self-management activities can be categorized into self-healing, self-optimization, self-configuration, self-protection, self-awareness, self-knowledge, self-maintenance and self-service (Lanthaler, 2000). According to Lanthaler (2000), there are four important tasks that an autonomic sensor node should perform for self-healing purpose. The functions are: i) to monitor and to collect data from the system, ii) to analyze whether something is going wrong in the system, iii) to plan or to create desired changes in order to execute the changes and iv) to perform the desired actions (Figure 1).

Self-Healing Behavior in Wireless Sensor Networks

Self-healing in the context of wireless sensor network is the capability of sensor network to recover from any types of behavior such as malicious attack, disruption or failure. In terms of failure, there exist two natures of failure: the failure of node and the failure of system. Failure in sensor network is due to many reasons such as failure in software and failure in hardware; unavailability of communication link, medium access contention and failure in routing. This failure in WSN is considered one of the main challenges and yet opens to many areas of research and discussions. Routing failure occurs due to routing holes, node failure, energy depletion, natural disruption in sensor area and many more. Self-healing in the context of routing failure is the capability of sensor nodes to monitor, detect, analyze, plan and execute events in order to recover or heal from the phenomena (Lanthaler, 2000). Moreover, self-healing by means of energy efficient routing is one of the main research focuses. There are various research projects going on that trying to tackle the self-healing problem in the context of software systems. The model proposed here can be used as guideline to perform self- healing system in terms of software, hardware, systems or

Figure 1. Autonomic computing architecture

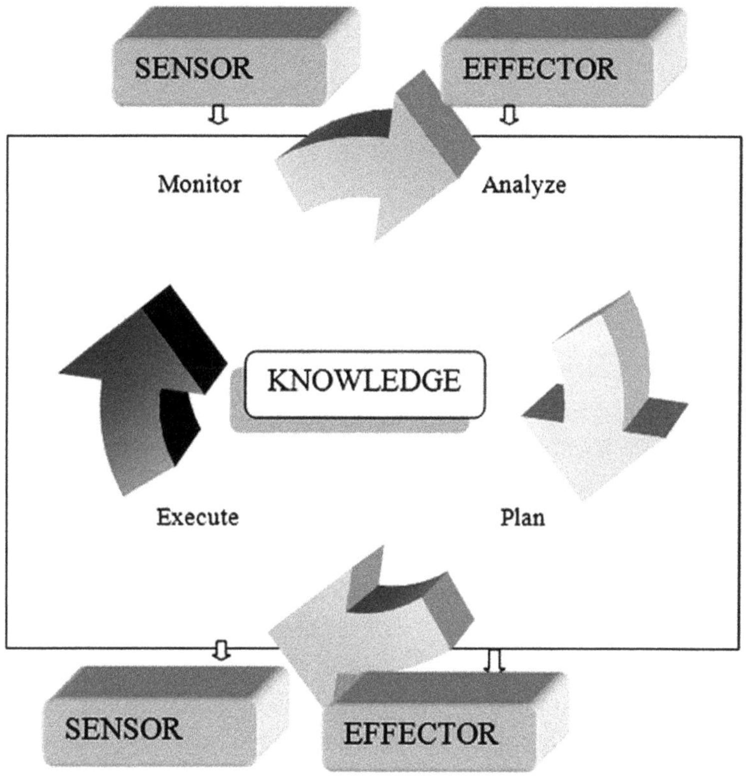

Figure 2. Self-healing architecture

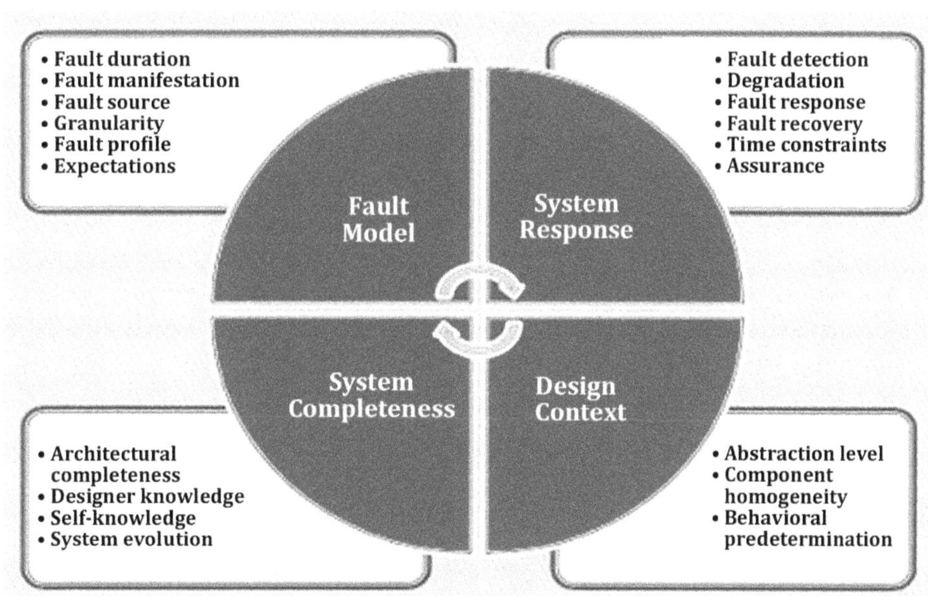

Figure 3. Sensor Network Overall View

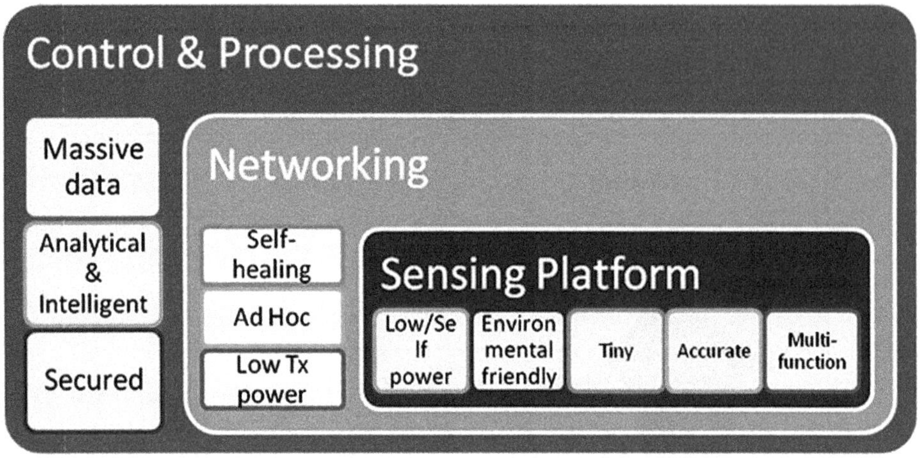

network. In conclusion, significant features of self-healing schemes are fault model, system response, system completeness and design context. Self-healing systems research is relatively interesting and getting better consideration, but it is still not finely defined in the aspects of scope and areas to be focused into self-healing. Koopman (2003) summarizes some of the different views researchers have in terms of self-healing aspects. Exists vast amount of knowledge in terms of dependable system and specific techniques considered for self-healing. As shown in Figure 2, there exist four common classifications to the self-healing system problem space: fault-model, system response, system completeness, and design context. These are the aspects that have been proposed by various researchers and have been combined into one single model. Figure 3 shows the context in which self-healing aspects falls in sensor network research.

Fault Model

Self-healing systems are actually considered as dependable systems. And all dependable systems have to be defined for some kind of fault tolerant scheme. Following fault hypothesis section will explain what kind of fault the system is about to tolerate. Therefore, self-healing system should have a fault tolerant model in order to specify what kind of fault the system is to self-heal from. Without such particular model it is difficult to assess how the system is going to heal from such behavior. Fault model is categorized into fault duration which is to detect whether the fault is permanent, intermittent or transient where it is essential to state the length of fault to self-heal, fault manifestation is to identify whether severe, fail-fast, fail-silent or others, fault source is the source where the fault comes from, granularity is the size of the fault and fault profile expectations are the profile of fault occurrences.

System Response

The most important thing of this model is to first detect the fault and way of responding to the types of faults. Reliability and safety or security is to be considered in order to decide on potential responses. System response is categorized into fault detection, either by internal component, peer-to-peer checking or supervisory checking, degradation is the degree of degraded operation, fault response is response mechanism, recovery is recovery operation to self-heal, constant time is to determine what kind of self-healing required and assurance is assurance for correctness.

System Completeness

Self-healing system should deal with limits in terms of knowledge, incomplete specifications, and incomplete design. System completeness is categorized into architectural completeness, designed with first implementation, designer understanding is how sound the fault model is categorized, system self-knowledge is the knowledge about themselves and environment in order to heal and system evolution should deal with changes over time.

Design Context

Several other factors need to be identified to formulate the system design. Design context is categorized into abstraction level which is self-healing to application, software or hardware, component homogeneity whether homogeneous or heterogeneous, behavioral predetermination is predetermined or deterministic behavior, user participation in healing, system linearity and system space is how big the system is. There are various research projects going on that trying to tackle the self-healing problem in the context of software systems. The model proposed here can be used as guideline to perform self- healing system in terms of software, hardware, systems or network. In conclusion, relevant aspects of self-healing systems are fault model, system response, system completeness and design context.

SURVEY ON SELF-HEALING WIRELESS SENSOR NETWORKS

This section comprises some of the related works in the area of self-healing, energy efficient routing and some of the bio-inspired mechanisms used for self-healing as well as energy efficient routing. The problems, suggestions and future works discussed in this paper helps in further research and enhancement of energy efficient self-healing and particularly using bio-inspired mechanisms. This survey is particularly useful to researchers looking into new areas of self-healing systems in networks. Following section starts by looking into various types of research papers followed by its protocol approach and results obtained. The results are useful in terms of making comparison with other protocols and future enhancements.

Immune-Inspired Algorithm

Atakan et. al (2007) argues distributed node and rate selection (DNRS) systems that is nature inspired immune system. The system forms a B-cell centered designated node (DN) to sense anticipated incident and to encounter the incident signal at the sink node. This theory is very similar to immune system whereby when the system spots a pathogen, B-cells are aggravated to create some antibodies at dissimilar densities to lessen the result of the pathogen. The most important phenomenon in this protocol is the selection of DNs to ensure that they are positioned to the event source as closer as possible and they are positioned farther from each other as much as possible. There are three factors used for the selection of DNs: i) the similarity between sensor node and event source, ii) affinity between sensor node and uncorrelated neighbor nodes and iii) the affinity between sensor node and correlated neighbor nodes. Upon selection of DNs, when an event occurs, the DNs sense the event and send sensed information with sampling frequency to sink node to construct the event signal. Natural immune system based mathematical model and equations are used to tackle the sampling frequency parameter and the congestion.Besides the determination of the DNs, the algorithm also selects the most suitable sampling frequencies of sensor nodes according to the congestion. Besides the selection of the minimum number of designated nodes, the algorithm offers substantial saving of energy to sensor network.

Bokareva et. al (2005) proposes a new mechanism of immunology inspired method of self-healing in wireless sensor network. This method is based on some inherent characteristics of human immune system with five main goals in mind. The first is where there are any failures on individual component there should not be any adverse effects on the whole sensor network. In human-inspired immune system, any cells are adept to replacing their basic cells thus do not affect the entire human system. The second implies that in a large deployed sensor network, there should be some degree of independence where sensor nodes would be capable of managing themselves without the support of outside object as in human immune system. In the third approach, several layers of functionality are combined into one single layer for protection and detection of possible faults. Fourth, similar to immune system where just a small number of detector cells are needed to detect pathogens, in a resource constrained sensor network, only limited sensor nodes with limited resource are needed in order to detect failure and threat. Diversity and adaptability should be the fifth mechanism where sensor nodes should have the capability to detect possible threat in advance as similar to human system where it can detect large number of pathogens and detecting unknown pathogens and remembering them in the future. SASHA is an architecture (Figure 4) proposed for self-healing sensor network and to support the resource constrained sensor networks. Several different network objects manage to deliver resourceful and effective fault-tolerance, automatic fault recognition, network monitoring and response.

As in the natural immune system, its main role is to identify the “self” and the removal of non-self proteins. Using the same concept in sensor network, it is important to understand what is meant by self and non-self set. One of the main focuses of sensor network is to monitor unknown environments and report the correct sensor readings to the base station. One main challenge in this case is what constitutes correct

Figure 4. SASHA architecture

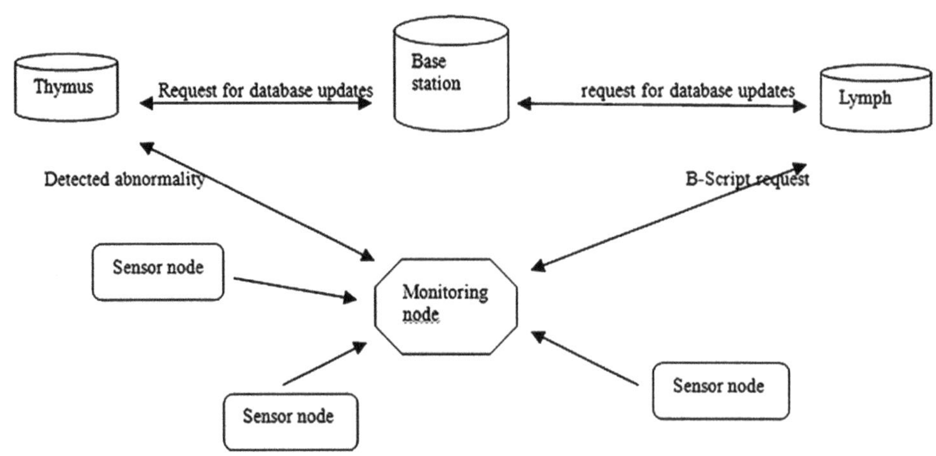

sensor readings. The paper focuses on detecting defective sensor readings using some pattern recognition techniques. The architecture in SASHA comprises numerous managing components, a great number of sensor nodes, numerous monitoring nodes, base station, Thymus and Lymph (database) machines.

Sensor nodes are resource constrained tiny nodes to monitor and report events to the base station. Monitoring nodes have additional capability of sensing, processing and communication capability. Each monitoring node covers a certain portion of the network in the form of a tree network rooted with the monitoring node. Lymph consists of B cells in human system is responsible for detecting anomalies (infected pathogens) and to destroy them. As in sensor network, the anomalies can be detected by running mobile scripts known as B-Script. These scripts are dynamically generated to act as filter for the abnormal behaviors of the forest. Lymph machine stores databases of previous attacks, severity of damage and the need for response and the solution. Thymus as similar to human immune system is a representation of self. There are three specific aims that SASHA is looking into: i) automatic fault detection in order to achieve robustness in sensor network, it is important to identify faulty sensor readings. The fault detection mechanism comprises trivial disseminated algorithm to detect faulty readings. ii) adaptive monitoring; this process is achieved by distributing mobile scripts onto the sensor network. This process consists of generation, maturing and migration of mobile monitoring scripts. These monitoring scripts are created automatically without intervention of human. and iii) corresponding reaction by returning to various types of faults. Coordinated response to malicious behavior is handled between monitoring nodes, Lymph and Thymus.

Faulty sensor readings can be achieved by firstly identifying what constitutes correct readings. This is performed by using Self-Organizing Neural Network with Competitive Learning (SONN). The main advantage is that SONN does not need earlier knowledge of the incident being observed. In order to classify sensor readings as faulty or correct, all monitoring nodes would run SONN that is acquired from the Lymph machine. An immunology inspired self-healing architecture is proposed for wireless sensor networks. It is a holistic approach to network design for automatic fault recognition and response over possible faults. It is adaptive in nature that can evolve over observing and inference abilities.

Ant-Inspired Algorithm

Camilo et. al (1996) proposes an energy optimized protocol called the EEABR protocol based on Ant Colony Optimization heuristics by Di Caro and Dorigo (1998) for wireless sensor networks. The general idea of this algorithm is based on basic ant-inspired routing algorithm. In ant-inspired routing algorithm (Dorigo et. al, 1996), at every interval, a forward ant is initiated with the intention of discovering a path from every network until reaching the final node. The identifier of node crossed is kept in memory of the ant. At every node, a forward ant selects the advancing node using some probabilistic values. The pheromone trail of node and routes visited are kept in a routing table. This task is performed by the forward ant which is taking its route towards the destination. The same concept is applied when backward ant returns back to its origin where it was created. Upon returning, the pheromone trail is again updated in the routing table and this allows each node to identify its best neighbor to forward the data to.

This section discusses about the improved version of the ant based routing algorithm. There are two improvements made in the algorithm to reduce the quantity of energy used by sensor nodes and decide on the energy quality path identified by the ants. In the general algorithm, forward ants do not have a specific location where it is sent and due to that the amount of information kept relating to identification of neighboring nodes at the routing table is huge. But this is not suitable for large networks since a huge amount of memory is needed to keep the neighborhood information. Improvement is made to save memory in which if the forward node is sent directly to the sink node, then the routing table is only going to store information about neighboring nodes that are in the same direction with the sink node. This actually reduces the size of routing table as well as the memory needed by the nodes. In this protocol the quality of a given path is determined not only by the distance but also based on the energy level of the path. This is done by proposing a new function to determine the pheromone trail that the backward ant would take during its journey back to origin.

There are more improvements made to this protocol by trying to reduce the communication load within nodes and the energy spent for communications. It is preferred that most of the data processing are done at the ant node itself instead of sending it to the sink to be processed. In order to perform this, the memory of the ant is kept at a low rate by only keeping information of only the last two nodes visited. So in this case the path followed by the ant is no more in the memory and therefore nodes should create memory to keep records of the nodes visited. When the node receives a forward ant, the ant's identification, previous node, next node and timer value are kept in the memory. The ant is then forwarded to the next node and this continues with other nodes as well. If the record of the ant is found in the memory, the ant is then eliminated to avoid possible looping. When the backward ant reaches the node, the node inspects its memory to know whether the ant's identification is found in the memory together with timer value and finds the next node where the ant should be forwarded to. The timer value is used to ensure that the records are deleted if it is found that the ant could not find the way back to the source.

The results are collected based on the three different versions of ant based routing algorithms discussed: the basic ant-based routing algorithm (BABR), the improved ant-based routing algorithm (IABR) and energy-efficient ant-based routing algorithm (EEABR). There are four metrics, average energy, minimum energy, standard deviation, the variance between energy level on all nodes and energy efficiency are used to compare the performance of the algorithm. Based on the results obtained, EEABR and IABR perform better results as compared to BABR as both algorithms are energy conscious. EEABR lessens the energy spent in relations to communication and drop in overhead of message exchanged.

Autonomic Computing: Self-Healing

This work (Natalija and Nelson, 2007) generally looks at one specific problem which is combating routing holes in sensor networks by deploying a single mobile node. As discussed in other research papers and literature reviews, self-healing by means of mobile nodes in WSN is still not widely explored. So this paper looks at combating the routing hole through deployment of a mobile node using a simple distributed algorithm for determining the Optimal Placement of a Mobile Node (OPlaMoN) proposed. According to the author, the easiest way to achieve energy-efficient mechanism is by minimizing the number of transmissions and by utilizing optimized routing path for transmission. However achieving optimized routing path is a great challenge as this tiny-battery powered sensor nodes are scattered randomly in an irregular geographic area. Due to these natures, the network topology would eventually lead into a routing hole. The routing hole becomes a challenging area to achieve optimized routing path. And the techniques to handle these issues are considered as self-healing in WSN. This paper is generally to be the first attempt in investigating the energy aspect of self-healing by using mobile robots to handle routing holes. There are two main contributions made by this authors as to: i) use 'energy' as the main criteria for placing a mobile node within a routing hole and ii) propose an algorithm for energy-efficient optimal location of mobile node within a routing hole.

In this protocol, the optimal path between the source and the sink may come across an anomaly known as routing hole. The node at which the routing hole is identified is known as node (1) and the node that exits the hole boundary is known as node (n). Only one mobile node is used to bridge the boundary hole and the node is placed at distance r from the hole in the direction of the exit node. There are two specific types of holes considered in this protocol: circular and rectangular shaped holes.

For the circular shaped scenario, the sensor nodes on the boundary of the hole would actually form a convex polygon with radius r. As the size of the hole increases, more sensor nodes would fit into the boundary of the hole, increasing the radius r and the degree of polygon. The experiment was first done with no mobile node around the hole and the energy required to transmit a bit from the source to the sink is calculated. Then by placing a mobile node within the hole, the energy required to transmit a bit from the source to the sink through the mobile node is calculated. And based on the results, the energy required to transmit through the mobile node is not really justifiable. The same situation is applied to a rectangular shaped scenario, and produces the same result as above. If energy is used as the main criterion for this case, routing around the hole would produce better results as compared to bridging the hole through a mobile node. So rather than using energy as the main criteria, an algorithm to find the optimal position of mobile node is proposed (OPlaMoN). This algorithm uses distributed decision-making by ensuring minimum exchange of information among effected nodes. Each effected boundary node would calculate a bidding-value, assumed to be a value gained by the effect if a mobile node were to be placed next to it. Each node would locally exchange its bidding-value with other nodes, and the highest node with bidding-value to be elected as the bidder. Placing a mobile node within the vicinity of the node with the highest value would assume to be energy gained. The algorithm has proven that placing a mobile node within the routing hole could possibly be a solution to combat the hole problem in WSN. However energy may not be the sole of the criterion in finding optimal solution for self-healing. Looking into other parameters such as overall transmission delay and static node failure should also be taken into consideration. So through simulation results, a distributed energy-efficient placement of mobile node within a routing hole is proposed.

The main focus of the work by (Branch et. al, 2007) is to provide a fault-tolerant and efficient routing protocol in WSN where nodes can automatically find the shortest route without deciding which neighbor to forward the data to. A transfer node announces the message to all the nodes within i transmission reach. The receiver then autonomously decides where to forward by using the knowledge of hop distance and a prioritized transmission back-off delay that ensures that the data is sent by only a single node. A time slotting mechanism is used to ensure that the available shortest route is always used and reduces the probability of packet collision. One of the main advantages of SHR is where it dynamically determines the shortest path even when topology changes occur.

This protocol is an extension of self-selective routing (SSR) where each node knows in advance its distance (hops) to the destination. In SSR, the sensor node would automatically broadcast its data to all the available neighbors within the range. Upon receiving data, the next node to send the data would be decided by using a back-off delay scheme. In this mechanism, once a node receives data it sets a random delay where the average is relative to distance from the endpoint. But based on the calculations, the algorithm ensures that only nodes nearest to the endpoint have the greatest chance of re-broadcasting the data. If a node hears the broadcast of the same packet that it is waiting to broadcast, it would automatically abandon the operation. All other nodes would be notified via an explicit acknowledgement of the broadcast so that they do not need to resend the same packet.

Though SSR has some benefits of low overhead and fault-tolerance, it has several weaknesses as well. First the back-off delay mechanism creates a probability of packet collision and due to this reason packets might need to travel along longer route while shorter routes are available in realistic cases. Moreover, SSR does not provide route healing mechanism for sending packets around damaged routes. So SHR introduces some new contributions to SSR in order to address the weaknesses discussed above.

The first improvement SHR is looking into is to introduce a time-slotted back-off delay scheme. This back-off delay is measured in transmission slots rather than using continuous time. This ensures that nodes closer to the destination would first send the packet than any other available nodes as in SSR. The second improvement would be route repair mechanism for propagating packets around severed routes. Severed route occurs when a sending node does not have any neighbor node that is closer to the destination. So SHR ensures that rerouting is performed on the shortest available routes by adjusting the hop distance of the sending node. SHR also ensures that node's hop distances are adjusted when departing node comes back to the network.

SHR protocol is operated by using a data structure of a cost table which consists of ID of the source and destination, the sequence number of the last packet and the hop distance from the destination to the current node. There are three phases uses in this protocol: i) destination request, ii) destination reply and iii) data transmission. When a node sends data to a destination in which no distance information is available, it floods a destination request packet (DREQ) to the network. The hop count value is set to 1 initially and when an intermediate node receives the DREQ packet, it creates a cost table entry if no table has been created previously. Otherwise the table is updated with packet ID, hop count field and sequence number. The intermediate node then would broadcast the packet with some random back-off delay to avoid collisions.

When the destination receives the DREQ packet, it will not rebroadcast it thus resulting in all nodes knowing their distance from the source. Destination node would broadcast a destination reply (DREP) packet. The same data field as in above is used and at this point of time all nodes would have already known the hop distance from the source node. When an intermediate node receives DREP packet, it updates its cost table with the actual hot count distance to the destination. The same process continues with any intermediate node that receives the DREP packet and the process continues until it reaches the source node where the source knows in advance that the destination node exists. Upon receiving DREP packet, the source starts sending DATA packet to the destination along the route previously used as each node knows in advance its distance to the destination.

The effect of node density is evaluated by changing the mean number of neighbors per node. As the mean value increases, more collisions occur as more neighboring nodes compete to forward the packet. But at the same time end-to-end delay decreases as network density increases and the performance of route repair mechanism also increases in SHR. Delivery ratio also improves as density increases. Moreover, with the route repair mechanism the networks rate of failure can be handled effectively by adjusting the number of packets required to invoke route repair.

Autonomic Computing: Self-Optimizing

Mobile Ad Hoc Network (MANET) is formed by nodes communicating through a shared multihop communication network. And each node acts as a host as well as a router. Mobile Ac Hoc network is becoming a popular area since nodes in this network can be easily deployed and do not need any infrastructure to set-up the network. Recent advancement in mobile wireless devices increases the potential for mobile ad hoc networks. But some of the major challenges faced by routing in MANET are frequent changing network topology, low transmission power and low available bandwidth. Routing protocols should be able to cope with network topology changes without incurring too much of overheads of control messages.

Normal Internet routing in wired network is not suitable for ad hoc networks, thus increases the need for further research in the routing of ad hoc networks. Since there have been many routing algorithms proposed for MANET, the work done in this paper does not propose a new algorithm. But this work, by any routing algorithm provides path between source and destination, so the path can be optimized dynamically to enhance performance and consume less energy. This work (Gui & Mohapatra 2003) proposes a Self-Healing and Optimizing Routing Technique (SHORT) for ad hoc networks. There are two categories being looked at in this technique: a) Path Aware (PA) Short for optimizing the healing path and b) Energy Aware (EA) Short for conserving power usage in MANET.

The shortest path created by on-demand ad hoc routing protocols may diverge from the shortest path and due to the movement of the nodes in ad hoc networks; the shape of routing route may vary while the connectivity still remains intact. Most of the earlier proposed protocols do not initiate a new mechanism for path discovery till link failure occurs. The variations in form can be corrected by developing improved routing paths by escaping significant overheads which will be discussed in the later part of the discussion.

For extending the network life time, there is a need to look into energy efficiency of routing protocols. Many studies have been conducted at energy efficient in routing protocols such as handling minimum cost path along multihop links as well as using dissimilar power aware metrics for determining routes. The goal of the framework in this paper is to enhance power consumption through route optimization. Energy aware routing looks at extending network and nodes lifetime by routing packets to node that have sufficient energy as well as avoiding routing packets to nodes that are low on battery. Routing protocols should be able to self-heal itself by allowing nodes to come out from situations of being over-used while other nodes are less utilized. The main goal in ad hoc networking is to increase performance as well as to monitor power conservation. This paper looks at improving both measures without incurring additional overhead on the network.

Based on Figure 5, for a routing path from node A to node J, the initial routing path is discovered through some path discovery processes with the shortest path between two nodes. In this case it would take ten hops to reach from node A to node J in which, ten hops are considered the shortest path between

Figure 5. SHORT node connections

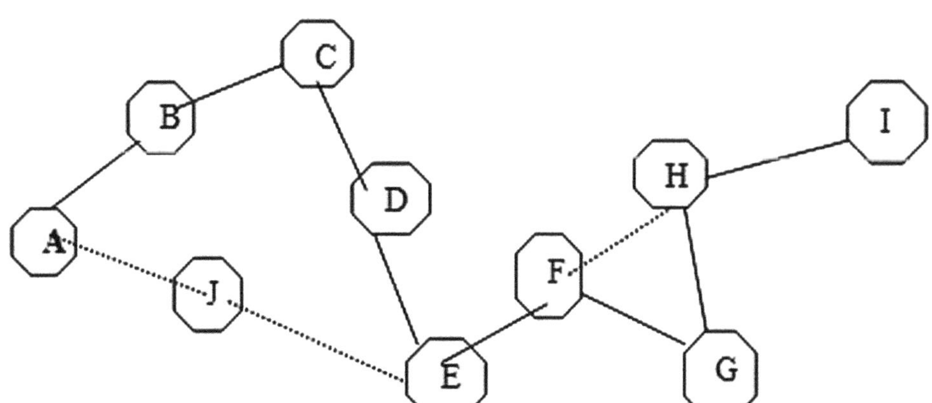

nodes. During the execution of the process the mobility nature of the nodes may cause changes in routing path as shown in Figure 4. In this case, node I is within the reach of node A and node J is within the reach of node F. With the new path available through node I and J, there should be a possibility of identifying the shortest path which is less than ten hops in this case. But since there is no such mechanism to update the routing table, it may not be possible for nodes to self-heal to identify such paths. So the main focus of this paper is to look into identifying such possibilities and allowing nodes to self-heal and optimize the paths available by modifying routing tables dynamically. The basic approach taken in this new protocol is known as (n,2) reduction where n hops along the path can be reduced to 2 hops only.

Autonomic Computing: Energy Aware

The main goal of energy aware load balancing is to balance the energy at each node by performing load balancing of traffic among all nodes. For example, for a connection between A and D, if the traffic continuously follows the route thus A to D ignore other available paths, then the load is just focused on this route. Nodes in other routes are left alone without incurring any energy. For example, nodes B and D can also participate in the routing process, but they are being eliminated due to no proper load balancing performed in order to distribute the traffic. This will cause nodes in the earlier route eventually be drained out of battery power. Whereas if it is to be an energy aware routing, then traffic is evenly loaded in order to consume equal amount of energy at each node. EA-Short basically looks at such mechanism to divert the traffic to other nodes.

There are two possible ways how this energy aware mechanism can be performed. In the first approach, when a node (G for example) which is closer to the current route of traffic between A, B and C overhears the message for consecutive times, it may take over the role of the intermediate node (node B in this case). Node G can perform this by comparing the energy level at node B and node G, and if the remaining energy at G is more than B, then G might take over in order to forward the messages to C. In the second approach, node G detects that node B needs help which could be due to low energy level at node B and G informs B that it may take over the role to divert traffic from A-B-C to A-G-C. This can be further explained by introducing other additional nodes along the path which may become helpers to help other nodes which are in trouble of low energy.

In this algorithm, there are two more additional fields to be included in the IP header. They are "Hop Counter" and "Residual Energy Level" together with other IP fields such as source, destination, sender and sequence number. Each node maintains an overhear table consists of source-destination pair, sequence number and "overhear list". When a node overhears a packet, it checks its source-destination field to know if the packet belongs to this new flow. If it belongs to this flow then a new entry is added into the overhear table. The residual energy field keeps the residual energy level at each transmitting node of the packet.

PA-Short algorithm is developed on both AODV and DSR with the enhanced name called AODV-SHORT and DSR-SHORT. The goal of this approach is to see how Short algorithm performs over other on demand routing protocols. Based on the experiment, DSR performs better than AODV and AODV-SHORT performs better than AODV in terms of path optimality. In terms of delivery rate, both AODV-SHORT and DSR-SHORT achieve a stable performance for all mobility rates. As for energy aware, Short is implemented on AODV called AODV-EA. Based on the experiment, it is evident that AODV-EA performs better load balancing thus performs better energy usage at all participating nodes rather than concentrating on certain nodes only. Short algorithm achieves routing optimality by con-

Table 1. A Summary of Protocols

No	Protocol	Advantages and Disadvantages
1	Yu et. al (2001)	Uniform and non-uniform Geographic energy aware routing Delivers more packets than GPSR (Intanagonwiwat et. al, 2000) For non-uniform traffic - longer network lifetime than GPSR Hole problems Picks neighbor closer to sink
2	Intanagonwiwat et. al (2000)	Geographic non-energy aware routing Avoids communication holes by delivering planar graph out of original network graph Hole problem – by using the shortest path, it enlarges hole Tends to focus traffic on the edge when it routes around holes therefore exhausting nodes on the edge
3	Wendi et. al (2000)	Uses a clustering mechanism whereby every node rotates the role as cluster head to well balance the energy distribution and to extend the network lifespan Cluster head communicates with member sensor nodes and forwards data to the base station Not appropriate for extended networks since direct communication between the cluster head and the base station incurs additional energy
4	Sha et al. (2005)	Uses right-hand rule (locating and bypassing routing holes) Guaranteed delivery and more energy efficient Avoids routing to the holes Two important notions – weight and hole
5	Youssef et al. (2002)	Uses distance to measure transmission energy and propagation delay Cluster-based approach Looks at maintaining reasonable end-to-end interval and throughput output
6	Vlajic & Moniz (2007)	Not only looking at energy but overall transmission delay and static node failure
7	Akkaya & Younis (2004)	Mobility of the gateway-dynamic maintenance of multihop Continuous adjusted network topology and update Gateway picks the best node based on its proximity to the next position it moves Notifies the last node on the current position to extend path to the new location
8	Shah & Rabaey (2002).	Not only single path but having many alternative paths for communication Energy at single node not depleted easily
9	Bokareva et. al (2005)	A naturally inspired system for hybrid sensor network by providing self-healing capabilities Automatic fault recognition and response Adaptive architecture to learn and evolve with unknown faults – self-healing Looking at faulty sensor readings which need self-healing Co-ordination between monitoring nodes
10	Gui & Mohapatra (2003)	Self-healing and optimizing techniques for mobile ad hoc networks Network performance is enhanced in relations to bandwidth and latency by not incurring extra cost Path determination can be done without using much energy Metrics such as path length, energy consumption and energy aware load balancing
11	Dressler et. al (2005)	Bio-inspired strategy in terms of self-organization and adapt to environmental changes Looking into energy efficient operation, quality of service, optimized task allocation and communication Looking into cell biology methods and mimics them into networking aspects mainly for self-organization context. Feedback loop adapted from cell biology Mobile robot systems(communication relay) and stationary sensor network
12	Branch et. al (2007)	Energy effective self-healing routing for WSN The shortest possible path is used, route repair and collision avoidance Future research-more energy efficient since random selection of neighbors which may cause more energy to be consumed.
13	Ganesan et. al (2001)	Propose two types of multipath to provide energy efficient recovery from failure Failure of primary path can be recovered without using flooding- flooding reduces network lifetime

continued on following page

Table 1. Continued

No	Protocol	Advantages and Disadvantages
14	Al-Fares et. al (2009)	Proposes a hierarchical routing protocol for WSN by looking into network survivability and redundancy features. Looks into node or link reliability problem and fault tolerance by introducing redundant cluster head Shows positive and better results compared to low energy adaptive clustering hierarchy (LEACH Wendi et. al (2000)).
15	Camilo et. al (2006)	Bio-inspired mechanism Uses ant pheromone trail Energy conscious. Energy consumption due to communication is reduced Lessening the overhead in terms of message interchange.
16	Tong et. al (2003)	A new concept of introducing mobile agent Sensor nodes are less burdened by reducing the load Hierarchical architecture More energy effective
17	Atakan & Akan (2007)	Biologically inspired mechanism Maps B-cell to designated nodes

Table 2. Comparison of protocols

	Biologically Inspired	Stationary Node	Energy Efficient	Hole Problem	Shortes t Path	Fault Tolerance	Hierarchical
Wendi et. al (2000)		√	√				√
(Gui & Mohapatra 2003)		√	√		√		
Branch et. al (2007)		√	√		√	√	
Bokareva et. al,(2005)	√	√				√	
Camilo et. al (2006)	√	√	√				
Sha et al. (2005)		√	√	√	√	√	
Yu et. al (2001)		√	√	√	√		
Intanagonwiwat et. al (2000)							
Akkaya and M. Younis (2004)		Mobile Base station	√				
Al-Fares et. al (2009)		√	√			√	√
Shah and Rabaey, (2002)		√	√				
Dressler et. al (2005)	√	√	√				
Youssef et al. (2002)		√	√		√		
Ganesan et. al (2001)		√	√		√		
Vlajic and N. Moniz (2007)		Partial mobile	√	√			
Atakan and Akan (2007)	√	√	√			√	
Tong et. al (2003)		Partial mobile	√			√	√

tinuously observing routing paths and redirecting paths to a more optimal one. Based on the simulation results, higher delivery rate and longer network lifetime are achieved by adopting SHORT algorithm.

Summary

Table 1. and Table 2 outlines a summary of protocols in the literature by considering its advantages and disadvantages. This section is discussed in brief to give an outline of self-healing protocols and their weaknesses to look for possible improvement in future research

CONCLUSION

The chapter is concluded by presenting a review of self-healing routing protocol in wireless sensor network. The protocols discussed in this chapter are categorized as energy efficient and self-healing, so that a reasonable review can be conducted. The review was conducted by analyzing the strengths and weaknesses of the protocols. The protocol is also presented in two different contexts as in bio-inspired and non bio-inspired mechanisms. Also the comparison in section 3 and 4 is done based on various metrics useful in sensor network performance. The comparative study done in terms of energy efficient, self-healing and bio-inspired methods of the protocols in the literature provides as insight for future better self-healing protocols. Moreover, the static and single layer of hierarchy should be given further thought to look into other alternative approaches such as n-tier and mobile nodes. The given bio-inspired mechanism provides a self-healing mechanism for energy efficient protocol to increase the lifetime of wireless sensor networks. Up to now, many self-healing routing protocols have been proposed but not many have focused on biologically-inspired mechanisms for autonomic behavior.

REFERENCES

Akkaya, K., & Younis, M. (2004). *Energy-Aware Routing to a Mobile Gateway in Wireless Sensor Networks*. Paper presented at the IEEE Globecom Wireless Ad Hoc and Sensor Networks Workshop, Dallas, TX.

Akkaya, K., & Younis, M. (2005). A survey on routing protocols for wireless sensor networks. *Ad Hoc Networks*, *3*(3), 325–349. doi:10.1016/j.adhoc.2003.09.010

Al-Fares, M. S., Sun, Z., & Cruickshank, H. (2009). *A hierarchical routing protocol for survivability in wireless sensor network (WSN)*. Paper presented at the International Multi Conference of Engineers and Computer Scientists.

Al-Karaki, J. N., & Kamal, A. E. (2004). Routing techniques in wireless sensor networks: a survey. *Wireless communications, IEEE, 11*(6), 6-28.

Assunção, H. P., Ruiz, L. B., & Loureiro, A. A. (2006). A service management approach for self-healing wireless sensor networks. In *Autonomic Networking* (pp. 215–228). Springer Berlin Heidelberg. doi:10.1007/11880905_18

Atakan, B., & Akan, Ö.B. (2007). Immune system-based energy efficient andreliable communication in wireless sensor networks. In Advances in Biologically Inspired InformationSystems – Models Methods and Tools, Studies in Computational Intelligence (SCI). Springer.

Bokareva, T., Bulusu, N., & Jha, S. (2005). *SASHA: towards a self-healing hybrid sensor network architecture*. Paper presented at the Second IEEE International Workshop on Embedded Networked Sensors. doi:10.1109/EMNETS.2005.1469101

Branch, J. W., Lisee, M., & Szymanski, B. K. (2007). *SHR: Self-Healing Routing for wireless ad hoc sensor networks*. Paper presented at the Intern. Symp. Performance Evaluation of Computer and Telecommunication Systems SPECTS.

Camilo, T., Carreto, C., Silva, J. S., & Boavida, F. (2006). An energy-efficient ant-based routing algorithm for wireless sensor networks. In Ant Colony Optimization and Swarm Intelligence. Springer Berlin Heidelberg. doi:10.1007/11839088_5

Di Car, G., & Dorgio, M. (1998). Antnet: distributed stigmergetic control for communication networks. *Journal of Artificial Intelligence Research*, (9), 317-365.

Dorigo, M., Maniezzo, V., & Colomi, A. (1996). Ant system: Optimization by a colony of cooperating agents. *IEEE Transactions on Systems, Man, and Cybernetics. Part B, Cybernetics*, *26*(1), 29–41. doi:10.1109/3477.484436 PMID:18263004

Dressler, F. (2006). Self-organization in ad hoc networks: Overview and classification. University of Erlangen, Dept. of Computer Science.

Dressler, F., & Carreras, I. (2007). *Advances in Biologically Inspired Information Systems*. Springer-Verlag Berlin Heidelberg.

Dressler, F., Kuger, B., Fuchs, G., & German, R. (2005). *Self-Organization in Sensor Networks using Bio-Inspired Mechanisms*. Paper presented at the 18th ACM/GI/ITG International Conference on Architecture of Computing Systems-System Aspects in Organic and Pervasive Computing (ARCS'05): Workshop Self-Organization and Emergence, Innsbruck, Austria.

Ganesan, D., Govindan, R., Shenker, S., & Estrin, D. (2001). Highly-resilient, energy-efficient multipath routing in wireless sensor networks. Paper presented at the ACM SIGMOBILE. *Mobile Computing and Communications Review*, *5*(4), 1125. doi:10.1145/509506.509514

Gui, C., & Mohapatra, P. (2003). *SHORT: Self-healing and optimizing routing techniques for mobile adhoc networks*. Paper presented at the 4th ACM international symposium on Mobile ad hoc networking & computing.

Heinzelman, W. R., Chandrakasan, A., & Balakrishnan, H. (2000). *Energy-efficient communication protocol for wireless microsensor networks*. Paper presented at the 33rd annual Hawaii international conference on. doi:10.1109/HICSS.2000.926982

Intanagonwiwat, C., Govindan, R., & Estrin, D. (2000). *Directed diffusion: a scalable and robust communication paradigm for sensor networks*. Paper presented at the 6th annual international conference on Mobile computing and networking. doi:10.1145/345910.345920

Karp, B., & Kung, H. T. (2000). *GPSR: Greedy perimeter stateless routing for wireless networks.* Paper presented at the 6th annual international conference on Mobile computing and networking. doi:10.1145/345910.345953

Koopman, P. (2003). Elements of the self-healing problem space. In *ICSE Workshop on Architecting Dependable Systems.*

Lanthaler, M. (n.d.). *Self-Healing Wireless Sensor Network.* Available at: http://74.125.155.132/scholar?q=cache:qoVzFSeebusJ:scholar.google.com/+selfhealing+wireless+sensor+netowrk+lanthaler&hl=en&as_sdt=2000

Sha, K., Du, J., & Shi, W. (2005). *Wear: A balanced, fault-tolerant, energy-aware routing protocol for wireless sensor networks.* Technical Report MIST-TR-2005-001, Wayne State University.

Shah, R. C., & Rabaey, J. (2002). *Energy Aware Routing for Low Energy Ad Hoc Sensor Networks.* Paper presented at the IEEE Wireless Communications and Networking Conference (WCNC), Orlando, FL. doi:10.1109/WCNC.2002.993520

Sohrabi, K., Gao, J., Ailawadhi, V., & Pottie, G. J. (2000). Protocols for Self-Organization of a Wireless Sensor Network. *IEEE Personal Communications*, *7*(5), 16–27. doi:10.1109/98.878532

Tong, L., Zhao, Q., & Adireddy, S. (2003). *Sensor networks with mobile agents.* Paper presented at the Military Communications Conference.

Vlajic, N., & Moniz, N. (2007). *Self-healing wireless sensor networks: Results that may surprise.* Paper presented at the Globecom Workshops, 2007 IEEE.

Youssef, M., Younis, M. F., & Arisha, K. (2002). *A constrained shortest-path energy-aware routing algorithm for wireless sensor networks.* Paper presented at the Wireless Communications and Networking Conference, 2002. doi:10.1109/WCNC.2002.993370

Yu, Y., Govindan, R., & Estrin, D. (2001). *Geographical and energy aware routing: A recursive data dissemination protocol for wireless sensor networks. Technical report ucla/csd-tr-01-0023.* UCLA Computer Science Department.

Chapter 12
Green Energy in Data Centers

Kanahavalli Mardamutu
Quest International University Perak, Malaysia

Vasaki Ponnusamy
Universiti Tunku Abdul Rahman, Malaysia

Noor Zaman
King Faisal University, Saudi Arabia

ABSTRACT

Green energy paradigm has been gaining popularity in the computing system from the software, hardware, infrastructure and application perspectives. Within that concept, data center greening is of utmost importance at the moment since data centers are one of the most energy conserving elements. Data centers are seen as the technology era's black energy-swallowing secret. Reducing energy consumption at data centers can reduce carbon footprint effect tremendously. Not addressing the issue immediately will lead to significant energy usage by data centers and will hinder the growth of data centers. The call for sustainable energy efficient data center leads to venturing into data center green computing. The green computing concept can be achieved by using several methods adopted by researchers including renewable energy, virtualization through cloud computing, proper cooling system, identifying suitable location to harvest energy whilst reducing the need for air-conditioning and employing suitable networking and information technology infrastructure. This paper focuses into several approaches used by researches to reduce energy consumption at data centers while deploying efficient database management system. This paper differs from others in the literature by giving some suitable solutions by looking into a hybrid model for green computing in data centers.

INTRODUCTION

The increase in need of data processing has led to a demand in obtainability for cheaper, faster, efficient and larger data management system. This has further introduced a significant problem in terms of energy consumption and energy usage as mentioned in (Barroso. L.A, 2005). Since the trend of Internet Infrastructure is moving towards services-based computing, data center plays a key role in this new

DOI: 10.4018/978-1-4666-9792-8.ch012

computing architecture. It is used widely in variety of services independent encapsulating any vendor, product and technology including web hosting, application services, electronic markets, outsourced storage, online-monitoring system, mashups, digital nervous system and other network services. Fostering renewable energy production, developing better renewable technologies and improving the environmental performance of conventional electricity benefits heightened through improving the long-term strategy of data center performance. Environmental Protection Agency (EPA) shows that the data center and database management system in the US consume power doubled between 2000 and 2006 and may multiply again within the next decade. According to EPA, it is projected that energy consumption by data centers for the year 2008 is more than 80 billion kWh. (US Department, 2013). Report to the Congress, EPA's Energy Star Program estimates that increase efficiency in data centers could yield energy savings as the burgeoning efforts to make data centers greener is the energy and the word 'green' adds multiple meaning when applied to the data centers. (Koomey J, 2011).

Many works have been done to produce efficient result in reducing the energy usage for the purpose of data center and data management for example by Cisco System Inc., Google Inc., and many more (Barroso.L.A,2005). Some of the approaches were to design highly efficient systems at the peak performance point with maximum utilization of resources. But very few studies have addressed the energy problem comprehensively (US Department, 2013).

PROBLEM STATEMENT

In today's technological world, it is proven that there is a significant relationship between data center and investment in capital overlay and outgoing costs. Table 1 shows the detailed breakdown on the distribution of the cost from data center (Greenberg et al., 2008).

The design of computing system is being optimized based on the execution time and operates at low consumption due to entrapment of resources and fragmentation of data. No concern was made for the issue of energy consumption by the system. Uncontrolled usage of energy in data centers has negative effects on the reliability, density, scalability of information processing and the environment. Need and awareness are raised across nation through multiple solution and disciplines to optimize and save energy use in data management system (Barroso. L.A, 2005). System or even its components like CPUs, memory and disk are hardly distributing energy efficiently. It trades power for performance (US Department, 2013). Power management has become a critical issue due to large energy usage and all over the world government has started imposing the taxes on the carbon emission, ultimately aiming to reduce the emission and help make computing into green computing (Bansal et al., 2011).

Table 1. Guide on the distribution of data center costs

Amortized Cost	Component	Sub-Component
~ 45%	Servers	CPU, memory, storage systems.
~ 25%	Infrastructure	Power distribution and cooling.
~15%	Power draw	Electrical utility costs.
~15%	Network	Links, transit, equipment.

a. Source of data: A. Greenberg et al, "The cost of a cloud: research problems in data center networks", ACM SIGCOMM Computer Communication Review, 39(1), pp. 68, 2008

Although research was conducted in designing power aware applications, an effort for implementing has not been taken place. Large expenses and allocations need to be augmented when data center reaches maximum provisioned power (Bansal et al., 2011). In the very near future, energy efficiency is expected to be one of the key procuring arguments in the society. Research should focus in building power-aware database management systems (DBMS) or GreenDB.

Analysis shows that giant companies are also working towards energy efficiency using green database system. Huge volume of energy usage goes to the data center. The global business market intelligence is a billion dollar market and still growing to see double-digit growth rates (US Department, 2013). With the growth of the internet data center (IDC), data systems are also running its management workload such as search engines, storage of multimedia like video and audio, streaming, web analytics, and cloud computing. Dataset are growing rapidly and workload is also increasing. Together with it, the energy usage has increased. Companies like Google are working towards reducing the energy usage excessively in data centers (US Department, 2013). Data center has become the major immobilizer in the topic of energy wastage.

Clients particularly are searching for high scalability setup, lower unit price and simplicity of management. As to maintenance costs, the second largest bill for large database centers comes from the power company. Statistic shows that in US alone, data centers consumed about 60 billion kilowatt-hours of electricity in 2006 which equaling to 1.5% of electricity consumption in US (US Department, 2013). Demand for a solution is in the high urge and is being researched by many giant companies

OBJECTIVES

This research is not to reinvent the wheel to conduct a critical analysis on the existing research to integrate the idea proposed to produce wholesome elucidation. Modules from many different researches have been integrated and evaluated to enhance its components to be fully functional with fewer flaws. Therefore, it is a better solution to show the benefits of energy efficiency through green database. The main objectives of this research are as follows;

1. To initiate and analyse energy-aware research conducted on database management in a way that improves energy efficiency of a data center.
2. To propose solution in implementing the energy efficiency through green computing and green database system.

Many companies, organizations and individuals have proposed methodology to resolve this problem based on the categories shown in Figure 1. This research paper analyzes similar papers which has solution but needs minor integration to propose new innovation with better and efficient energy saving. Analysis focuses on power consumption patterns and identification of power-saving opportunity in GreenDB. Critical analysis done based on the following observation.

- Power-aware database.
- Data management through software and hardware.
- Data management through disk power management.
- Green computing architecture.

Figure 1. Critical analysis flow model

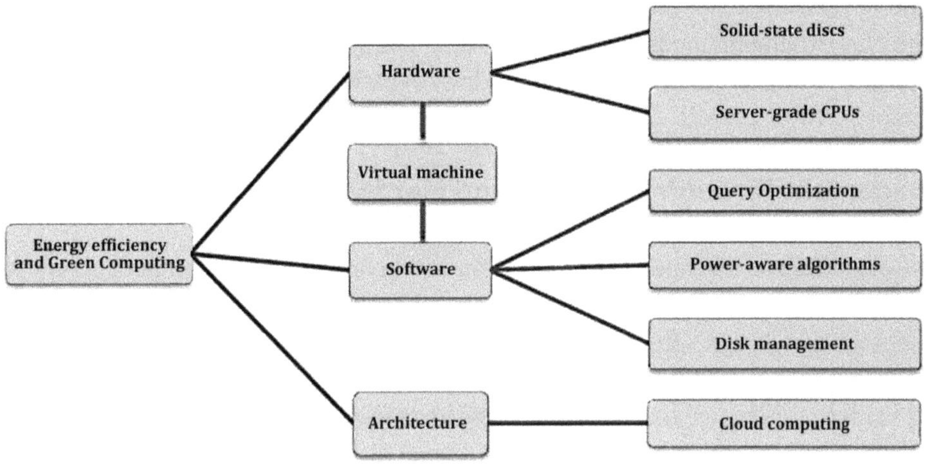

- Renewable energy.

Power-Aware Database

Power-aware computing research at the application level in the aspect of hardware and operating system level found to be synergistic for energy proficiency in the core systems. Xu. Z., 2011 describes on creating a power-aware database management (P-DBMS) and initial ideas on the design as such systems, with the focus on a power-aware query optimization module in the DBMS. This particular paper has proposed and created a power model to accurately measure the energy costs of query executions and plan selection. It is in contrast with the heuristic-based adaptive solutions that rely on manual timing and extensive empirical evaluation. It is meaningless without considering the database performance. It is either to make the query run faster by optimizing the query execution or let the system run in lower power state while sacrificing the performance to some extent. They proposed that the throughput is improved while maintaining the power line. It indirectly reduces the power cost while not sacrificing the throughput performance. Implementation has been taken place through initial design in algorithm based software named PostgreSQL rather than power-saving through hardware and operating system levels

Table 2. Critical overview on PostgreSQL system

Attribute	PostgreSQL System	
	Description	**Analysis**
Hardware	Hardware Optimisation	x
	Power Management	x
Software	Software Optimisation	/
	Query management and optimisation	/
Architecture	Architecture Manuveuring	x
	Distributed System	x

a. Evaluation based on the result of proposed research paper by Author modification. The proposed solution resolves the power aware of the database as seen in Table 2. Only setback would be database management system needs to be revisited or redesigned and initial efforts be invested. The plan evaluation model has a stumbling block where the system is stable and foreseeable but the database is barely stable and predictable. Thus, modeling error occurs and estimation per operation cost cannot be 100% correct.

Data Management through Software and Hardware

Synergy between database management through algorithm and hardware architecture are promising in energy efficiency computing. Harizopoulus, S. et al., 2009 recommends that adjusting existing system-wide configuration and design algorithm for energy proficiency is not the same as designing algorithm based on the throughput and performance. Deploying an energy-efficient data management platform is by choosing hardware components with good performance-per-watt characteristics by choosing high-performance components (server-grade CPUs, solid-state disks etc.). Installing an energy-efficient system was also done through software-based approaches for reducing wasted energy in data management systems. This paper proposes to power down the unused hardware components, revisit and redesign data structures and algorithm from a wider perspective. The 'push-to' design components offers more control over power performance trade-offs for example usage of multicore chips which can be activated at any time. The increase of hardware heterogeneity offers technology refresh cycle like the virtual machine. The disadvantage here is that some workloads may be able to use these additional resources while others will underutilize them and create a waste of power. Table 3 shows that hardware configuration and application-agnostic power management is only part of the solution proposed in this paper.

Data Management through Disk Power Management

Reducing energy consumption through hardware depends on many aspects. Earlier research shows that the result from monitoring data center is that the average idle period for a server disk is very small compared to the time taken in spinning up and down. This significantly limits the conservation of energy. Zhu, Q. et al., 2004 proposed reducing energy consumption of disk storage using power-aware cache management with algorithm which provides more opportunities for causal disk power management structures in energy efficiency. So far only a few studies (D. Colarelli and D. Grunwald, 2002 ; Gurumurthi et al.,

Table 3. Critical overview on Stavros review

Attribute	Synergistic System	
	Description	Analysis
Hardware	Hardware Optimisation	/
	Power Management	x
Software	Software Optimisation	/
	Query management and optimisation	/
Architecture	Architecture Manuveuring	x
	Distributed System	x

Table 4. Critical Overview On Power-Aware Cache Management Algorithm

Attribute	Synergistic System	
	Description	Analysis
Hardware	Hardware Optimisation	x
	Power Management	/
Software	Software Optimisation	/
	Query management and optimisation	x
Architecture	Architecture Manuveuring	x
	Distributed System	x

a. Evaluation based on the result of proposed research paper by Author

2003 ; Gurumurthi et al. 2003 ; Carrera et al., 2003) addressed the problem in energy efficiency for the storage systems at the data centers. Based on these studies, multi-speed disks or multiple rotational speed disks shown promising results by using synthetics workloads. Therefore, power-aware cache management algorithm was designed whereby it is aware of the base disk energy ingestion and power management plan for read and writes accesses.

It is an energy-optimal cache replacement algorithm which uses the dynamic programming in polynomial time leading to 16% less energy and lead to 50% better average response time. But this study has its limitation such as it is only designed for multiple disks and focuses on high-end storage systems. Single disk may not work with this algorithm. Also it has low effect on prefetching as well. It does not consider the storage cache energy consumption and focuses on smaller part of the overall management as shown in Table 4.

Green Computing Architecture

Virtual machine (VM) has been globally used in the data center environment due to its important feature and architecture. Liu. L. et al. 2009 uses the latest technology trend which is the cloud computing as a guaranteed performance and designed to reduce the data center energy consumption. It is effective and enables the widespread online-monitoring, live VM migrating and placement optimization. It utilizes the live migration feature of VM Technology such as XEN (Bansal et al.,2011) to implement the architecture for example Amazon Elastic Compute Cloud (Zu. Z. 2011) provides a virtualized computing environment hosted with different kind of Linux-based service. Evaluation results show that 27% energy can be saved when applying Green computing architecture. VM Migration allows lesser usage of computer. This architecture assures real-time performance requirement as well as saves the consumption of energy by Internet Data Center (IDC). Green computing architecture with the integration of VM technology is an effective method in reducing server power consumption.

The problem may arise in the angle of automatic scheduling decision on dynamic migration or consolidating the VMS among physical servers to meet the workload requirements meanwhile saving energy, especially for performance sensitive. The aspect of performance trade-off between the operating system and applications or middleware has not been tested which can be a flaw. But based on Table 5, this technology alone may not be sufficient to provide efficiency.

Table 5. Critical Overview On Green Computing and VM Technology

Attribute	Green Computing and VM Technology System	
	Description	**Analysis**
Hardware	Hardware Optimisation	/
	Power Management	/
Software	Software Optimisation	/
	Query management and optimisation	x
Architecture	Architecture Manuveuring	/
	Distributed System	x

a. Evaluation based on the result of proposed research paper by Author

Renewable Energy

Exploiting green energy sources is another opportunity but it is variable and intermittent. So far, there has been less work on integrating renewable energy with database system. Chen, C., et al., (2012) uses the hardware optimization to save energy and retaining the execution part of the hardware. They proposed through integration of renewable energy which is the solar to power the execution of database management system. This research bridges the gap between renewable energy and database management system with ReinDB (Renewable Energy Integrated Database). The purpose would be to minimize the usage on carbon intensive energy consumption on a database server with both renewable energy and carbon intensive energy supplies. This works on Supply Driven Execution (SDE) on a database server with two mechanisms.

1. Workload scheduling to meet workload deadlines.
2. Dynamic configurations on hardware setting.
3. Brown-aware workload shifting.
4. Battery-aware workload shifting.

Although it increases the usage of renewable energy (solar usage) but there are still some disadvantages to take into consideration. Studies show that the mismatch between workloads and green energy supplies challenges in the utilization of energy. Divergence in the context of data center from brown energy and green causes excessive usage of energy (Goiri et al,. 2011; Goiri et al,. 2012 ; Krioukov et al 2012 ; Liu et al. 2011). Clients are required to change the hardware to new system which may need initial investment that may cost high. Also the latency time (slack time) is also another issue whereby the clients have to tolerate a large slack time to complete a task with the green techniques. In the economical aspect, the cost involves in manufacturing solar panel and battery provision. Table 6, shows that, there are many other aspect to be incorporated.

Table 6. Critical overview on integration of renewable energy

Attribute	Integration of Renewable Energy	
	Description	**Analysis**
Hardware	Hardware Optimisation	/
	Power Management	x
Software	Software Optimisation	/
	Query management and optimisation	x
Architecture	Architecture Manuveuring	/
	Distributed System	x

a. Evaluation based on the result of proposed research paper by Author

Table 7. Physical usage

Project/ Attribute	Hardware Modification	Software Modification	Solar Integration
Zichen Xu	No	No	No
Cheng Chen	Yes	Yes	Yes
Liang Liu	No	Yes	No
Stavros Harizopoulos	Yes	Yes	No
Qingbo Zhu	No	Yes	No

ANALYSIS AND PROPOSED GREEN SOLUTIONS

The concept of green database and energy efficiency is emerging high and gaining more popularity and everyone wants to be on top of it and they have started researches on the concept of energy saving in the DBMSs. Companies like IBM, Oracle that develops their own DBMSs want to decrease the energy consumption. All researches are focusing on individual area only. Integration of success methods from each research can result in wholesome solution.

Evaluation and analysis done to multiple proposal and project paper resulting in integration of the best modules for many different sources tosses excellent solution. Comparisons shown in Table 7, 8, 9, 10 and 11 define the strength and weakness based on the critical analysis done.

Virtualization and Green Cloud

Virtualization allows software and hardware virtualization by providing virtual environment. This reduces the need for overwhelming use of servers. More processing can be done with lesser energy consumption. Every single component of IT can be virtualized from the servers, routers, switches, input-output, applications, operating systems and desktop. Virtual machine allows more than one operating system to be operated in one computer. Each operating system runs its own virtual machine and resources like I/O, storage and processor can be assigned to each virtual machine. It is to automatically make the scheduling decision on dynamically migrating or consolidating Virtual Machines among physical servers to meet the workload requirements while saving energy, especially for performance sensitive. The heuristic

Table 8. Power coverage

Project/ Attribute	Power State	Power Cost	Power efficiency
Zichen Xu	Low	Reduced	30%
Cheng Chen	Very low	Reduced	28-53%
Liang Liu	Low	Reduced	27%
Stavros Harizopoulos	Low	Reduced	45%
Qingbo Zhu	Low	Reduced	16-35%

Table 9. Query and throughput

Project/Attributes	Query Execution Speed	Query Optimizer	Throughput time
Zichen Xu	Faster	Increased	Maintain
Cheng Chen	No change	Scheduled Query	peak performance
Liang Liu	No change	Optimized	Guaranteed real time
Stavros Harizopoulos	No change	Optimized	Peak performance
Qingbo Zhu	Faster	Optimized	Limited

Table 10. Other Aspects

Project/Attributes	Scalability	Reliability	Slack time
Zichen Xu	Inefficient	Inefficient	Retained
Cheng Chen	Efficient	Efficient	Retained
Liang Liu	Efficient	Efficient	Retained
Stavros Harizopoulos	Efficient	Efficient	Retained
Qingbo Zhu	Efficient	Inefficient	Retained

Table 11. Application and Remark

Project/Attributes	Application Level	Remark/ New Invention
Zichen Xu	Very little	Cost evaluation model and power evaluation model
Cheng Chen	Very little	ReinDB – brown and green energy usage with additional equipment to be installed
Liang Liu	Increasing	VM, Heuristic search algorithm and performance sensitive
Stavros Harizopoulos	Low	Algorithm based and redesign the system component
Qingbo Zhu et al.	Low	Works only for multiple disk and focuses on high-end storage system.

algorithm is used in virtual machine migration to achieve the optimization. States of server power is adjusted through the dynamic voltage or frequency scaling.

Renewable Energy Sources

Datacenters can be equipped with renewable energy sources such as solar powered energy and hydroelectric energy supply. Solar powered energy can be good source of energy for data centers located at locations that has wide availability of solar energy. The solar panel nodes can be charged during the day with the sunlight and stored in backup sources to be supplied during the absence of sunlight. Although installation of such facility is comparatively costly, but the outcome can be very promising as these are renewable energy sources. Moreover energy generation through hydroelectric source is another promising factor. This is applicable in areas where there plenty of water sources for hydro energy generation. A good example would be the Facebook datacenter located in Sweden which is supplied with hydroelectric power without interruption due to the abundant resources of hydroelectric power. In conclusion, the choice of renewable energy is very much dependent on the geographical areas where the datacenters are located (1). More research work can be carried on vibration energy generation from the vibration caused by the servers and the other high end equipment present in the datacenter. This requires extensive research on the strength of vibration generated and amount of energy can be drawn from that.

Disk Power Model and Management

Managing the multi disks speed to increase the amount of energy saved with data center workloads by identifying the active, idle and standby mode of the disk. During the active mode, read and write mode is actively processing in the platters spin and the head. Likewise, during the idle, no disk is required whereby energy can be saved by idling the rotation and spinning. Mathematically, a lower rotational speed mode consumes lesser energy compare to higher rotations. Reduction in Input/ Output Operations per Second (I/Ops) directly leads to reduction in the energy consumption which is shown in (1). An average latency is the time taken by the disk that is accessed to rotate to the position under read/ write head. Meanwhile the average seek time would be the time (ms) taken for the hard drive's read/write head to locate itself over the read/ write track.

I/Ops Calculation:

1 / ((σεεκ τιμε / 1000) + (λατενχψ / 1000) = I/Oπσ (1)

Ενεργψ Χονσυμπτιον= Ποωερ (Ωαττ) ξ I/Oπσ

Power-Aware Database Management System

It monitors the query optimizer, buffer manager, storage manager and many other database query optimization modules.

Cost can be reduced by improving the query optimization time whereby the processing time can be improved and obtained through the setup of the server which is shown in (2).

Cost estimation, C:

X = (Θυερψ Οπτιμιζατιον ξ τιμε) ξ Προχεσσινγ Τιμε (2)

Datacenter Cooling System

Cooling system in the datacenter is very important aspect as it guarantees proper operation of datacenter equipment such as servers and storage devices. A suitable cooling method to maintain the thermal condition of the CPU's can provide efficient energy usage. The Facebook datacenter in northern Sweden provides an energy efficient cooling system as compared to other datacenters. Besides that, the arctic climate in the northern region allows the Facebook datacenter to be in a colder temperature condition requiring less cooling needs. The river along Lulea provides a lot of resources for hydroelectricity supply for the entire datacenter without the need for generators. The datacenter depends on the climate chilling system rather than depending on air conditioning system. An intelligent heat, ventilation and air conditioning (HVAC) system can be designed with proper cooling system (Parisio et al.; Mousavi et al.) Datacenters with central air systems, direct liquid cooling and humidification can allow for proper cooling system.

Power Management Using Wireless Sensor Network

One way to make datacenters energy efficient is by providing good network infrastructure. The servers, routers, switches and other equipment linked with an efficient networking facility can decide the energy consumption of the datacenter. Congested network linkage allows unnecessary communication among devices and allows for additional energy consumption. One potential networking infrastructure that could facilitate with proper energy management is through employing wireless sensor networks (WSN). WSN provides intelligent services by deploying various types of sensors to monitor the environment in the datacenter to make it a smart datacenter. The smart datacenter is intended to be designed to create a system that performs energy management in terms of energy consumption and optimizing power consumed by datacenter devices. The framework of the system is shown in Figure 2. The system is categorized into fault detection and energy management. Fault detection stage should identify malfunctioned datacenter equipment and resources that is unnecessarily consuming power and wasting energy. The energy management stage should be able to manage energy consumption, optimize power consumption and perform automated notification. The fault detection system consists of data acquisition from various sensors and information processing using fuzzy logic engine. The energy management system performs event detection and action mechanism based on changes in the sensor data and motion detection. In this system, datacenter environment parameters such as temperature, humidity, light and sound are detected for deciding whether electrical appliances and lights to be switched on or off for energy management in buildings. Besides that, motion sensors are used for detecting movement so that lights can be switched on and off based on occupancy. Motion sensors can also be used to detect heat emitted from the equipment so that notification alarm is triggered to take necessary action. The temperature and humidity sensors can be used to decide whether the datacenter cooling system to be turned on or off. This system can be created using some machine intelligent capabilities to make the system self-autonomic. Moreover WSN is a cost effective solution as it does not require extensive IT infrastructure and data communication.

Figure 2. WSN data center monitoring system

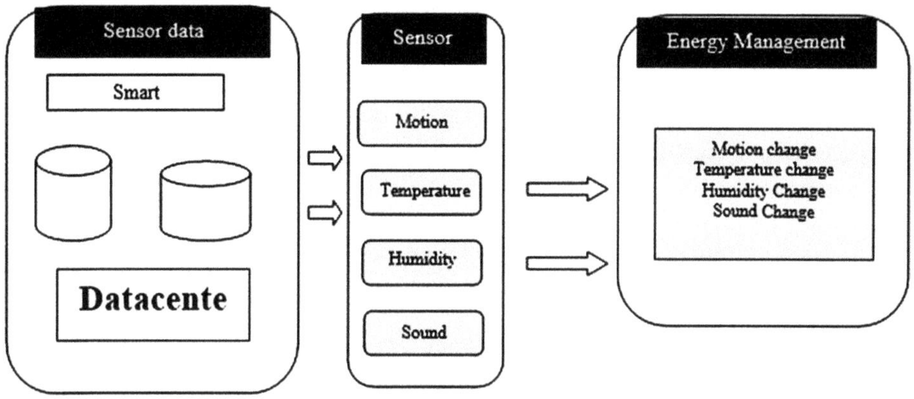

Hybrid Model

A hybrid model (Figure 3) can take advantage of a combination of energy efficient solutions including virtualization by cloud, WSN monitoring system and renewable energy system (Gholamhosseinian, and Khalifeh, 2012). Machine learning systems such as Case Based Reasoning (CBS) or decision trees can be integrated together with the virtualization concepts so that lesser servers are needed for data storage and processing (Pawlish et al., 2014). These intelligent systems allow servers to make automatic decisions and require lesser tasks. Moreover intelligent agents such as pattern recognition, fuzzy logic and neural network can be integrated with WSN so that energy usage at datacenter can be monitored intelligently. A proper selection hardware that requires less cooling system is also an added advantage. For example the Open Computer hardware has been designed in a simplified way for energy efficiency. It is just a piece of metal sheet with the motherboard that allows for proper air flow. A casing free system can allow for faster cooling process in datacenters. In conclusion the hybrid model should provide the following advantage:

Figure 3. Datacenter hybrid model

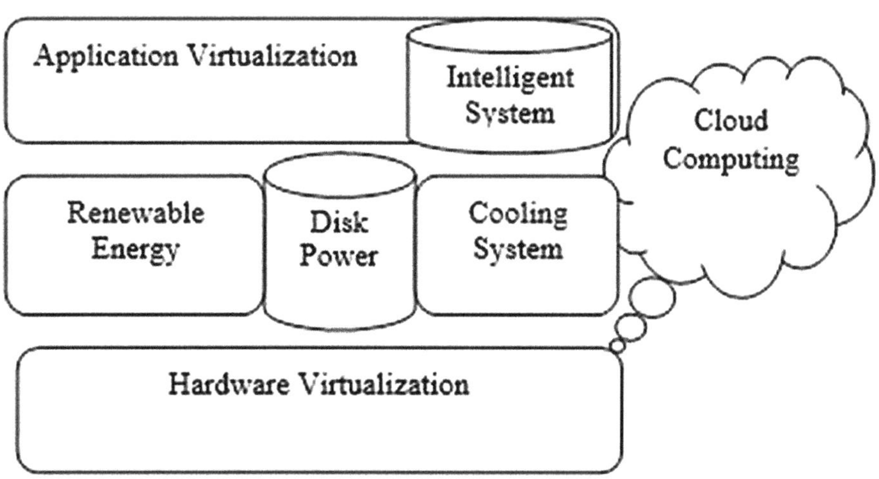

1. Reducing energy consumption compared to the traditional datacenter model
2. More utilization of virtualization by employing green cloud architecture by allowing lesser usage of storage, machines and processing without compromising the number of applications can be operated
3. Utilizing natural resources for renewable energy for datacenter. Since natural resources are ample depending on the geographical location, a proper energy harvesting method could solve energy usage at datacenters

FUTURE WORK

Looking at the hybrid model proposed, a complete framework and simulation of the model will prove the feasibility of the model. Taking renewable energy into consideration, future work in constructing data centers should focus on natural resources available. More emphasis should be given to locations near the northern poles to reduce the need for cooling system. Focus on areas that have plenty of sunlight to harvest energy using solar system or locating data centers near hydroelectric power generators. Virtualization should be highest importance by deploying systems using cloud computing infrastructure. Provide an integrated system operation with software, hardware, controls and sensors to be coupled together with the grid electricity. An automation system using self-healing energy management can be considered in this case. The system stores huge amount of energy in the storage via renewable energy and constantly monitor the level of energy at the grids and allowing power to be supplied during energy crisis.

CONCLUSION

The green energy at data centers will help to reduce the serious energy consumption issues at data centers to reduce the critics from authorities. Total energy consumption of data centers in the US for example can be reduced from 2% to even lower. The suggested solutions may not only reduce the greenhouse effect but offer a more cost effective data center setup. In this paper, several approaches of reducing energy consumption at data centers have been proposed. The proposals are validated by adopting some of the solutions addressed in the literature providing reasonable metrics. Ultimately the aim and objectives are to minimize the wastage of energy and to retain the efficiency. It is foreseen that implementation of technology collaborated with the green energy will ultimately enhance the technology to be ecofriendly.

ACKNOWLEDGMENT

This paper is a revised and expanded version of a paper entitled 'A Critical Analysis of Green Database System for Energy Efficiency and Green Computing' presented at the International Symposium on Mathematical Sciences and Computing Research 2013 (iSMSC 2013) 6-7 December 2013, Perak, MALAYSIA

REFERENCES

Amazon Elastic Compute Cloud (EC2). (n.d.). Retrieved July, 2008, from http://www.amazon.com/cc2/

Vance, A. (2013). *Inside the Arctic Circle, Where Your Facebook Data Lives*. Retrieved June, 21, 2013, from http://www.bloomberg.com/bw/articles/2013-10-03/facebooks-new-data-center-in-sweden-puts-the-heat-on-hardware-makers

Bansal, K., Goel, H., & Shrivastava, V. K. (2011). A Critical Review on Concept of Green Databases. *Global Journal of Business Management and Information Technology*, *1*(2), 113–118.

Barroso, L. A. (2005). The price of performance. *Queue*, *3*(7), 113–118. doi:10.1145/1095408.1095420

Carrera, E. V., Pinheiro, E., & Bianchini, R. (2003, June). Conserving disk energy in network servers. In *Proceedings of the 17th annual international conference on Supercomputing* (pp. 86-97). ACM. doi:10.1145/782814.782829

Chen, C., He, B., Tang, X., Chen, C., & Liu, Y. (2013). Green Databases through Integration of Renewable Energy. In CIDR.

Colarelli, D., & Grunwald, D. (2002, November). Massive arrays of idle disks for storage archives. In *Proceedings of the 2002 ACM/IEEE conference on Supercomputing* (pp. 1-11). IEEE Computer Society Press. doi:10.1109/SC.2002.10058

Data Center Knowledge. (n.d.). *Inside Facebook's Lulea Data Center*. Retrieved June, 28, 2013, from http://www.datacenterknowledge.com/inside-facebooks-lulea-data-center/

Crouch, D. (2014). *UK firm takes the cloud to chillier climes with Swedish data centre*. Retrieved June, 21, 2013, from http://www.theguardian.com/technology/2014/oct/22/cloud-storage-data-centre-sweden-arctic-hydro66

Gholamhosseinian, A., & Khalifeh, A. (2012). *Cloud computing and sustainability: Energy efficiency aspects*. Academic Press.

Greenberg, A., Hamilton, J., Maltz, D. A., & Patel, P. (2008). The cost of a cloud: Research problems in data center networks. *Computer Communication Review*, *39*(1), 68–73. doi:10.1145/1496091.1496103

Goiri, Í., Le, K., Haque, M. E., Beauchea, R., Nguyen, T. D., Guitart, J., & Bianchini, R. et al. (2011, November). Greenslot: scheduling energy consumption in green datacenters. In *Proceedings of 2011 International Conference for High Performance Computing, Networking, Storage and Analysis* (p. 20). ACM. doi:10.1145/2063384.2063411

Goiri, Í., Le, K., Nguyen, T. D., Guitart, J., Torres, J., & Bianchini, R. (2012, April). GreenHadoop: leveraging green energy in data-processing frameworks. In *Proceedings of the 7th ACM european conference on Computer Systems* (pp. 57-70). ACM. doi:10.1145/2168836.2168843

Gurumurthi, S., Zhang, J., Sivasubramaniam, A., Kandemir, M., Franke, H., Vijaykrishnan, N., & Irwin, M. J. (2003, March). Interplay of energy and performance for disk arrays running transaction processing workloads. In *Performance Analysis of Systems and Software, 2003. ISPASS. 2003 IEEE International Symposium on* (pp. 123-132). IEEE.

Gurumurthi, S., Sivasubramaniam, A., Kandemir, M., & Franke, H. (2003, June). DRPM: dynamic speed control for power management in server class disks. In *Computer Architecture, 2003. Proceedings. 30th Annual International Symposium on* (pp. 169-179). IEEE. doi:10.1109/ISCA.2003.1206998

Harizopoulos, S., Shah, M., Meza, J., & Ranganathan, P. (2009). *Energy efficiency: The new holy grail of data management systems research*. arXiv preprint arXiv:0909.1784.

Krioukov, A., Goebel, C., Alspaugh, S., Chen, Y., Culler, D. E., & Katz, R. H. (2011).. . *Integrating Renewable Energy Using Data Analytics Systems: Challenges and Opportunities. IEEE Data Eng. Bull., 34*(1), 1–9.

Koomey, J. (2011). *Growth in data center electricity use 2005 to 2010*. A report by Analytical Press, completed at the request of The New York Times, 9.

Liu, Z., Lin, M., Wierman, A., Low, S. H., & Andrew, L. L. (2011, June). Greening geographical load balancing. In *Proceedings of the ACM SIGMETRICS joint international conference on Measurement and modeling of computer systems* (pp. 233-244). ACM. doi:10.1145/1993744.1993767

Liu, L., Wang, H., Liu, X., Jin, X., He, W. B., Wang, Q. B., & Chen, Y. (2009, June). GreenCloud: a new architecture for green data center. In *Proceedings of the 6th international conference industry session on Autonomic computing and communications industry session* (pp. 29-38). ACM. doi:10.1145/1555312.1555319

Mardamutu, K., Joon, R. A. L. A., Jegatheesan, J., Ponnusamy, V., & Mei, Y. (2008). A Critical Analysis of Green Database System for Energy Efficiency and Green Computing. *Computer Communication Review*, *39*(1), 68.

Mohammadi, E., Karimi, M., & Heikalabad, S. R. (2011). A novel virtual machine placement in cloud computing. *Australian Journal of Basic and Applied Sciences*, *5*(10), 1549–1555.

Mousavi, A., Vyatkin, V., Berezovskaya, Y., & Zhang, X. (2015, September). Towards energy smart data centers: Simulation of server room cooling system. In *Emerging Technologies & Factory Automation (ETFA), 2015 IEEE 20th Conference on* (pp. 1-6). IEEE.

Parisio, A., Molinari, M., Varagnolo, D., & Johansson, K. H. (n.d.). Control of HVAC Systems in Sweden. *Current Status and Future Directions. Academic Press.*

Pawlish, M., Varde, A. S., Robila, S. A., & Ranganathan, A. (2014). A call for energy efficiency in data centers. *SIGMOD Record*, *43*(1), 45–51. doi:10.1145/2627692.2627703

Powers, C. (2011). *Leadership in Green IT (Brochure) (No. DOE/GO-102011-3295). National Renewable Energy Laboratory*. Golden, CO: NREL.

Ranganathan, P., Rivoire, S., & Moore, J. (2009). Models and Metrics for Energy-Efficient Computing. *Advances in Computers*, *75*, 159–233. doi:10.1016/S0065-2458(08)00803-6

US Environment Protection Agency. Energy star. (2013). *Report to Congress on Server and Data Center Efficiency: Public Law 109-431*. Retrieved June, 21, 2013, from http://www.energystar.gov/ia/ partners/prod_development/downloads/EPA_Datacenter_Report_Congress_Final1.pdf

US Department of Energy, Pacific Northwest National Laboratory. (n.d.). Retrieved June, 28, 2013, from http://eioc.pnnl.gov/ research/sitawareness.stm

Xen Users' Manual v 3.3. (2008). Retrieved June, 3, 2013, from http://bits.xensource.com/Xen/docs/user.pdf

Xu, Z. (2010, June). Building a power-aware database management system. In *Proceedings of the Fourth SIGMOD PhD Workshop on Innovative Database Research* (pp. 1-6). ACM. doi:10.1145/1811136.1811137

Zhu, Q., David, F. M., Devaraj, C. F., Li, Z., Zhou, Y., & Cao, P. (2004, February). Reducing energy consumption of disk storage using power-aware cache management. In *Software, IEEE Proceedings* (pp. 118-118). IEEE.

Section 3

Exploring and Illustrating the Current and Future Trends in Combining Biologically-Inspired Computing with Energy Harvesting in WSN and Energy Transmission

Chapter 13
Energy Harvesting Models and Techniques:
A Review

Saira Muzafar
King Faisal University, Saudi Arabia

ABSTRACT

Wireless sensor networks (WSNs) has gain popularity due to their wide range of applications in almost all walks of life including industry controls, environmental monitoring, health, transportation, military, civil infrastructure, science, security and more. Wireless sensor nodes are cheap and tiny in size therefore its deployment is easier. They perform well in harsh environments where human intervention is almost difficult or not possible. However, wireless sensor networks are resource constraints and its power supply has been a big challenge to keep the sensor nodes functional for a longer period. Advancement in low power electronics helped a lot but the use and maintenance of conventional batteries with a limited life span cannot address the power supply problem effectively in a long run. Therefore, harvesting energy from ambient environment is an effective alternative both in terms of power and cost, which can help sensor networks to live longer. This chapter mainly focuses on different possible energy sources available in ambient environment and current technological mechanism to harvest energy for WSNs.

INTRODUCTION

Energy Harvesting

Harvesting energy from environment is not a new technology; it has been used since decades to scavenge energy from wind, sun and moving water. The new thing is to implement this idea in modern systems. It is a process of conversion of environmental energy into electrical energy. Energy is around us in many different forms such as sound, vibration, mechanical, thermal, chemical, electrical, solar, wind, RFetc. Figure1(Hi-globe,n.d.)shows the ambient energy sources in the environment. There are many types of energy harvesters each offering differing degrees of usefulness depending on the application (Alippi C.

DOI: 10.4018/978-1-4666-9792-8.ch013

Figure 1. Different sources of ambient energy

et al.,2009).Energy harvesting or energy scavenging has an important application in areas where wiring is not practical or it is difficult to supply energy, hence it has great scope in wireless sensor networks to keep them operational for almost infinite time.

Wireless sensor networks are suitable to deploy in areas with harsh weather condition and flexible enough to embed with any system of interest.However, constant power supply is a big challenge to keep sensor node functional. To address power problem in WSN, we broadly have two solutions:

- Reduce Power consumptionby using low power networks

Currently, low power consumption in WSNs is a hot research area to increase the life cycle of sensor networks. Research has been conducted to improve/investigatelow-power MAC protocols (Lu C. et al., 2010) (Filip N. et al., 2011), routing protocols (Hua W.S. et al., 2014)(Raghunathan V. et al., 2005) transport protocols (Nacef A. B. et al., 2011), and even operating systems (Dash. S et al., 2012).

- Extracting energy from ambient environment

Another feasible option is to extract energy from ambient environment as the total power cost of wireless sensor nodes has been reduced to low level by now (Zhou H.-Y. et al.,2008) with the development of microelectronics, low-power design, and network technology (Vullers R. J. M. et al., 2009), and this give rise to adapt energy harvesting techniques for the provision of energy in WSNs.

Traditionally, conventional batteries used for power supply in WSN. Although batteries are common and reliable source for energy supply but they have limitation of size and aging even they are not used at the fullest. Function of the wireless node is to sense/compute energy from the environment in the form of heat, sound, pressure and to communicate neighboring nodes. It is good to capture operational energy from the ambient environment. Wireless sensors are useless when their battery dies, also batteries can cause a serious environmental hazard because batteries contain heavy metals and toxic chemicals for example mercury and lead. Moreover, it is expensive to maintain and replace batteries especially for the places difficult to reach such as environmental monitoring, biomedical fields and military surveillance. Life of the network depends on the life of the sensor nodes, which requires continuous power supply. Using batteries as a power is not suitable in long run, because average battery life is 1-18 months and they discharge 0.1 to 5% a month, the discharge rate will be increased if they exposed to high temperature (>40C) constantly and consistently. Batteries can be a good choice for the applications of wireless sensor networks if they require less energy conversion overhead. Due to short life span of batteries and other limitations, it is important to find an alternative energy source. Energy harvesting is a promising technique to power sensor nodes continuously at lower cost. However, it is important to consider environment, size and life of energy harvesting system. Different type of energy harvesters are available, their usefulness depends on the type of application. Some of the potential applications are reviewed below:

Energy Harvesting Applications

WSN has been declared as top ten breakthrough technologies of 21stcentury that will change the world (MITTechnology Review 2003). It finds innovative applications in almost all walks of life.

Application in Industry

Sensors application in industry requires continuous power supply and often operated in harsh environment. Industry applications usually require continuous monitoring of pressure and temperature parameters, such as oil and gas fields.

Applications in Biomedical and Fitness Devices

Energy harvesting technique has useful and innovative applications for implantable body sensors in medical field, as they need regular battery replacement. Such as pacemakers, defibrillator, ear devices for monitoring biological activity in patient ear.

Environmental Monitoring Applications

Wireless sensors are widely used for environmental and habitat monitoring such as forests, pollution detection, weather monitoring, agriculture, waterways, cattle monitoring, glaciers in ocean (Padhy P.et al., 2005) and many more.

Military Applications

Wireless sensor networks are most suitable technology for activities related to security and surveillance purpose, as the sensors provide high spatial resolution and coverage. Such as perimeter and force protection, logistic information, exchange of critical information, command, intelligence, intrusion, monitoring sea and land area.

Table 1. Power density comparison from various energy scavenging technologies

Energy Harvesting Method	Power Density
Solar energy-Outdoor Solar energy-Indoor	15m 15mW/cm3—bright sunny day 0.15mW/cm3—cloudy day 10–100 μW/cm2
Vibrations (piezoelectric—shoe inserts) Vibrations (electrostatic conversion) Vibrations (electromagnetic conversion)	330 μW/cm3 0.021 μW/mm3—105Hz 306 μW/cm3—52Hz
Thermoelectric	5∘C gradient 40 μW/cm3
Wind flow	16.2 μW/cm3—5m/s
Acoustic noise	3 nW/cm3—75 dB 960 nW/cm3—100 dB
Magnetic field energy	130 μW/cm3—200 μT, 60Hz

Applications in Structural Maintenance

Structures such as buildings and bridges require regular inspection and maintenance in order to avoid any harmful incident. Sensors can indicate problems in advance this will reduce the cost of maintenance and failure.

Applications in Transport Management

Traffic congestion is a main problem in big cities all over the world. Intelligent transport management system consists of sensors help to prevent traffic congestion by synchronizing traffic lights and measuring vehicle flow.

Energy Harvesting Sources

Several methods can be used to harvest available free energies from the ambient environment. Table 1. (Zhou Get al., 2014) shows the comparison of power density of some of the potential energy scavenging technologies.

The power density expresses the captured energy per unit volume, area, or mass.

Common known energy sources include solar, magnetic, mechanical vibrations, thermal, wind and water. The captured energy can directly be used or stored in a storage device i.e. a capacitor or a battery depends on the output power impedance and magnitude, and requirement of the application.

Potential energy harvesting methods are discussed below:

1. Photonic Method

Generally, two methods are used for solar power extraction. First is solar thermal power generation which is suitable for large engineering systems and not applicable in WSNs for it is not available on small or nano scale (Tabbakh S. R. K. et al., 2010). The second method is to use photovoltaic solar cells and is suitable for WSNs. The energy harvesting by sun is cheaper and nonpolluting, but is not very efficient for dark areas, as this method require continuous light source. Therefore, it is not suitable for networks deployed in buildings and overcast areas. The other light such as fluorescent and infrared can also be used as the power source for solar cells (Tabbakh S. R. K. et al., 2010).

Photonic process is based on conversion of photons coming from indoor or outdoor light source. Photovoltaic cells are used to harness energy from solar or artificial light. Photovoltaic cells are solid state electrical device capable of converting solar/light energy into electrical energy by photovoltaic effect having the efficiency rate up to 30% (Stojcev M. K. et al., 2009).

Wireless sensor networks based on solar power source must consider following points

- Power supply requirements of the application
- Environmental parameters affecting the solar cells efficiency, such as solar radiation intensity, dense cloudy and snow weather, seasonal and climatically changes, latitude etc
- Angle and position at which the solar panel installed
- Characteristics and power features of the solar panel
- Chemistry and capacity of energy storage components used to store the harvested energy (Chalasani S. et al., 2008) (Morais R. et al., 2008) (Stojcev M. K. et al., 2009).

Broadly, we can divide solar cells into two categories

1. **Crystalline Solar cells:**
   a. **Monocrystalline Silicon Solar Cells:** Made up of pure silicon and considered as most efficient of all solar cells up to 30%
   b. **Polycrystalline Solar Cells:** First solar cells ever introduced to the industry, in 1981, they are less efficient than monocrystalline operated at 13-16% efficiency. Also, due to lower heat tolerance their efficiency decreased in high temperature.
2. **Thin Film Solar Cells:** Thin film solar cells are less efficient than Crystalline Solar cells and are usually suitable for smaller-scale applications, such as solar powered pocket calculators, travel lights, and camping gear used in remote locations. Thin- film solar cells consists of thin layers made of semiconductor material such as amorphous silicon (a-Si) and copper indium gallium deselenide (CIGS).The layers of semiconductor material used such as a-Si must be thin in order to absorb solar energy efficiently. On the other hand, sun exposure cause significant degradation in power output of these cells. This problem can be addressed by thinning a-Si layers, resulting less absorption of sunlight. Hence, they are not suitable for larger-scale applications.

Although photovoltaic energy harvesting is a mature technology, it can only generate power on availability of light. It is not a good choice if a network is deployed in overcast area.

Thermal Energy Harvesting

This method uses two techniques.

1. Thermo Energy Harvesting

In this method electricity is produced by temperature difference using thermoelectric generators. Thermoelectric generators based on Seebeck effect can generate electricity from heat flux, they are used for temperature sensing and to harvest waste heat. The performance of thermoelectric material is based on its electrical conductivity, thermal conductivity and the ability to produce voltage difference based on Seeback effect. There are many sources of waste heat which can be used around us, such as geothermal (Stokes C. D .et al. 2010), industrial waste heat, engine exhaust and the heat of sun (Smoker J.. et al., 2012).Energy can be harvested as long as difference in temperature is maintained and the temperature gradient is proportional to the efficiency of thermoelectric generator. Because it is difficult to maintain a significant temperature gradient on a small device (Yick J. et al. 2008), the small available temperature difference between the surfaces of the TEG, commonly no higher than 10◦C (Yong Y.-T. et al., 2009), and the micro size of the devices (Nacef A. B. et al., 2011)generate less output power resulting limited efficiency. Current range of conversion efficiency of thermoelectric generators is 6% to 11% and or the temperature gradient less than 40C, the thermoelectric generators are less efficient i.e. hardly 1%.Therefore, wireless sensor nodes based on thermoelectric harvesters are suitable in surroundings where temperature is high such as industry. They have wide application of converting waste heat energy of industrial processes into electricity. Using waste heat energy for energy harvesting helps to reduce pollution in environment, however due to lower rate of energy conversion and high cost their use is limited. Presently, TEGs are mainly used in high-tech fields such as medical treatment, military, submarines, and cosmic space (Zhou, G. et al., 2014).

2. Pyroelectric Energy Harvesting

Pyroelectric energy harvesting is efficient than thermoelectric energy harvesting. However, they require a varying temperature input which limits their application. In this process voltage is generated by cooling or heating pyroelectric materials. Charge can only be generated when the temperature of the material is changing; if temperature is constant charge will not be produced.

Energy Harvesting through Mechanical Vibration

Mechanical, vibration and kinetic energy can be generated by moving objects and is widely available all the time. These energies can be converted into electrical energy by using following mechanism.

1. Piezoelectric
2. Electrostatic
3. Electromagnetic

Depending on the type of force excitation vibration energy harvesting system converts the mechanical energy into electrical energy.Table2 (Zhou, G. et al., 2014) shows the vibration source and the corresponding energy harvesting method.

1. EH through Piezo

Piezo comes from Greek word "piezein" which means press. Mechanical energy from vibration, force or pressure can be converted into electricity by using piezoelectric materials. Piezoelectric materials are

Table 2. Energy Harvesting Methods corresponding to different vibration sources

Form of Vibration Source	Energy Harvesting Method
Force Excitation	Piezoelectric Energy Harvesting Method
Velocity Excitation	Electromagnetic Energy Harvesting Method
Displacement Excitation	Piezoelectric Electrostatic Energy Harvesting Method

crystals generally constructed of semiconductor materials such as Silicon and Germanium with added elements. They are capable of producing electricity when vibrated or compressed. Vibration based energy harvesting requires the frequency matching of source and harvesting device. Piezoelectric effect can be direct or indirect. For energy harvesting, direct piezoelectric effect is used. When a piezoelectric material is subject to mechanical stress it produces an electric charge. Piezoelectric devices have minimum moving parts and generate energy without an additional voltage source. The main disadvantage is piezoelectric materials are easily breakable and sometimes they allow current leakage. Several piezoelectric materials are available such as piezoelectric monocrystal, piezoelectric ceramics, piezoelectric polymers, and piezoelectric composites. Piezoelectric ceramic PZT is most commonly used material because of low cost, high efficiency rate and large electromechanical coupling constants. However, it is not capable of sustaining high strain, hence not suitable for the applications of high-frequency periodic load. An alternative is polyvinylidene fluoride (PVDF), although this piezoelectric material has smaller electromechanical coupling but offers more flexibility and higher mechanical strength which makes it suitable for high-frequency cyclic load.Figure2(Zhou, G. et al., 2014) shows two coupling modes of operation for piezoelectric materials. In thed33mode, the vibration force is applied in the same direction as the poling direction, while in thed31mode the force is applied perpendicular to the poling direction (SaadonS. et al., 2011).The d33 mode has a higher coupling coefficient than d31 mode, but d31 mode is able to produce a larger strain in a small external force at very low voltage source and small-sized devices compared to d33 mode, and it allows piezoelectric material to be strained on a bending cantilever (Lu F. et al., 2004), hence they

Piezoelectric transducer is one of the key structures in piezoelectric energy harvesting system and the selection of it had better be consistent with the environmental vibration so as to maximize the harvesting of mechanical energy. According to the different structures, piezoelectric transducers can be divided into unimorph, bimorph, rainbow type, cymbal type, moonie type, stack type, and so on (Erturk A.et al., 2009) (Erturk A. et al., 2010)(Saadon S. et al., 2011) (KaramiM.A.et al., 2011) (Li X. et al., 2011) (Palosaari J. et al., 2012).Table 3(Zhou, G. et al., 2014) shows the characteristics of different piezoelectric transducers.

The advantages of piezoelectric energy harvesting devices are simple structure, low cost, no fever, no electromagnetic interference, easiness to manufacture, relatively high output voltage level, electromechanical conversion efficiency, and so forth (Wang X. 2012).At present, the piezoelectric harvesters are capable of producing voltage from 2 to 10 V. In order to obtain better performance and improve the efficiency as well as practicability of piezoelectric energy harvester, the ongoing research mainly focus to improve the existing piezoelectric materials property, researching new piezoelectric materials with large piezoelectric constant strain, high electromechanical coupling coefficient and low loss, and improving the structure of piezoelectric energy harvester to increase the output power.

Figure 2. Illustration of (a) d31 mode; (b) d33 mode

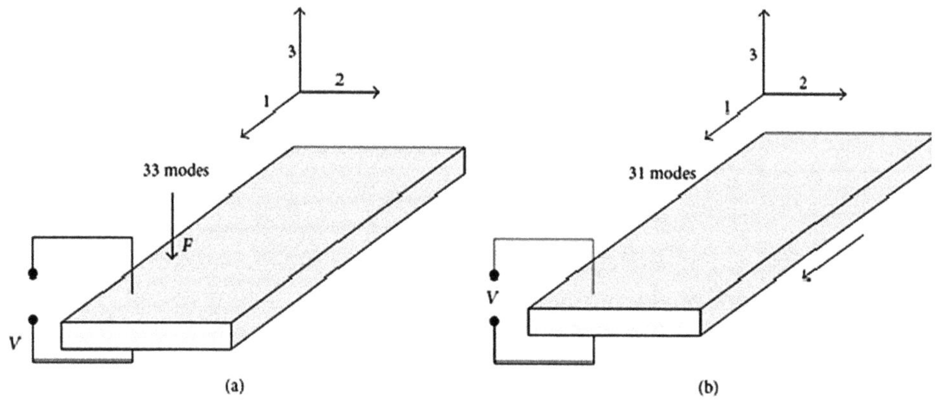

Table 3. Characteristics of different piezoelectric transducers

Piezoelectric Transducers	Characteristics
Unimorph	A highest power can be generated under lower excitation frequencies and load resistances, widely used in cantilever beam structure
Bimorph	Variable range of resonant frequency: 5Hz~100Hz, can be used in various configuration such as cantilever beam and end-end clamped beam
Rigid bimorph	Similar to bimorph but easier mounting, suitable for wide bandwidth, and widely used in cantilever beam structure
Macrofiber composite (MFC)	Flexible, both d31 and d33 modes practicable, suitable for low strain and high frequency application, large area coverage, and can be used as a bimorph component
Stack type	Suitable for the application of low frequency (~10Hz) and large uniaxial stress, easy Mounting
Rainbow type	Ability to generate a higher output voltage in the given stress level, can be stacked to amplify output voltage
Moonie and Cymbal type	Ability to withstand high stress, high output charge level, and adjustable resonance frequency

2. Electrostatic (Capacitive)Energy Harvesting

This technique based on changing the capacitance of a vibration dependent variable capacitor. Electrostatic harvesters can integrated with microelectronic devices but they need voltage source to charge the capacitor at initial stage.

3. Electromagnetic Energy Harvesting

This technique is based on Faraday's law of electromagnetic induction and generates electrical energy from mechanical energy by using inductive spring mass system. In this technique there is no mechanical contact between the parts of a separate voltage source, which results improvement in the reliability and reduction in the mechanical damping. However, they are not suitable for sensor nodes because of their larger size.

Wireless Energy Harvesting

Wireless energy harvesting can be divided broadly into two categories:

1. RF energy harvesting

Electromagnetic radiation is commonly available in environment today, and is a good option for continuous low power energy harvesting. In this method electromagnetic signals can be converted into electric signals by using rectenna. RF energy harvesting devices can completely replace the battery from sensor node. It has a wide application in military, agriculture, industry etc.

2. Resonant energy harvesting

This technique based on resonant inductive coupling. In this method electrical energy is transferred and captured between two highly resonant coils at the same frequency. Generally, two implementations are possible for resonant inductive coupling, weak and strong inductive coupling. In weak inductive coupling coils are distance apart by few centimeters only, if the receiving and external powered coil are matched strong coupling is possible between electromagnetic resonant devices for longer distance. Since both the coils have no physical connection, resonant inductive coupling is considered as a wireless energy harvesting technique. Some implementations for WSNs are discussed in (Reinisch H. et al., 2011) (Mandal S. et al., 2010) (Heer R. et al., 2010).

Dynamic Fluid Energy Harvesting

Dynamic fluid energy refers to the power of air and flowing water. Two methods can be used to harvest kinetic energy of wind or flowing water. The first method uses mechanical parts i.e. miniature wind turbines (Fei F. et al., 2012)and the second is based on non mechanical parts that works on the similar principal but induces mechanical vibration which can be converted into electricity by piezoelectric (Wang D.-A. et al., 2010)(Pobering S. et al., 2009), electrostatic (Bu L. et al., 2011), or electromagnetic principles (Wang D.-A. et al., 2010).

Harvesting Wind Energy

At present, four methods are in use to harvest wind energy, miniature wind turbines, piezoelectric wind harvesters, miniature wind belt and electromagnetic wind generators. Traditional wind turbines are the mature technique for harvesting wind energy, even though large-scale wind turbines are efficient; however small-scale wind turbines are less efficient because of high viscous drag on the blades at low Reynolds numbers (Matova S.P. et al., 2011).The micro wind belt generator has a higher output power at high wind speeds (>5.5m/s), but its output power reduces significantly at low wind speeds (<3.5 m/s),and the wind belt generator can be extremely noisy during Operation(Zhu D. et al., 2010).Wind energy harvesters made of piezoelectric materials are efficient, smaller in size, light weight, start instantly without any dead time, having compact structure, sensitive to low speed of wind. They almost have no heat dissipation and showing extremely low magnetic permeability(Tan Y. K. et al., 2007). However, they may generate very low output power due to low strain coefficient. Another limitation is piezoelectric material must vibrate at their resonant frequency for the generation of maximum power. The electromagnetic wind generators are reliable and have small mechanical damping, but the heavy magnet will result in the generator being not able to operate at low wind speed (Ramasur D. et al., 2012).Tan and Panda presented a piezoelectric wind harvester and utilized the bimorph piezoelectric actuator for a low power autonomous wind speed sensor (Tan Y. K. et al., 2007). Comparing with other wind mills, this harvester has smaller size and it is lighter in weight, but the harvesting power is of lesser magnitude and also the network is not capable of handling unexpected events as the duty cycle of sensor is too small. Hence, it is suitable for WSNs having insensitive delay and handling smaller amounts of information. Some examples of wireless sensor networks implemented with wind energy harvesting systems can be found in (Fei F. et al.,2012) (Sardini E. et al., 2011) (Tan Y. K. et al., 2011).Wind energy is infinitely available everywhere, it is non polluting and renewable. However, having low energy density, regional differences, and unstable factor for the wind speed is fluctuant and unpredictable which will resulting in equable electricity all the time (Ucar A. et al., 2009).Moreover, the wind harvester can be noisy as it has moving mechanical part (Wan Z. G. et al., 2011). Therefore, research is needed to find techniques for maximizing wind energy capture.

Harvesting Energy from Flowing Water

The flowing water contains kinetic energy due to the water pressure fluctuation, and it can be converted into electrical energy by energy harvesters (Wang D.-A. et al., 2011). Like wind flowing water is also renewable, pollution-free, continuous, and dependable energy source for wireless sensor nodes (Pobering S. et al., 2004). For hydropower can be developed in any size and any scale, it is applicable in WSNs (Tabbakh S. R. K. et al., 2010).Energy harvesting from flowing water is used for the WSNs implemented to monitor the water quality and hydraulic state (Ye G. et al., 2012)of rivers, canals and other water distribution systems. Flowing water is capable of generating enough power for WSNs, however the big size of the water turbine is a big challenge to install it in pipelines and integrate with the sensor nodes. While the flow-induced vibration energy harvester has a micro size, it needs further research to increase the conversion efficiency and optimize the generated power to be appropriate for powering wireless sensor nodes (Mohamed M. I. et al., 2011).

Acoustic Energy Harvesting

Sound is available and generated abundantly in environment such as noise of jet engines, train station; loud subways, during industrial processes, hydraulic systems, machine shops and even a sound produced by a crying baby can be harvested to power the sensor nodes. For this purpose, an acoustic Energy device is used to convert the noise into electrical energy. Acoustic Energy Harvesting device comprises of three main components a resonator, which oscillates at resonating frequency to produce mechanical oscillations (sound), a membrane capturing energy is attached to the one end of the resonator and a storage device. The membrane is connected to a piezoelectric material that converts the vibration caused by sound into electrical energy which is then stored by a storage device with signal conditioning circuit (Zhou, G. et al., 2014).An acoustic energy harvesting will not only satisfy the power demand of wireless sensors, but also improve the environmental conditions by reducing the noise (Filip N. et al., 2011) (Hua W.S. et al., 2014).

Hybrid Energy Harvesting

The amount of energy from a single energy harvester is typically small (Seah W. K. G. et al. 2009)and highly variable with time, its location, and working conditions of load (Li L. et al., 2008) (Li. Yet al., 2008),hence depending on a one energy source for some particular application may not help to keep the node functional in case of short supply of energy from environment. Many types of energy sources are available in the environment simultaneously. In order to make each sensor node obtain the required energy from environment as much as possible, we may need to design such a kind of energy harvesting system that it cannot only harvest energy from one type of power source (Shang Y. et al., 2007).Therefore, it is necessary to integrate a variety of energy harvesting modules in one sensor node (Silva R. et al., 2011).

HOW ENERGY HARVESTING WORKS

Energy is available abundantly in ambient environment; usually this available free energy reserve is remaining unused or wasted. Such as mechanical energy generating from vibration, stress and strain, thermal energy due to heat escaped from furnaces, combustion engines, biological, solar energy from all types of light sources; electromagnetic energy captured via inductors, coils and transformers; wind and fluid energy resulting from air and liquid flow; chemical energy from naturally recurring or biological processes; and RF energy due to radio transmitters and television broadcasting. The available sources provide energy in a form of very small packets and have unique characteristics depending on the application .Energy-harvesting devices scavenge and store these small energy packets and generate electricity for sensor nodes. Current energy harvesting circuits in use are designed particularly for the requirements of specific application.

However, in general a good harvesting system should possess following characteristics

- Harvesting circuit power consumption should be less than the amount of energy extracted from the surroundings.
- It must be in permanent active mode to capture available energy and capable of providing proper conditioned output as per application requirement.
- Harvester circuit must be capable to retain harvested energy for long periods of time with minimal leakage or loss.
- The energy management must be well-defined — and tolerate a wide range of voltage, current and waveform inputs, including over-voltage, overcharge and other irregular input conditions.

The energy harvesting system consists of two main parts:

Figure 3. Block diagram of energy harvesting system

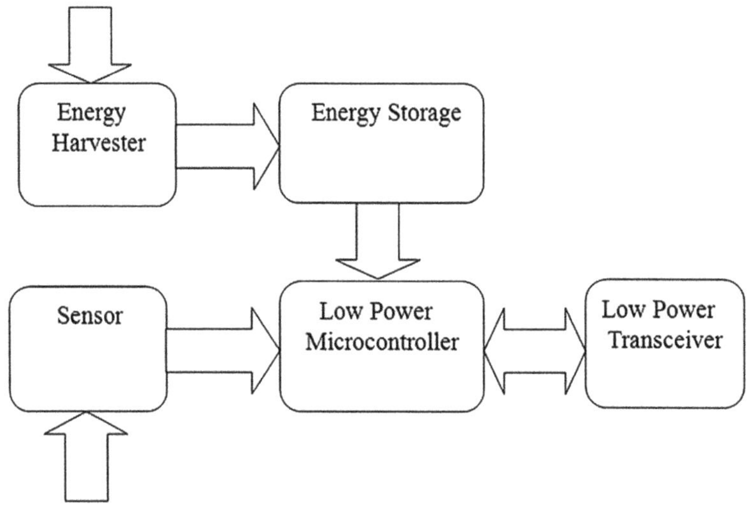

- Conversion of the ambient energy from the surrounding
- storing of captured energy for later use

Typically, energy-harvesting systems contain similar components and are ideal for sensor network applications. An MCU is responsible for the processing, sensing, and communication activities. The MCU can be integrated to any number of sensors to collect data from surroundings and will usually send or receive data via a wireless transceiver. A typical block diagram is shown in Figure 3.When system is in a state of collecting energy, the MCU, sensors and communication systems are put in a standby mode which is helpful to minimize their drain on the available power.

An energy buffer is used to store the power for the rest of the application. Size and technical specification of the buffer depends on the nature of the application. A large buffer is required if the application is discontinue to access available energy source for a long period of time. However, a small buffer will be enough if the application is adjacent to energy source and does not need to be active frequently especially in case of low duty-cycle applications. An ideal energy buffer must have efficient energy acceptance and delivery, negligible leakage and volume with unlimited capacity.

A typical energy harvester circuit consists of at least one transducer, a power conditioning circuit, and an energy storage unit. All these components work together to capture energy and deliver power to the sensor node. In general, transducer is a device which converts energy from one form to another, when it is used to convert any measureable quantity such as sound, pressure etc. to an electrical current or voltage it is termed as a sensor. Due to random and uncontrolled input energy, transducer may generate irregular output and some sensors may be very sensitive to the change of voltage supply, then the output voltage needs to be regulated by DC-DC converter (Raghunathan V. et al., 2005). Hence the power conditioning circuit is indispensable. A specialized DC-DC converter microchip takes in power from the transducer and outputs typically 3 to 5V, which can then be stored or used. Nano power comparators are used to support power consumption; they control and monitor the minimum and maximum limits of the input energy and the energy storage. The additional regulation may also be required between sensor and energy storage device to ensure that the sensor node works with a stable and safe voltage (Zhou, G et al., 2014).

Thus, energy harvesting process can be summarized as:

1. Detect energy source and harvest energy.
2. Harvested energy converted into electrical energy and saved in a storage device for example capacitor.
3. Saved electrical energy used by MCU and sensor to perform sensing, processing and communication activities.
4. Data delivered from the sensor through wireless transceiver.

CONCLUSION

Wireless Sensor Networks perform well in harsh environments where human intervention is almost difficult or not possible. However, they are highly resource constraints networks, advancement in low power electronics helped a lot but the use and maintenance of conventional batteries with a limited life span cannot address the power supply problem effectively for longer run. Therefore, harvesting energy from ambient environment is an effective alternative both in terms of power and cost, which can help sensor networks to live longer. Harvesting energy from an ambient environment is the promising technology to prolong the life of sensor node and most likely in future will replace the batteries to make WSNs independent of power supply.This chapter gives a brief intro to various possible ambient energy harvesting technologies and the selection of which should be suitable for the particular applications and working environments of WSNs. Besides of that it also enlightens different mechanism of harvesting energyto achieve the ultimate goal of permanent use of wireless sensors and expending the use of WSN application.

REFERENCES

Alippi, C., Anastasi, G., Di Francesco, M., & Roveri, M. (2009). Energy Management in Wireless Sensor Networks with Energy-hugry Sensors. *IEEE Instrumentation & Measurement Magazine*, *12*(2), 16–23. doi:10.1109/MIM.2009.4811133

Bu, L., Wu, X. M., Wang, X. H., & Liu, L. T. (2011). Liquid encapsulated energy harvester for low frequency vibrations. In *Proceedings of the 16th International Solid-State Sensors, Actuators and Microsystems Conference (TRANSDUCERS '11)*. doi:10.1109/TRANSDUCERS.2011.5969863

Chalasani, S., & Conrad, J. M. (2008). A survey of energy harvesting sources for embedded systems. In *Proceedings of IEEE Southeastcon.*

Dash, S., Swain, A. R., & Ajay, A. (2012). Reliable energy aware multi token based MAC protocol for WSN. In *Proceedings of the26th IEEE International Conference on Advanced Information Networking and Applications (AINA '12)*. doi:10.1109/AINA.2012.61

Erturk, A., & Inman, D. J. (2009). An experimentally validated bimorph cantilever model for piezoelectric energy harvestingfrom base excitations. Iopscience Smart Materials and Structures, 18(2).

Erturk, A., Vieira, W. G. R., DeMarqui, C. Jr, & Inman, D. J. (2010). On the energy harvesting potential of piezoaeroelastic systems. *Applied Physics Letters*, *96*(18), 184103. doi:10.1063/1.3427405

Filip, N., & Simu, I. (2011). About the noise energy conversion from agriculture tractor engines. In *Proceedings of the 39thInternational Symposiumon Agricultural Engineering.*

Heer, R., Wissenwasser, J., Milnera, M., Farmer, L., Opfner C. H, & Vellekoop, M. (2010). Wireless powered electronic sensors for biological applications. In Proceedings of IEEE EMBC 2010.

Hi-globe. (n.d.). Retrieved September 15, 2015, from http://hi-globe.com/?page_id=85

Hua, W. S., Li-Dong, D., De-Yi, K., Hao-Yue, P., Zhen, F., & Zhan, Z. (2014). Hybrid device for acoustic noise reduction and energy harvesting based on a silicon micro-perforated panel structure. *Chinese Physics B, 23*(4), 044302. doi:10.1088/1674-1056/23/4/044302

Karami, M. A., Bilgen, O., Inman, D. J., & Friswell, M. I. (2011). Experimental and analytical parametric study of single-crystal unimorph beams for vibration energy harvesting. *IEEE Transactions on Ultrasonics, Ferroelectrics, and Frequency Control, 58*(7), 1508–1520. doi:10.1109/TUFFC.2011.1969 PMID:21768034

Li, L., Zhang, Y., San, H., Guo, Y. I., & Chen, X. (2008). MEMS for vibration energy harvesting. In *MEMS/MOEMS Technologiesand Applications III, Proceedings of the SPIE.*

Li, X., Guo, M., & Dong, S. (2011). A flex-compressive-mode piezoelectric transducer for mechanical vibration/strain energy harvesting. *IEEE Transactions on Ultrasonics, Ferroelectrics, and Frequency Control, 58*(4), 698–703.

Li, Y., Yu, H., Su, B., & Shang, Y. (2008). Hybrid micropower source for wireless sensor network. *IEEE Sensors Journal, 8*(6), 678–681. doi:10.1109/JSEN.2008.922692

Lu, C., Park, S. P., Raghunathan, V., & Roy, K. (2010). Analysis and design of ultra low power thermoelectric energy harvesting systems. In *Proceedings of the ACM/IEEE International Symposiumon Low-Power Electronics and Design (ISLPED '10).* doi:10.1145/1840845.1840882

Lu, F., Lee, H. P., & Lim, S. P. (2004). Modeling and analysis of micro piezoelectric power generators for micro-electromechanical systems applications. *Smart Materials and Structures, 13*(1), 57–63. doi:10.1088/0964-1726/13/1/007

Mandal S., Turicchia L., &Sarpeshkar R. (2010). A low-power, battery-free tag for body sensor networks. *IEEE Pervasive Computing, 9*(1), 71-77.

Matova S. P., Elfrink R., Vullers R. J. M., & Schaijk R. V. (2011). Harvesting energy from airow with a michromachined piezoelectric harvester inside a Helmholtzresonator. *Journal of Micromechanics and Microengineering, 21*(10), 1-6.

MIT Technology Review. (2003). *10 Emerging Technologies That Will Change the World.* Retrieved September 15, 2015 from http://www2.technologyreview.com/featured-story/401775/10-emerging-technologies-that-will-change-the/2/

Mohamed, M. I., Wu, W. Y., & Moniri, M. (2011). Power harvesting for smart sensor networks in monitoring water distribution system. In *Proceedings of the International Conference onNetworking, Sensing and Control (ICNSC '11).* doi:10.1109/ICNSC.2011.5874897

Morais, R., Matos, S. G., Fernandes, M. A., Valente, A. L. G., Soares, S. F. S. P., Ferreirac, P. J. S. G., & Reis, M. J. C. S. (2008). Sun, wind and water flow as energy supply for small stationary data acquisition platforms. *Computers and Electronics in Agriculture*, *64*(2), 120–132. doi:10.1016/j.compag.2008.04.005

Nacef, A. B., Senouci, S.-M., Doudane, Y. G., & Beylot, A.-L. (2011). A cooperative low power Mac protocol for wireless sensor networks. In *Proceedings of the IEEE InternationalConference on Communications (ICC '11)*. doi:10.1109/icc.2011.5962416

Padhy, P., Martinez, K., Riddoch, A., Hart, J., & Ong, H. (2005). Glacialenvironmentmonitoring using sensor networks. In *Proceedings of the Conference on Real-World Wireless Sensor Networks*.

Palosaari, J., Leinonen, M., Hannu, J., Juuti, J., & Jantunen, H. (2012). Energy harvesting with a cymbal type piezoelectric transducer from low frequency compression. *Journal of Electroceramics*, *28*(4), 214–219. doi:10.1007/s10832-012-9713-8

Pobering, S., & Schwesinger, N. (2004). A novel hydropower harvesting device. In *Proceedings of the International Conference onMEMS, NANO and Smart Systems (ICMENS '04)*.

Pobering, S., & Schwesinger, N. (2009). Power supply for wireless sensor systems. In *Proceedings of the IEEE Sensors Conference(SENSORS '08)*.

Raghunathan, V., Kansal, A., Hsu, J., Friedman, J., & Srivastava, M. (2005). Design considerations for solar energy harvesting wireless embedded systems. In *Proceedings of the 4th InternationalSymposium on Information Processing in Sensor Networks(IPSN '05)*. doi:10.1109/IPSN.2005.1440973

Ramasur, D., & Hancke, G. P. (2012). A wind energy harvester for low power wireless sensor networks. In *Proceedings of the IEEE International Instrumentation and Measurement Technology Conference (I2MTC '12)*. doi:10.1109/I2MTC.2012.6229698

Reinisch H., Gruber S., Unterassinger H., Wiessecker M., Hofer G., Pribyl W., & Holweg G. (2011). An electro-magnetic energy harvesting system with 190 nW idle mode power consumption for a BAW based wireless sensor node. *IEEE Journal of Solid-State Circuits, 46*(7), 1728-1741.

Saadon, S., & Sidek, O. (2011). Vibration-based MEMS piezoelectric energy harvester (VMPEH) modeling and analysis for green energy source. In *Proceedings of the 4th International Conferenceon Developments in eSystems Engineering (DeSE '11)*. doi:10.1109/DeSE.2011.85

Saadon, S., & Sidek, O. (2011). A review of vibration-based MEMS piezoelectric energy harvesters. *Energy Conversion and Management*, *52*(1), 500–504. doi:10.1016/j.enconman.2010.07.024

Sardini, E., & Serpelloni, M. (2011). Self-powered wireless sensor for air temperature and velocity measurements with energy harvesting capability. *IEEE Transactions on Instrumentation and Measurement, 60*(5), 1838-1844.

Seah, W. K. G., Zhi, A. E., & Tan, H.-P. (2009). Wireless Sensor Networks Powered by Ambient Energy Harvesting (WSNHEAP)—survey and challenges. In *Proceedings of the 1stInternational Conference on Wireless Communication, Vehicular Technology, Information Theory and Aerospace and Electronic Systems Technology, Wireless (VITAE '09)*.

Shang, Y., Li, Y., Yu, H., Sun, H., & Su, B. (2007). Experimental investigation on thermoelectric generator of micro hybrid power source. In *Proceedings of the 3rd International Symposiumon Advanced Optical Manufacturing and Testing Technologies: Design, Manufacturing, and Testing of Micro- and Nano-OpticalDevices and Systems, Proceedings of the SPIE.*

Silva, R., Farinholt, K., & Park, G. (2011). Hybrid energy sources for embedded sensor nodes. In *Proceedings of the Industrialand Commercial Applications of Smart Structures Technologies.*

Smoker, J., Nouh, M., Aldraihem, O., & Baz, A. (2012). Energy harvesting from a standing wave thermoacoustic-piezoelectric resonator. *Journal ofApplied Physics, 111*(10), 104901. doi:10.1063/1.4712630

Stojcev, M. K., Kosanovic, M. R., & Golubovic, L. R. (2009). Power management and energy harvesting techniques for wireless sensor nodes. In *Proceedings of the 9th International Conference onTelecommunication in Modern Satellite, Cable, and Broadcasting Services (TELSIKS '09).* doi:10.1109/TELSKS.2009.5339410

Stokes, C. D., Duff, A., Mantini, M. J., Grant, B. A., & Venkatasubramanian, R. (2010). Nanostructured thermoelectric material and device technology for energy harvesting applications. In *Proceedings of the 4th IEEE Nanotechnology Materials and Devices Conference (NMDC '10).* doi:10.1109/NMDC.2010.5652061

Tabbakh, S. R. K., Maarefdoust, R., Kyun, N. C., & Ali, B. M. (2010). Environmental taxonomy of power scavenging techniques for autonomous self powered wireless sensors. In *Proceedings of theAsia Pacific Conference on Circuit and System (APCCAS '10).*

Tan, Y. K., & Panda, S. K. (2007). A novel piezoelectric based wind energy harvester for low-power autonomous wind speedsensor. In *Proceedings of the 33rd Annual Conference of the IEEE Industrial Electronics Society (IECON '07).*

Tan, Y. K., & Panda, S. K. (2011). Energy harvesting from hybrid indoor ambient light and thermal energy sources for enhanced performance of wireless sensor nodes. *IEEE Transactions on Industrial Electronics, 58*(9), 4424-4435.

Tan, Y. K., & Panda, S. K. (2011). Self-autonomous wireless sensor nodes with wind energy harvesting for remote sensing of wind-driven wild re spread. *IEEE Transactions on Instrumentation and Measurement, 60*(4), 1367-1377.

Ucar, A., & Balo, F. (2009). Investigation of wind characteristics and assessment of wind generation potentiality in Uludaˇg-Bursa, Turkey. *Applied Energy, 86*(3), 333–339. doi:10.1016/j.apenergy.2008.05.001

Vullers, R. J. M., Schaijk, R. V., Doms, I., Hoof, C. V., & Mertens, R. (2009). Micropower energy harvesting. *Solid-State Electronics, 53*(7), 684–693. doi:10.1016/j.sse.2008.12.011

Wan, Z. G., Tan, Y. K., & Yuen, C. (2011). Review on energy harvesting and energy management for sustainable wirelesssensor networks. In *Proceedings of the IEEE 13th International Conference on Communication Technology (ICCT '11).*

Wang, D.-A., & Chang, K.-H. (2010). Electromagnetic energy harvesting from flow induced vibration. *Microelectronics Journal, 41*(6), 356–364. doi:10.1016/j.mejo.2010.04.005

Wang, D.-A., & Ko, H.-H. (2010). Piezoelectric energy harvesting from flow-induced vibration. *Journal of Micromechanics and Microengineering*, *20*(2), ID025019. doi:10.1088/0960-1317/20/2/025019

Wang, D.-A., & Liu, N.-Z. (2011). A shear mode piezoelectric energy harvester based on a pressurized water flow. *Sensors andActuators, A. Physical*, *167*(2), 449–458.

Wang, X. (2012). Piezoelectric nanogenerators-Harvesting ambient mechanical energy at the nanometer scale. *Nano Energy*, *1*(1), 13–24. doi:10.1016/j.nanoen.2011.09.001

Ye, G., & Soga, K. (2012). Energy harvesting from water distribution systems. *Journal of Energy Engineering*, *138*(1), 7–17. doi:10.1061/(ASCE)EY.1943-7897.0000057

Yick, J., Mukherjee, B., & Ghosal, D. (2008). Wireless sensor network survey. *Computer Networks*, *52*(12), 2292–2330. doi:10.1016/j.comnet.2008.04.002

Yong, Y.-T., Chow, C.-O., Kanesan, J., & Ishii, H. (2009). A survey on design of self-powered wireless sensor network. In *Proceedings of the International Conference for Technical Postgraduates(TECHPOS '09)*. doi:10.1109/TECHPOS.2009.5412044

Zhou, G., Huang, L., Li, W., & Zhu, Z. (2014). Harvesting Ambient Environmental Energy for Wireless Sensor Networks: A Survey. *Journal of Sensors*, 815467.

Zhou, H.-Y., Wu, F., & Hou, K.-M. (2008). An event-driven multithreadingreal-time operating system dedicated to wirelesssensor networks. In *Proceedings of the International Conferenceon Embedded Software and Systems (ICESS '08)*.

Zhu, D., Beeby, S., Tudor, J., White, N., & Harris, N. (2010). A novel miniature wind generator for wireless sensing applications. In *Proceedings of the 9th IEEE Sensors Conference (SENSORS '10)*.

Chapter 14
Exploring Current Trends of Energy Harvesting

Shakeel Ahmed
King Faisal University, Saudi Arabia

ABSTRACT

There are numerous forms of energy that can gratify the needs of computing, these energies include solar, thermal, mechanical, wind, acoustic, and wave for different devices and applications. With ubiquitous computing requirements for self-powered devices such as MEMS (Micro-electromechanical systems) and these devices have narrow capacity of finite power sources and it is needed that these systems are needed to be supplied with energy for the lifetime of the system. Mostly Batteries are the main source of energy for many of these mobile systems, embedded systems and remote system applications. There is a requirement for Energy harvesting which is a method of extracting energy from the nearby environment called Environmental energy which can be used as supernumerary to electrochemical battery. Studies are being conducted to harvest energy from solar, wind, thermal, tidal and other mechanical sources which have limited lifespan. This paper serves as a study to identify various sources of energy available for harvesting centered on various technical papers available and present the work carried out by investigators in identifying them.

INTRODUCTION

With the widespread improvements in the field of MEMS (Micro-electromechanical systems) with tiny mechanical devices such as sensors, valves, gears, mirrors, and actuators embedded in semiconductor chips and as sensors are required in these devices and these devices need to have a long lifespan. In MEMS using Batteries are not always beneficial since they need human involvement to exchange them. Hence, obtaining the electrical power desired to operate these devices is the main worry. To overcome this alternate type of energy source to conventional batteries must be well-thought-out. Energy utilization is swiftly growing day by day and hence novel and proficient energy resources are need to be discovered. Researchers are discovering new ways of utilizing abundant renewable energy from different sources and it is an ongoing research topic.

DOI: 10.4018/978-1-4666-9792-8.ch014

Table 1. Energy Harvesting Sources (Atwood B, et al 2001)

Harvesting Technique	Power Density
Thermoelectric	$40\mu W/cm^3$
Vibration	$116\mu W/cm^3$
Piezoelectric	$330\mu W/cm^3$
Solar Cells	$15mW/cm^3$

The various kinds of typical ambient energies are thermal energy, sunlight, RF energy and mechanical energy. The energy harvesting sources can be used to surge the lifetime and capability and to run these sensor devices by moreover switching or increasing the battery usage (Beeby et al., 2006; Kansal et al., 2003; Kansal et al., 2004; Kansal et al., 2005; Meninger, 1999; Rahimi et al., 2003). The energy that can be obtained from different energy sources by selecting the thermal, light, or mechanical energies accessible in the surrounding locations. This process aids in providing unlimited energy for the lifetime of the electronic device. Hence, the process of extracting energy from the ambient environment and altering it into expendable electrical energy is identified as energy harvesting or power scavenging. The energy obtained from numerous sources and used in growing technological demand. Energy harvesting is capturing energy from environment and storing the energy for gathering the imminent needs. Most popular several energy harvesting traditional mechanisms are such as energy produces form wind farms, energy generated from solar farms, energy generated from geothermal farms, tidal energy farms and so on. Advancement in technology has improved the way by which these sources can be used to power electronic systems. All stated popular energy harvesting mechanisms have been well established which leads to the development of self-powered systems which require very less maintenance and can be deployed in unreachable locations anywhere easily. The main target of energy harvesting is to power this self-powered system.

The classification of energy harvesting is made on the base of the method of energy scavenging the power. Renewable sources of energy comprise are characterized on the basis of the numerous sources which are Solar energy, tidal energy wind energy, and thermal energy for energy harvesting which contain from photovoltaic cells, wind turbines, thermoelectric generators and mechanical vibration devices such as piezoelectric devices, electromagnetic devices. However different methods impose varied restrictions on the working environment of electronic devices based on the type of energy they belong to and the source of energy they gain power from unlike all these sources of energy, mechanical energy can be found in examples when others become inadequate. The different source of mechanical energy can be gained vibrations, heat, motion, noise, flow induced vibrations etc., Mechanical energy can be used to power electronic systems by harvesting energy from it. Piezoelectric materials, pyro electric materials, electrostatic materials are used to convert the mechanical energy into useful energy form. For example piezoelectric harvesting devices scavenge into operational electrical energy from mechanical energy. Table 1 shows several harvesting methods by their power generation capability (Atwood et al., 2001).

The wide-range of properties to be considered to describe a movable energy supplier is described (Fry et al., 1997). They different properties that include are the physical, electrical and environmental properties, electrical properties like maximum voltage and current, power density, physical properties such as size, shape and weight; environmental properties such as water resistance and operating temperature range; as well as operational and maintenance properties are to be considered. The important and necessary protection measures must be taken when using the energy harvesters in the embedded systems to increase the performance and lifetime of the system.

Mostly the embedded systems to increase the energy for a longer time depend on the on multiple power sources (MPS). The embedded systems harvest energy for the ambient sources such as solar power or by using chargeable batteries. The energy harvested from these battery sources is utilized to power the system and the rechargeable battery (Park & Chou, 2004).

ENERGY HARVESTING ARCHITECTURES

Generally, energy harvesting can be divided into two architectures— (i) Harvest-Use: Energy is harvested just-in-time for use and (ii) Harvest-Store-Use: Energy is harvested whenever possible and stored for future use.

Harvest-Use Architecture

Figure 1 (a) shows the Harvest-Use architecture. In this architecture, the harvesting system directly powers the sensor node and thus results in the node to be operational; the power output generated from the harvesting system has to be continuously above the minimum operating point. If adequate energy is not available, the node will be disabled. Abrupt variations in harvesting capacity close to the minimum power point will cause the sensor node to oscillate in ON and OFF states. A Harvest-Use system can be built to use mechanical energy sources like pushing keys/buttons, walking, pedaling, etc. For example, the push of a key/button can be used to deform a piezoelectric material, thereby generating electrical energy to send a short wireless message. Similarly, piezoelectric materials strategically placed within a shoe may deform to different extents while walking and running. The harvested energy can be used to transmit RFID signals, used to track the shoe-wearer.

Harvest-Store-Use Architecture

Figure 1 (b) illustrates the Harvest-Store-Use architecture. The architecture consists of a storage component that stores harvested energy and also powers the sensor node. Energy storage is useful when the

Figure 1. Energy harvesting architecture with and without storage capability

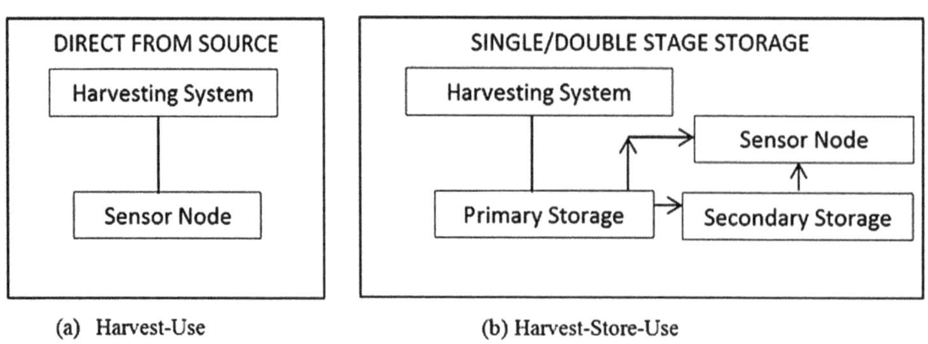

harvested energy available is more than the current usage. Excess energy is stored for later use when either harvesting opportunity does not exist or energy usage of the sensor node has to be increased to improve capability and performance parameters. The storage component itself may be single-stage or double-stage. Secondary storage is a backup storage for situations when the Primary storage is exhausted. As an example, a Harvest-Store-Use system can use uncontrolled but predictable energy sources like solar energy. During the daytime, energy is used for work and also stored for later use. During night, the stored energy is conservatively used to power the sensor node.

SOURCES OF ENERGY HARVESTING

Solar Energy Harvesting

The energy derived from the Sun is defined as solar energy which is joined by several technologies like that of photovoltaic panels and solar thermal collectors. Solar power is the amendment of sunlight into electricity using photovoltaic (PV) or concentrated solar power (CSP) moreover directly or indirectly respectively. PV uses photoelectric effect to transform light into electric current. In CSP sunlight in big space is concentrated into lesser beam using lenses or mirrors.

Photovoltaic Energy Harvesting

In photovoltaic technique, semiconductors displaying photoelectric effect are used to acquire direct current electricity from solar radiations. Photoelectric effect is generated by using Photovoltaic cell or solar cells that convert light into electric current. The novel PV technology Dye Sensitized Solar Cell is used to harvest energy as dyes absorb light like to that of photosynthesis process in plants using chlorophyll. The device which is used to convert light energy in to electrical energy is a photovoltaic cell. The form of energy is characteristically light energy gained usually from sunlight.

Photovoltaic cells are a convenient solution for places where the accessibility of light and usage of batteries and further means of power supply are not possible or expensive. Examples include marine locations and roadway signs. Various factors are to be considered during designing sources which are generally scavenge solar energy, are availability of day light, effects of operation at higher latitudes, snow cover, periods of dense cloud and characteristics of the photovoltaic cell used, the power supply requests, intensity of the incident light, are to be considered (Callaway et al., 2004; Raghunathan et al., 2005). Silicon-based cells are the most prevalent photovoltaic cells which are more subtle to light, and are easily available which offer a realistic price to performance ratio.

Concentrated Solar Power Harvesting

With the usage of lenses or mirrors the Sunlight from a large area is concentrated into small beam, which could be additionally used as a heat source for conventional power plant. Numerous techniques like solar power tower, Stirling dish, Fresnel reflector are used to track the sun light and focus light. The working fluid in Concentrated Solar Power (CSP) systems is produced power by heating with the help of concentrated sunlight.

Wind Energy Harvesting

For more than a century wind power has been considered as a feasible source of "free" energy. Since the early 1980s Wind turbines have been a very important player as a renewable energy. Wind power covers widespread variety of applications and can be coupled by large wind turbine farms providing up to 800 MW of power, and small domestic wind turbines providing 3 kW to a home. Wind turbine generator transforms the mechanical energy into electrical energy. These Wind turbine generators are a bit infrequent, compared to other producing units usually found attached to the electrical grid, in that the generator has to work with a power source which deliveries fluctuating mechanical power.

Thermal Energy Harvesting

The process of converting heat energy into electricity with very high efficiency process is known as thermionic conversion. Investigators have been trying for many long years to develop a beneficial thermionic generator since the thermal energy harvesting has projecting results. Heat is generated by steam, and by using this steam the turbines spin and these turbines drive the generator, which harvests electricity. This organization requires a reasonable volume of equipment as well as space is very common. Thermal Energy (TE) energy-harvesting system takes benefit of slight temperature difference amid its two surfaces. Temperature pitches are all over the place and Current is produced when there is a temperature difference among two junctions of a conducting material. Thermal energy harvesting uses temperature differences or gradients to generate electricity the essential electrical energy is generated by thermoelectric generators which follow the principle of thermo electricity. The concept of creating electric potential with a temperature difference and vice-versa can be termed as thermoelectricity. Here, the thermal energy is foraged to obtain electrical energy to power the electronic devices. Thermoelectric devices are mainly used in space and earthly applications.

Bio-Energy Harvesting

The energy derived from material of the biological sources is called Bio-energy and is renewable energy. Chemical energy in the form of Biomass which is an organic material. The patent of Voltree Power's for bio-energy harvester changes living plant metabolic energy to operational electricity, providing exclusive battery replacement alternative for ultra-low power sensors and ultimately a power source in nanotechnology. These naturally-occurring occurrences can serve as a power source in environmental sensing networks for climatic and wild life observation in areas with limited accessibility and where battery replacement is not a possibility. Future applications that are depended on Self-sustaining power sources such as tree power will play acute role. An alternate of bio-Energy harvesting is using of trees which can be used to power circuits, but as found out the voltage generated by these trees to too small to charge batteries.

Tidal Energy Harvesting

Oceans are considered as the best source of potential energy. The energy is transmitted in the form of waves, currents and tides of which certain amount of the energy can be used to perform work. The energy is obtained by taking the energy from the moving water mass due to tides from the tidal power. Tides can generate energy in different forms and types, two types of tidal energy are popular where energy can be extracted include kinetic energy and potential energy which are achieved by current between ebbing and surging tides and potential energy of currents between high and low tides simultaneously. Producing energy through tidal is easy and possible enough, it is comparable with generating energy through water by building different dams and other mechanisms. Currently this source is considered as a common and cheap source, numerous coastal sites around the world are being inspected for their suitability to produce tidal energy. Power based on tidal is more reliable and predictable.

Mechanical Energy Harvesting

Mechanical energy is found suitable in some occurrences where solar, wind, thermal, or other form of renewable energy is not found. Mechanical energy is formed with the assistance of vibrating arrangement, motion of living creatures, flow induced vibration and noise. A piezoelectric material is used in changing this mechanical energy into electric energy. Piezoelectric materials have capacity of producing electrical charge once a mechanical load is applied on them; mechanical strain into electric current or voltage is converted by the direct piezoelectric effect. The energy created by a device when it is set into tremor, due to which an inertial mass that be used to generate movement and this measure can be adapted to electrical energy using three mechanisms: piezoelectric, electrostatic and electromagnetic. The form of energy consumed here is the mechanical energy. Piezoelectric materials have essential ability to detect vibrations and that is the reason they have become a viable energy scavenging source. The choices of using piezoelectric materials are many and they are selected based on different criteria for selection among the choice for sensing, actuating or harvesting energy. Some of these are naturally occurring materials such as quartz. Polycrystalline ceramic is a common piezoelectric material. Lead Zirconate Titanate (PZT) is being considered since it shows a high efficiency of mechanical to electrical energy conversion (Joseph, 2005). Different properties of piezoelectric materials are considered by the researchers to develop various piezoelectric harvesters' application in order to power different applications (Amirtharajah et al., 1998; Cavallier et al., 2005; Cornwell et al., 2005; Li et al., 2000; Lit et al., 2005; Mitcheson et al., 2004; White et al., 2001; Williams et. Al., 1995) [13-20].

Energy Harvesting from Machinery Vibrations

To convert kinetic energy into electrical energy to harvest energy from machinery vibrations a linear generator is designed to convert kinetic energy into electrical energy to harvest energy from machinery vibrations. Numerous mechanisms for transduction such as piezoelectric transducers, electromagnetic and electrostatic are used. One of these three transducers which is electromagnetic energy harvesters have maximum power density, further these are not appropriate for MEMS applications. Both the piezoelectric energy and power density harvester have same as that of electromagnetic energy and has simple structure. Out of these three Electrostatic energy harvesters having low power density is suitable for MEMS applications.

Energy Harvesting from Flow Induced Vibrations

Many of the researches have focused there attention on Energy harvesters from flow-induced vibration. Existing devices are capable to generate energy, the start flow speed for them has been reduced at minimum as possible as 2.5m.s-1. Most of the devices which are capable to produce beneficial energy are normally huge in volume if we compare them to the other available vibration energy harvesters. Thus, it is a great challenge and tough to integrate these devices into wireless sensor nodes easily or with other available wireless electronic systems. Future the work should mainly emphasis on shrinking these energy harvesters while these devices have to maintain current power level. In addition, researches should consider to reduce the start flow speed and to allow these technologies to be accessible for wider applications.

Energy Harvesting from Human Movement

Random, broadband non-linear or non-resonant devices are preferred in order to harvest energy from human movement which is Feet which is the main part in the human body that generates energy. The force produced by the human body is concentrated mainly on the feet due to rigorous movement. Using this mechanism certain amount of useful energy is produced for movable electronic devices by the energy harvesters. Additional attention must be taken in order to avoid discomfort in human body due to energy harvesting.

Electromagnetic Energy Harvesting

The manner of producing voltage in a conductor by altering the magnetic field round the conductor is defined as Electromagnetic induction. By applying the principle of electromagnetic induction Electromagnetic Energy Harvesting can be attained. Permanent magnets, a coil and a resonating cantilever beam are used in producing electromagnetic induction for energy harvesting (Beeby et al., 2006). According to (El-Hami et al., 2001) where he defines a vibration-based electromechanical power generator that consisted of a cantilever beam and a pair of magnets. Various researchers in the literature have (Amirtharajah et al., 1997; Beeby et al., 2005; Huang et al., 2003; Kulah et al., 2004; Mizuno et al., 2003; Shearwood et al., 1997; Snarski et al., 2004; Umeda et al 1997; Williams et al., 1998) [23-31] recognized the procedures employed to produce power from electromagnetic resources. The electromagnetic generators intended can be protected from the outside environment which is advantage. Actuators and sensors are made using these magnetostrictive materials where they have the ability of altering magnetic energy into kinetic. These materials are suitable to high frequency vibration and overwhelmed the limitations of other vibrational sources and are extremely flexible.

Table 2 highlights the different form of Harvesting Energy the mechanism involved and the drawback of using these forms of energy.

Table 2. Comparative Analysis

Energy	Mechanism	Drawback
Light	Photo Voltaic cell	Connected when it is daytime and sunny
Wind	Wind turbine generator	Noise, wind cannot be forecasted
Wave	Ocean thermal energy conversion	Changeable Intensity of sea wave and loss to the power generation units.
Heat	Thermo electric devices	Spoil atmosphere due to discharge of fumes and smoke
Mechanical	Electromagnetic, electrostatic, piezoelectric transducers	Probable only in high power machines

PREVIOUS WORK USING PIEZOELECTRIC CRYSTAL

Power Generating Sidewalk

The piezoelectric being used to generate power with different ways and generating power through side walk using piezoelectric is one of the common way where piezoelectric crystal arrays are usually placed under the roadways, footpaths and other high traffic areas like speed breakers, freeways for voltage generation at maximum. The energy harvested through these ways and resources can be used to charge the capacitors, chargeable Lithium batteries etc. later these charged batteries can be used as per the need.

Power Generating Boots or Shoes

DARPA used a unique idea to generate energy by involving soldiers as they considered that the number of soldiers is great and if the idea gets success to generate energy using some technique through their shoes than it will be a huge impact. That idea was practiced by DARPA by using piezoelectric generators to place them in the solders shoes. That idea was being used for Energy Harvesting successfully for a little amount. However, it was affecting the health of the solders constantly while wearing the shoes and generating the power.

Gyms and Workplaces

Energy harvesting through piezoelectric crystals as of DAPA used through solders shoes, research was also conducted on other different angles to place them in Gyms with different exercise machines specially to place them under the treadmill. Researchers are focusing that most of the machines are remain active for longer period due to different exercise sessions so it can become a good source of energy harvesting. The same idea can be supported for other related offices and stores where energy can be harvested in a way from daily office routine. Also, the studies are being carried out to utilize the vibrations in a vehicle, like at clutches, gears, seats, shockups, and footrests.

Mobile Keypad and Keyboards

Researchers are working to generate power through different easy ways which belongs to our daily life, where peoples may not feel burden or duty. One of the recent idea to generate power through smart phones nowadays where piezoelectric crystals can be placed under the keypad of mobile while user will use it for texting or different purpose it will become source of power generating and the same energy can be harvested to charge the cell phone batteries.

People Powered Dance Clubs

In the light of stated research advanced countries already used piezoelectric crystals to generate the power and then harvest thc generated energy in different ways such as in European countries piezoelectric crystals are place beneath the dance club floors in different dance house. The dance floors become source of vibration due to dance parties and that become source for piezoelectric crystals to generate the power which lately can be harvested as per the need.

CONCLUSION

There are numerous forms of energy that can gratify the needs of computing, these energies include solar, thermal, mechanical, wind, acoustic, and wave for different devices and applications. With ubiquitous computing requirements for self-powered devices such as MEMS (Micro-electromechanical systems) and these devices have narrow capacity of finite power sources and it is needed that these systems are needed to be supplied with energy for the lifetime of the system. Mostly Batteries are the main source of energy for many of these mobile systems, embedded systems and remote system applications. To address this issue different Energy harvesting mechanisms based on environment are being well-thought-out and work out as possible feasible prospect to switch the current power supplies for energy constrained in embedded systems. The need to use power independent devices drives to attain vast development in the arena of energy harvesting. The researchers are working towards generating new methods to overcome the few boundaries such as the lower amount of power generated by using the power harvesters. These new methods would help in retaining the energy harvesters as one of the finest bases to power portable devices in the field of wireless sensor network in particular while in general for wireless technology.

REFERENCES

Amirtharajah, R., & Chandrakasan, A. P. (1997). Self-powered low power signal processing. *IEEE Symposium, VLSI Circuits, Digest of Technical Papers.*

Amirtharajah, R., & Chandrakasan, A. P. (1998). Self-powered signal processing using vibration-based power generation. *IEEE Journal of Solid-State Circuits*, *33*(5), 687–695. doi:10.1109/4.668982

Atwood, B., Warneke, B., & Pister, K. S. J. (2001). Smart Dust mote forerunners. *Proceedings of 14th Annual International Conference on Microelectr mechanical Systems.*

Beeby, S. P., Tudor, M. J., Koukharenko, E., White, N. M., O'Donnell, T., Saha, C., Kulkarni, S. & Roy, S., (2005). Micromachined silicon generator for harvesting power from vibration. *Proceedings of Transducers.*

Beeby, S. P., Tudor, M. J., & White, N. M. (2006). Energy harvesting vibration sources for Microsystems applications. *Journal of Measurement Science and Technology*, *17*(12), 175–195. doi:10.1088/0957-0233/17/12/R01

Beeby, S. P., Tudor, M. J., & White, N. M. (2006). Energy harvesting vibration sources for Microsystems applications. *Journal of Measurement Science and Technology*, *17*(12), 175–195. doi:10.1088/0957-0233/17/12/R01

Callaway, Jr., & Edgar, H. (2004). *Wireless Sensor Networks.* Boca Raton, FL: Auerbach Publications, CRC Press LLC.

Cavallier, B., Berthelot, P., Nouira, H., Foltête, E., Hirsinger, L., & Ballandras, S. (2005). Energy harvesting using vibrating structures excited by shock. *IEEE Ultrasonics Symposium.* doi:10.1109/ULTSYM.2005.1603006

Cornwell, P. J., Goethal, J., Kowko, J., & Damianakis, M. (2005). Enhancing power harvesting using a tuned auxiliary structure. *Journal of Intelligent Material Systems and Structures*, *16*(10), 825–834. doi:10.1177/1045389X05055279

El-Hami, M., Glynne-Jones, P., James, E., Beeby, S. P., White, N. M., Brown, A. D., & Hill, M. et al. (2001). Design and fabrication of a new vibration-based electromechanical power generator. *Sensors and Actuators*, *92*(1-3), 335–344. doi:10.1016/S0924-4247(01)00569-6

Fry, D. N., Holcomb, D. E., Munro, J. K., Oakes, L. C., & Maston, M. J. (1997). Compact Portable Electric Power Sources, Report, Oak Ridge National Laboratory. *ORNL. U. S. Atomic Energy Commission*, TM-13360.

Huang, W. S., Tzeng, K. E., Cheng, M. C., & Huang, R. S., (2003). Design and fabrication of a vibrational micro-generator for wearable MEMS. *Proceedings of Eurosensors XVII.*

Joseph, A. D. (2005). Works in Progress-Energy Harvesting Projects. *IEEE Pervasive Computing / IEEE Computer Society [and] IEEE Communications Society*, 69–71.

Kansal, A., Potter, D., & Srivastava, M. B. (2004). Performance Aware Tasking for Environmentally Powered Sensor Networks. *Proceedings on the Measurement and Modeling of Computer Systems in Joint International Conference*. ACM Press. doi:10.1145/1005686.1005714

Kansal, A., & Srivastava, M. B. (2003). An Environmental Energy harvesting framework for Sensor Networks. *Proceedings of International Symposium on Low Power Electronics and Design (ISLPED '03).* ACM Press. doi:10.1145/871619.871624

Kansal, A., & Srivastava, M. B. (2005). Distributed Energy Harvesting for Energy Neutral Sensor Networks. *IEEE Pervasive Computing / IEEE Computer Society [and] IEEE Communications Society*, 4.

Kulah, H., & Najafi, K. (2004). An electromagnetic micro power generator for low-frequency environmental vibrations. *MicroElectro Mechanical Systems—17th IEEE Conference on MEMS.*

Li, W. J., Chan, G. M. H., Ching, N. N. H., Leong, P. H. W., & Wong, H. Y. (2000). Dynamical modeling and simulation of a lasermicromachined vibration-based micro power generator. *International Journal of Nonlinear Science Simulation*, *1*, 345–353.

Liu, Y., Ren, K. L., Hofmann, H. F., & Zhang, Q. (2005). Investigation of Electrostrictive Polymers for Energy Harvesting. *IEEE Transactions on Ultrasonics, Ferroelectrics, and Frequency Control*, *52*(12), 2411–2417. doi:10.1109/TUFFC.2005.1563285 PMID:16463508

Meninger, S. (1999). *A low power controller for a MEMS based energy converter.* M Sc Massachusetts Institute of Technology.

Mitcheson, P. D., Green, T. C., Yeatman, E. M., & Holmes, A. S. (2004). Architectures for Vibration-Driven Micro-Power Generators, IEEE. *Journal of Microelectromechanical Systems*, *13*(3), 429–440. doi:10.1109/JMEMS.2004.830151

Mizuno, M., & Chetwynd, D. (2003). Investigation of a resonance microgenerator, Micromech. *Microeng*, *13*(2), 209–216. doi:10.1088/0960-1317/13/2/307

Park, C., & Chou, P. H. (2004). Power utility maximization for multiple-supply systems by a load-matching switch. *Proceedings of ACM/IEEE International Symposium on Low Power Electronics and Design*. doi:10.1145/1013235.1013281

Raghunathan, V., Kansal, A., Hsu, J., Friedman, J., & Srivastava, M. (2005). Design Considerations for Solar Energy Harvesting Wireless Embedded Systems. *Fourth IEEE/ACM International Conference on Information Processing in SensorNetworks*. doi:10.1109/IPSN.2005.1440973

Rahimi, M., Shah, H., Sukhatme, G. S., & Estrin, D. (2003). Studying the feasibility of Energy harvesting in a mobile sensor network. *Proceedings of IEEE International Conference on Robotics and Automation*. doi:10.1109/ROBOT.2003.1241567

Shearwood, C., & Yates, R. B. (1997). Development of a resonant electromagnetic micro-generator. *Electronic Letters, 33*(22), 1883-1884.

Snarski, S. R., Kasper, R. G., & Bruno, A. B. (2004). Device for electromagnetohydrodynamic (EMHD) energy harvesting. *Proceedings of the Society for Photo-Instrumentation Engineers*, *5417*, 147–161. doi:10.1117/12.545170

Umeda, M., Nakamura, K., & Ueha, S. (1997). Energy Storage characteristics of a piezo-generator using impact induced vibration. *Journal of Applied Physics*, *36*(Part 1, No. 5B), 3146–3151. doi:10.1143/JJAP.36.3146

White, N. M., Glynne-Jones, P., & Beeby, S. (2001). A novel thickfilm piezoelectric micro-generator. *Smart Materials and Structures*, *10*(4), 850–852. doi:10.1088/0964-1726/10/4/403

Williams, C. B., Pavic, A., Crouch, R. S., & Wood, R. C. (1998). Feasibility study of vibration-electric generator for bridge vibration sensors. *IMAC-Proceedings 16th International Modal Analysis Conference*.

Williams, C. B., & Yates, R. B. (1995). Analysis of a micro-electric generator for microsystems," in Proceedings of Solid-StateSensors and Actuator and in Eurosensors IX. *Transducers*, *1*, 369–372.

Compilation of References

Aarts, E., & Wichert, R. (2009). *Ambient intelligence*. Springer Berlin Heidelberg.

Aelterman, P., Rabaey, K., Clauwaert, P., & Verstraete, W. (2006). Microbial fuel cells for wastewater treatment. *Water Science and Technology: A Journal of the International Association on Water Pollution Research, 54*(8), 9–15. Retrieved from http://www.ncbi.nlm.nih.gov/pubmed/17163008

Ahn, Y., & Logan, B. E. (2013a). Altering Anode Thickness To Improve Power Production in Microbial Fuel Cells with Different Electrode Distances. *Energy & Fuels, 27*(1), 271–276. doi:10.1021/ef3015553

Ahn, Y., & Logan, B. E. (2013b). Domestic wastewater treatment using multi-electrode continuous flow MFCs with a separator electrode assembly design. *Applied Microbiology and Biotechnology, 97*(1), 409–416. doi:10.1007/s00253-012-4455-8 PMID:23053104

Akhtar, M., Epps, J. & Ambikairajah, E. (2008). Advances in Eukaryotic Gene Prediction. *Selected Topics in Signal Processing, 2*(3), 310 – 321.

Akhtar, M., Ambikairajah, E., & Epps, J. (2008). Optimizing period-3 methods for eukaryotic gene prediction. *IEEE International Conference on Acoustics, Speech and Signal Processing*. doi:10.1109/ICASSP.2008.4517686

Akkaya, K., & Younis, M. (2004). *Energy-Aware Routing to a Mobile Gateway in Wireless Sensor Networks*. Paper presented at the IEEE Globecom Wireless Ad Hoc and Sensor Networks Workshop, Dallas, TX.

Akkaya, K., & Younis, M. (2005). A survey on routing protocols for wireless sensor networks. *Ad Hoc Networks, 3*(3), 325–349. doi:10.1016/j.adhoc.2003.09.010

Alajmi, B. N., Ahmed, K. H., Finney, S. J., & Williams, B. W. (2011). Fuzzy-Logic-Control Approach of a Modified Hill-Climbing Method for Maximum Power Point in Microgrid Standalone Photovoltaic System. *IEEE Transactions on Power Electronics, 26*(4), 1022–1030. doi:10.1109/TPEL.2010.2090903

Albino, V., Garavelli, A., & Schiuma, G. (2001). A metric for measuring knowledge codification in organizational learning. *Technovation, 1*(2), 413–422. doi:10.1016/S0166-4972(00)00058-4

Al-Fares, M. S., Sun, Z., & Cruickshank, H. (2009). *A hierarchical routing protocol for survivability in wireless sensor network (WSN)*. Paper presented at the International Multi Conference of Engineers and Computer Scientists.

Alippi, C., Anastasi, G., Di Francesco, M., & Roveri, M. (2009). Energy Management in Wireless Sensor Networks with Energy-hugry Sensors. *IEEE Instrumentation & Measurement Magazine, 12*(2), 16–23. doi:10.1109/MIM.2009.4811133

Alippi, C., & Galperti, C. (2008). An Adaptive System for Optimal Solar Energy Harvesting in Wireless Sensor Network Nodes. *IEEE Transactions on Circuits and Systems, 55*(6), 1742–1750. doi:10.1109/TCSI.2008.922023

Al-Karaki, J. N., & Kamal, A. E. (2004). Routing techniques in wireless sensor networks: a survey. *Wireless communications, IEEE, 11*(6), 6-28.

Allen, R., Douence, R., & Garlan, D. (1998). *Specifying and analyzing dynamic software architectures*. Proc. of Fundamental Approaches to Software Engineering.

Alves, A. S., Paquete, C. M., Fonseca, B. M., & Louro, R. O. (2011). Exploration of the "cytochromome" of Desulfuromonas acetoxidans, a marine bacterium capable of powering microbial fuel cells. *Metallomics : Integrated Biometal Science, 3*(4), 349–353. doi:10.1039/c0mt00084a PMID:21298162

Amazon Elastic Compute Cloud (EC2). (n.d.). Retrieved July, 2008, from http://www.amazon.com/cc2/

Amirtharajah, R., & Chandrakasan, A. P. (1997). Self-powered low power signal processing. *IEEE Symposium, VLSI Circuits, Digest of Technical Papers*.

Amirtharajah, R., & Chandrakasan, A. P. (1998). Self-powered signal processing using vibration-based power generation. *IEEE Journal of Solid-State Circuits, 33*(5), 687–695. doi:10.1109/4.668982

Andersen, P. H., Mathews, J. A., & Rask, M. (2009). Integrating private transport into renewable energy policy: The strategy of creating intelligent recharging grids for electric vehicles. *Energy Policy, 37*(7), 2481–2486. doi:10.1016/j.enpol.2009.03.032

Angenent, L. T., Karim, K., Al-Dahhan, M. H., Wrenn, B. A., & Domíguez-Espinosa, R. (2004). Production of bioenergy and biochemicals from industrial and agricultural wastewater. *Trends in Biotechnology, 22*(9), 477–485. doi:10.1016/j.tibtech.2004.07.001 PMID:15331229

Arends, J. B. A., & Verstraete, W. (2012). 100 years of microbial electricity production: Three concepts for the future. *Microbial Biotechnology, 5*(3), 333–346. doi:10.1111/j.1751-7915.2011.00302.x PMID:21958308

Aristeidis, K., Joannopoulos, J. D., & Marin, S. (2008). Efficient wireless non-radiative mid-range energy transfer. Elsevier.

Arveni. (n.d.). *Arveni*. Retrieved from http://www.arveni.fr/en/

Asif, M., & Muneer, T. (2007). Energy supply, its demand and security issues for developed and emerging economies. *Renewable & Sustainable Energy Reviews, 11*(7), 1388–1413. doi:10.1016/j.rser.2005.12.004

Assunção, H. P., Ruiz, L. B., & Loureiro, A. A. (2006). A service management approach for self-healing wireless sensor networks. In *Autonomic Networking* (pp. 215–228). Springer Berlin Heidelberg. doi:10.1007/11880905_18

Atakan, B., & Akan, Ö.B. (2007). Immune system-based energy efficient andreliable communication in wireless sensor networks. In Advances in Biologically Inspired InformationSystems – Models Methods and Tools, Studies in Computational Intelligence (SCI). Springer.

Atlas Scientific. (n.d.). *Atlas Scientific ORP Kit*. Retrieved January 15, 2014, from https://www.atlas-scientific.com/product_pages/kits/orp-kit.html

Atwood, B., Warneke, B., & Pister, K. S. J. (2001). Smart Dust mote forerunners. *Proceedings of 14th Annual International Conference on Microelectr mechanical Systems*.

Baccelli, F., & Liu, Z. (1992). Comparison properties of stochastic decision free Petri nets. *IEEE Transactions on Automatic Control, 37*(12), 1905–1920. doi:10.1109/9.182477

Bain, R. L. (2007). World Biofuels Assessment, Worldwide Biomass Potential: Technology Characterizations" NREL/MP-510-42467. *National Renewable EnergyLaboratory, Golden, CO, USA, 140*, 144–150.

Balasubramaniam, R. (2013). Wireless energy transfer using resonant inductive coupling. In *Proceedings of Sixth IRAJ International Conference* (pp. 44-47). Pune, India: Power Publishers.

Balato, M., & Vitelli, M. (2014). A new control strategy for the optimization of Distributed MPPT in PV applications. *International Journal of Electrical Power & Energy Systems*, *62*, 763–773. doi:10.1016/j.ijepes.2014.05.032

Balconi, M. (2002). Tacitness, codification of technological knowledge, and the organization of industry. *Research Policy*, *31*(3), 357–379. doi:10.1016/S0048-7333(01)00113-5

Bandivadekar, A., Bodek, K., Cheah, L., Evans, C., Groode, T., Heywood, J., & Weiss, M. (2008). *On the road in 2035: Reducing transportation's petroleum consumption and ghg emissions*. MIT Laboratory for Energy and the Environment.

Bansal, K., Goel, H., & Shrivastava, V. K. (2011). A Critical Review on Concept of Green Databases. *Global Journal of Business Management and Information Technology*, *1*(2), 113–118.

Bardeen, J., Cooper, L. N., & Schrieffer, J. R. (1957). Theory of superconductivity. *Physical Review*, *108*(5), 1175–1204. doi:10.1103/PhysRev.108.1175

Barnes, J., Bessant, J., Dunne, N., & Morris, M. (2001). Developing manufacturing competitiveness within South African industry: The role of middle management. *Technovation*, *21*(5), 293–309. doi:10.1016/S0166-4972(00)00047-X

Barroso, L. A. (2005). The price of performance. *Queue*, *3*(7), 113–118. doi:10.1145/1095408.1095420

Bayar, T. (2011). *Malaysian Explores Its Renewable Options.* Retrieved from http://www.renewableenergyworld.com/rea/news/article/2011/09/malaysia-explores-its-renewables-options

Beeby, S. P., Tudor, M. J., Koukharenko, E., White, N. M., O'Donnell, T., Saha, C., Kulkarni, S. & Roy, S., (2005). Micromachined silicon generator for harvesting power from vibration. *Proceedings of Transducers.*

Beeby, S. P., Tudor, M. J., & White, N. M. (2006). Energy harvesting vibration sources for microsystems applications. *Measurement Science & Technology*, *17*(12), R175–R195. doi:10.1088/0957-0233/17/12/R01

Benitez, L. E., Benitez, P. C., & Van Kooten, G. C. (2008). The financials of wind power with energy storage. *Energy Financials*, *30*(4), 1973–1989.

Bennetto, H. (1990). Electricity generation by microorganisms. *Biotechnology Education*. Retrieved from http://www.ncbe.reading.ac.uk/NCBE/MATERIALS/METABOLISM/PDF/bennetto.pdf

Bertsimas, D., & Tsitsiklis, J. (1993). Simulated Annealing. *Statistical Science*, *8*(1), 10–15. doi:10.1214/ss/1177011077

Bettmann, O. (1974). *The Good Old Days They Were Terrible*. Academic Press.

Bidar, M., & Rashidy, K. H. (2013). Modified firefly algorithm using fuzzy tuned parameters. In *Fuzzy Systems (IFSC),13th Iranian Conference* (pp. 1 - 4). Qazvin, Iran: IEEE.

Blanco, M. I. (2009). the financials of wind energy. *Renewable & Sustainable Energy Reviews*, *13*(6-7), 1372–1382. doi:10.1016/j.rser.2008.09.004

Boaventura, A., Collado, A., Carvalho, N. B., & Georgiadis, A. (2013, March-April). Optimum Behavior. *IEEE Microwave Magazine*, *14*(2), 26–35. doi:10.1109/MMM.2012.2234631

Boghani, H. C., Kim, J. R., Dinsdale, R. M., Guwy, A. J., & Premier, G. C. (2013). Control of power sourced from a microbial fuel cell reduces its start-up time and increases bioelectrochemical activity. *Bioresource Technology*, *140*, 277–285. doi:10.1016/j.biortech.2013.04.087 PMID:23708786

Boghani, H. C., Papaharalabos, G., Michie, I., Fradler, K. R., Dinsdale, R. M., Guwy, A. J., & Premier, G. C. et al. (2014). Controlling for peak power extraction from microbial fuel cells can increase stack voltage and avoid cell reversal. *Journal of Power Sources*, *269*, 363–369. doi:10.1016/j.jpowsour.2014.06.059

Bokareva, T., Bulusu, N., & Jha, S. (2005). *SASHA: towards a self-healing hybrid sensor network architecture*. Paper presented at the Second IEEE International Workshop on Embedded Networked Sensors. doi:10.1109/EMNETS.2005.1469101

Boonma, P., & Suzuki, J. (2007). BiSNET: A biologically-inspired middleware architecture for self-managing wireless sensor networks. *Computer Networks*, *51*(16), 4599–4616. doi:10.1016/j.comnet.2007.06.006

Borghino, D. (2014, May 23). *Gizmag*. Retrieved Feb 3, 2015, from MIT finds new way to harvest energy from heat: http://www.gizmag.com/thermogalvanic-harvesting-waste-heat/32212/

Bou, E., Alarcon, E., & Gutierrez, J. (2012). A comparison of analytical models for resonant inductive coupling wireless power transfer. In *Progress In Electromagnetics Research Symposium Proceedings* (pp. 689-693). Moscow, Russia: The Electromagnetics Academy.

Branch, J. W., Lisee, M., & Szymanski, B. K. (2007). *SHR: Self-Healing Routing for wireless ad hoc sensor networks*. Paper presented at the Intern. Symp. Performance Evaluation of Computer and Telecommunication Systems SPECTS.

Brancheau, J., & Wetherbe, J. (1986). Information architectures: Methods and practice. *Information Processing & Management*, *22*(6), 453–463. doi:10.1016/0306-4573(86)90096-8

Brian, B. (2009). Using the theory of habitus to move beyond the study of barriers to technology integration. *Computers & Education*, (52): 353–364.

Bu, L., Wu, X. M., Wang, X. H., & Liu, L. T. (2011). Liquid encapsulated energy harvester for low frequency vibrations. In *Proceedings of the 16th International Solid-State Sensors, Actuators and Microsystems Conference (TRANSDUCERS '11)*. doi:10.1109/TRANSDUCERS.2011.5969863

Bullen, R., Arnot, T., Lakeman, J., & Walsh, F. (2006). Biofuel cells and their development. *Biosensors and ...*. Retrieved from http://www.sciencedirect.com/science/article/pii/S0956566306000406

Bullough. (n.d.). Several Views of Metal Hailde and High Pressure Sodium Lighting for Outdoor Application. *Lighting Research and Technology*.

Buonanno, A., D'Urso, M., & Pavone, D. (2011). An ultra wide-band system for RF energy harvesting. in *antennas and propagation (EUCAP)*. In *Proceedings of the 5th European Conference* (pp. 388 - 389). Rome, Italy: IEEE.

Burrus. (2014). *The Internet of Things Is Far Bigger Than Anyone Realizes*. Retrieved from Wired web site: http://www.wired.com/2014/11/the-internet-of-things-bigger/

Callaway, Jr., & Edgar, H. (2004). *Wireless Sensor Networks*. Boca Raton, FL: Auerbach Publications, CRC Press LLC.

Camilo, T., Carreto, C., Silva, J. S., & Boavida, F. (2006). An energy-efficient ant-based routing algorithm for wireless sensor networks. In Ant Colony Optimization and Swarm Intelligence. Springer Berlin Heidelberg. doi:10.1007/11839088_5

Campos, J., Chiola, G., & Silva, M. (1991). Properties and performance bounds for closed free choice synchronized monoclass quering networks. *IEEE Transactions on Automatic Control*, *36*(12), 1368–1382. doi:10.1109/9.106153

Caponetto, R. (2008). Power Consumption reduction in a remote controlled street lighting system. *Power Electronics, Electrical Drives, Automation and Motion*, 428-433.

Carannante, G., Fraddanno, C., Pagano, M., & Piegari, L. (2009). Experimental Performance of MPPT Algorithm for Photovoltaic Sources Subject to Inhomogeneous Insolation. *IEEE Transactions on Industrial Electronics*, *56*(11), 4374–4380. doi:10.1109/TIE.2009.2019570

Carrera, E. V., Pinheiro, E., & Bianchini, R. (2003, June). Conserving disk energy in network servers. In *Proceedings of the 17th annual international conference on Supercomputing* (pp. 86-97). ACM. doi:10.1145/782814.782829

Cavallier, B., Berthelot, P., Nouira, H., Foltête, E., Hirsinger, L., & Ballandras, S. (2005). Energy harvesting using vibrating structures excited by shock. *IEEE Ultrasonics Symposium*. doi:10.1109/ULTSYM.2005.1603006

CDC. (2013, July 23). *Centers for Disease Control and Prevention*. Retrieved March 9, 2015, from Division for Heart Disease and Stroke Prevention: http://www.cdc.gov/dhdsp/atlas/index.htm

Chalasani, S., & Conrad, J. M. (2008). A survey of energy harvesting sources for embedded systems. In *Proceedings of IEEE Southeastcon*.

Chan, D. S. H., & Phang, J. C. H. (1987). Analytical methods for the extraction of solar-cell single- and double-diode model parameters from I-V characteristics. *IEEE Transactions on Electron Devices*, *34*(2), 286–293. doi:10.1109/T-ED.1987.22920

Chang, K.-C., Lin, W.-M., Lee, T.-S., & Chung, K.-M. (2011). Subsidy programs on diffusion of solar water heaters: Taiwan's experience. *Energy Policy*, *39*(2), 563–567. doi:10.1016/j.enpol.2010.10.021

Chao, K.-H. (2014). An extension theory-based maximum power tracker using a particle swarm optimization algorithm. *Energy Conversion and Management*, *86*, 435–442. doi:10.1016/j.enconman.2014.05.018

Charikleia, K., Haris, D., & John, P. (2010). Technology transfer through climate change: Setting a sustainable energy pattern. *Renewable & Sustainable Energy Reviews*, *14*(6), 1546–1557. doi:10.1016/j.rser.2010.02.001

Charlier, R. H. (2003). Sustainable co-generation from the tides: A review. *Renewable & Sustainable Energy Reviews*, *7*(3), 187–213. doi:10.1016/S1364-0321(02)00011-4

Chattopadhyay, G., Schlecht, E., Maiwald, F., Dengler, R. J., Pearson, J. C., & Mehdi, I. (2002). Frequency multiplier response to spurious signals and its effects on local oscillator systems in millimeter and submillimeter wavelengths. *Proceedings - Society of Photo-Optical Instrumentation Engineers*, *4855*, 480–488.

Chen, C., He, B., Tang, X., Chen, C., & Liu, Y. (2013). Green Databases through Integration of Renewable Energy. In CIDR.

Chen, G., Wei, B., Logan, B. E., & Hickner, M. A. (2012). Cationic fluorinated polymer binders for microbial fuel cell cathodes. *RSC Advances*, *2*(13), 5856. doi:10.1039/c2ra20705b

Chen, G., Zhang, F., Logan, B. E., & Hickner, M. A. (2013). Poly(vinyl alcohol) separators improve the coulombic efficiency of activated carbon cathodes in microbial fuel cells. *Electrochemistry Communications*, *34*, 150–152. doi:10.1016/j.elecom.2013.05.026

Cheng, D. K. (1989). *Field and Wave Electromagnetics*. Addison Wesley.

Chen, H., Gu, F., & Liu, F. (2005). Predicting protein secondary structure using continuous wavelet transform and Chou-Fasman method. *27th Annual International Conference of the Engineering in Medicine and Biology Society*.

Chen, Z., Kawasaki, S., & Carvalho, N. (2013, March-April). Wireless power transmission-The last cut of wires. *IEEE Microwave Magazine*, *14*(2), 22–24. doi:10.1109/MMM.2012.2234361

Choi, J., Cho, J. K., & Seo, C. (2011). Analysis on Transmission efficiency of wireless energy transmission resonator based on magnetic resonance. In Microwave Workshop Series on Innovative Wireless Power Transmission: Technologies, Systems, and Applications (IMWS), 2011 IEEE MTT-S International (pp. 199-202). IEEE. doi:10.2528/PIERM11050903

Christidis, A., Koch, C., Pottel, L., & Tsatsaronis, G. (2012). The contribution of heat storage to the profitable operation of combined heat and power plants in liberalized electricity Markets. *Energy*, *41*(1), 75–82. doi:10.1016/j.energy.2011.06.048

Christopher, W., Craighead, G., Tomas, M., & David, J. (2009). The effects of innovation–cost strategy, knowledge, and action in the supply chain on firm performance. *Journal of Operations Management*, (27): 405–421.

Ciufudean, C. (2000). Modeling the reliability of the interaction man-machine in railway transport. The Annals of the "Stefan cel Mare" University of Suceava, (13), 80-84.

Ciufudean, C. (2003). Discrete event control of railway systems. In *Proc of International Conference AQTR Cluj Napoca*.

Ciufudean, C., & Larionescu, A. (2002). Safety criteria for production lines modeled with Petri nets. Advances in Electrical and Computer Engineering, 2(2), 15-20.

Ciufudean, C., Filote, C., & Amarandei, D. (2008). Control Charts of Workflows. *Advances in Data Mining, 8th Industrial Conference ICDM 2008*, (pp. 330-344). Leipzig, Germany: Springer Verlag Berlin-Heidelberg.

Ciufudean, C. (2002). *Discrete Event Systems for Modelling the Railway Traffic*. Bucharest: Matrix Rom Publishing House. (in Romanian)

Clemons, E., & Row, M. (1991). Sustaining IT advantage: The role of structural differences. *Management Information Systems Quarterly*, *15*(3), 275–292. doi:10.2307/249639

Cohendet, P., & Steinmueller, W. (2000). The codification of knowledge: A conceptual and empirical exploration. *Industrial and Corporate Change*, *9*(9), 195–209. doi:10.1093/icc/9.2.195

Colarelli, D., & Grunwald, D. (2002, November). Massive arrays of idle disks for storage archives. In *Proceedings of the 2002 ACM/IEEE conference on Supercomputing* (pp. 1-11). IEEE Computer Society Press. doi:10.1109/SC.2002.10058

Cook-Chennault, K. A., Thambi, N., & Sastry, A. M. (2008). Powering MEMS portable devices—a review of non-regenerative and regenerative power supply systems with special emphasis on piezoelectric energy harvesting systems. *Smart Materials and Structures*, *17*(4), 043001. doi:10.1088/0964-1726/17/4/043001

Cooper, L. N. (1956). Bound electrons pairs in a degenerate Fermi gas. *IEEE Transactions on Microwave Theory and Techniques*, *104*, 1189–1190.

Cornwell, P. J., Goethal, J., Kowko, J., & Damianakis, M. (2005). Enhancing power harvesting using a tuned auxiliary structure. *Journal of Intelligent Material Systems and Structures*, *16*(10), 825–834. doi:10.1177/1045389X05055279

Cottone, F., Vocca, H., & Gammaitoni, L. (2009, February). Nonlinear energy harvesting. *Physical Review Letters*, *102*(8), 806–811. doi:10.1103/PhysRevLett.102.080601 PMID:19257728

Cowan, R., & Foray, D. (2002). The economics of codification and the diffusion of knowledge. *Industrial and Corporate Change*, (6): 595–622.

Crouch, D. (2014). *UK firm takes the cloud to chillier climes with Swedish data centre*. Retrieved June, 21, 2013, from http://www.theguardian.com/technology/2014/oct/22/cloud-storage-data-centre-sweden-arctic-hydro66

Da Silva, T. L., Roseiro, J. C., & Reis, A. (2012). Applications and perspectives of multi-parameter flow cytometry to microbial biofuels production processes. *Trends in Biotechnology*, *30*(4), 225–232. doi:10.1016/j.tibtech.2011.11.005 PMID:22257766

Daft, L., & Lengel, H. (1986). Organizational information requirements, media richness, and structural design. *Management Science*, *32*(5), 554–571. doi:10.1287/mnsc.32.5.554

Dagdeviren, C., Yang, B. D., Su, Y., Tran, P. L., Joe, P., Anderson, E., & Rogers, J. A. et al. (2014). Conformal piezoelectric energy harvesting and storage from motions of the heart, lung, and diaphragm. *Proceedings of the National Academy of Sciences of the United States of America*, *111*(5), 1927–1932. doi:10.1073/pnas.1317233111 PMID:24449853

Das, R. (2009, November). *Energy Harvesting Journal*. Retrieved from http://www.energyharvestingjournal.com/articles/1842/batteryless-infrared-remote-control-from-arveni

Dash, S., Swain, A. R., & Ajay, A. (2012). Reliable energy aware multi token based MAC protocol for WSN. In *Proceedings of the26th IEEE International Conference on Advanced Information Networking and Applications (AINA '12)*. doi:10.1109/AINA.2012.61

Data Center Knowledge. (n.d.). *Inside Facebook's Lulea Data Center*. Retrieved June, 28, 2013, from http://www.datacenterknowledge.com/inside-facebooks-lulea-data-center/

Datta & Asif. (2004). DFT based DNA splicing algorithms for prediction of protein coding regions. *Conference Record of the Thirty-Eighth Asilomar Conference on Signals, Systems and Computers*.

Datta, S., Asif, A., & Wang, H. (2004). Prediction of protein coding regions in DNA sequences using Fourier spectral characteristics. *IEEE Sixth International Symposium on Multimedia Software Engineering Proceedings*. doi:10.1109/MMSE.2004.63

Datta, S., & Asif, A. (2005). A fast DFT based gene prediction algorithm for identification of protein coding regions. *IEEE International Conference on Acoustics, Speech, and Signal Processing*. doi:10.1109/ICASSP.2005.1416388

Davis, F., & Higson, S. P. J. (2007). Biofuel cells--recent advances and applications. *Biosensors & Bioelectronics*, *22*(7), 1224–1235. doi:10.1016/j.bios.2006.04.029 PMID:16781864

Davis, J. B., & Yarbrough, H. F. (1962). Preliminary Experiments on a Microbial Fuel Cell. *Science*, *137*(3530), 615–616. doi:10.1126/science.137.3530.615 PMID:17836545

De Dominicis, C. M., F. A. (2011). On the development of a wireless self localizing streetlight monitoring system. *Sensors Application SymposiumIEEE*. doi:10.1109/SAS.2011.5739806

Debabov, V. G. (2008). Electricity from microorganisms. *Microbiology*, *77*(2), 123–131. doi:10.1134/S002626170802001X PMID:18522314

Degrenne, N., Allard, B., & Buret, F. (2011). Comparison of 3 self-starting step-up DC: DC converter topologies for harvesting energy from low-voltage and low-power microbial fuel cells. *Proceedings of the ....* Retrieved from http://ieeexplore.ieee.org/xpls/abs_all.jsp?arnumber=6020291

Degrenne, N., Buret, F., Allard, B., & Bevilacqua, P. (2012). Electrical energy generation from a large number of microbial fuel cells operating at maximum power point electrical load. *Journal of Power Sources*, *205*, 188–193. doi:10.1016/j.jpowsour.2012.01.082

Deicon Dynamics & Controls. (n.d.). *Tuned Mass Dampers*. Retrieved March 5, 2015, from http://www.deicon.com/tuned-mass-dampers/

DeMeo, E. A., & Galdo, J. F. (1997). *Renewable Energy Technology Characterizations. TR-109496*. Washington, DC: U.S. Department of Energy and Electric Power Research Institute.

Deterre, M. (2013). *Toward an energy harvester for leadless pacemakers.* (PhD Thesis). Universite Paris Sud - Paris XI. Retrieved from https://tel.archives-ouvertes.fr/tel-00868838

Dewan, A., Ay, S. U., Karim, M. N., & Beyenal, H. (2014). Alternative power sources for remote sensors: A review. *Journal of Power Sources*, *245*, 129–143. doi:10.1016/j.jpowsour.2013.06.081

Dewan, A., Beyenal, H., & Lewandowski, Z. (2009). Intermittent Energy Harvesting Improves the Performance of Microbial Fuel Cells. *Environmental Science & Technology*, *43*(12), 4600–4605. doi:10.1021/es8037092 PMID:19603683

Di Car, G., & Dorgio, M. (1998). Antnet: distributed stigmergetic control for communication networks. *Journal of Artificial Intelligence Research*, (9), 317-365.

Dimroth, F. (2014, Dec 1). *Fraunhofer Institute for Solar Energy Systems ISE.* Retrieved January 14, 2015, from Fraunhofer ISE: Press Releases 2014: http://www.ise.fraunhofer.de/en/press-and-media/press-releases/press-releases-2014/new-world-record-for-solar-cell-efficiency-at-46-percent

Donelan, J. M., Li, Q., Naing, V., Hoffer, J. A., Weber, D. J., & Kuo, A. D. (2008, February8). Biomechanical Energy Harvesting: Generating Electricity During Walking with Minimal User Effort. *Science*, *319*(5864), 807–810. doi:10.1126/science.1149860 PMID:18258914

Dorigo, M., Maniezzo, V., & Colomi, A. (1996). Ant system: Optimization by a colony of cooperating agents. *IEEE Transactions on Systems, Man, and Cybernetics. Part B, Cybernetics*, *26*(1), 29–41. doi:10.1109/3477.484436 PMID:18263004

Dowhan, L., Wymyslowski, A., & Urbanski, K. (2009). Simulated annealing as global optimization algorithm used in numerical prototyping of electronic packaging. In *Thermal, Mechanical and Multi-Physics simulation and Experiments in Microelectronics and Microsystems, 10th International Conference* (pp. 1-5). Delft, Netherlands: IEEE.

Dressler, F. (2006). Self-organization in ad hoc networks: Overview and classification. University of Erlangen, Dept. of Computer Science.

Dressler, F., & Carreras, I. (2007). *Advances in Biologically Inspired Information Systems.* Springer-Verlag Berlin Heidelberg.

Dressler, F., Kuger, B., Fuchs, G., & German, R. (2005). *Self-Organization in Sensor Networks using Bio-Inspired Mechanisms.* Paper presented at the 18th ACM/GI/ITG International Conference on Architecture of Computing Systems-System Aspects in Organic and Pervasive Computing (ARCS'05): Workshop Self-Organization and Emergence, Innsbruck, Austria.

Du, Z., Li, H., & Gu, T. (2007). A state of the art review on microbial fuel cells: A promising technology for wastewater treatment and bioenergy. *Biotechnology Advances*, *25*(5), 464–482. doi:10.1016/j.biotechadv.2007.05.004 PMID:17582720

EE Times. (2009, November 6). Retrieved Jan 20, 2015, from Connecting the Global Electronics Community: http://www.eetimes.com/document.asp?doc_id=1253967

Egelhoff, G. (1982). Strategy and structure in multinational corporations: An information-processing approach. *Administrative Science Quarterly*, *27*(2), 435–458. doi:10.2307/2392321

EIA. (2009). *Energy Consumption by Manufacturers—Data Tables.* EIA.

Elefsiniotis, A., Kokorakis, N., Becker, T., & Schmid, U. (2014, February1). A thermoelectric-based energy harvesting module with extended operational temperature range for powering autonomous wireless sensor nodes in aircraft. *Sensors and Actuators. A, Physical*, *206*, 159–164. doi:10.1016/j.sna.2013.11.036

El-Hami, M., Glynne-Jones, P., James, E., Beeby, S. P., White, N. M., Brown, A. D., & Hill, M. et al. (2001). Design and fabrication of a new vibration-based electromechanical power generator. *Sensors and Actuators*, *92*(1-3), 335–344. doi:10.1016/S0924-4247(01)00569-6

Emile, A., Jan, K., & Wil, M. (2005). *Search methodologies*. Springer. doi:10.1007/0-387-28356-0

Enrique, J. M., Durán, E., Sidrach-de-Cardona, M., & Andújar, J. M. (2007). Theoretical assessment of the maximum power point tracking efficiency of photovoltaic facilities with different converter topologies. *Solar Energy*, *81*(1), 31–38. doi:10.1016/j.solener.2006.06.006

Erturk, A., & Inman, D. J. (2009). An experimentally validated bimorph cantilever model for piezoelectric energy harvestingfrom base excitations. Iopscience Smart Materials and Structures, 18(2).

Erturk, A., Vieira, W. G. R., DeMarqui, C. Jr, & Inman, D. J. (2010). On the energy harvesting potential of piezoaeroelastic systems. *Applied Physics Letters*, *96*(18), 184103. doi:10.1063/1.3427405

Esram, T., & Chapman, P. L. (2007). Comparison of Photovoltaic Array Maximum Power Point Tracking Techniques. *IEEE Transactions on Energy Conversion*, *22*(2), 439–449. doi:10.1109/TEC.2006.874230

ETSAP. (2010). *Marine Energy Technology Brief E13 - November, 2010, Energy Technology Systems Analysis Programme*. Paris, France: International Energy Agency.

European Commission. (2014, Dec 11). Retrieved January 6, 2015, from European Commission: Climate Action: http://ec.europa.eu/clima/policies/brief/eu/

European Editors. (2013). *Digi-Key Corporation*. Retrieved 2 24, 2015, from Article Library-Smaller, Smarter, Ultra-Low-Power Sensors Raise Potential for Energy Harvesting in Medical Implants: http://www.digikey.com/en/articles/techzone/2013/sep/smaller-smarter-ultralowpower-sensors-raise-potential-for-energy-harvesting-in-medical-implants

Eu, Z. A., Tan, H. P., & Seah, W. K. G. (2009). *Routing and relay nodeplacement in wireless sensor networks powered by ambient energy har-vesting. In Proc. of the IEEE Wireless Communications & Networking Conference*. WCNC.

Faheem, A., & Luiz, F. (2008). The software product line architecture: An empirical investigation of key process activities. *Information and Software Technology*, (50): 1098–1113.

Falcione, A., & Krogh, B. H. (1993). Design recovery for relay ladder logic, *IEEE Trans. on Cont. Syst. Mag.*, *24*(4), 90–98.

Farook, S., & Raju, P. S. (2013). Evolutionary hybrid genetic-firefly algorithm for global optimization. *International Journal of Computational Engineering & Management*, *16*(3).

Fernando, E., García, M., Eva, P., José, E., & Navas, L. (2009). Knowledge codification and technological innovation success: Empirical evidence from Spanish biotech companies. *Technological Forecasting and Social Change*, *76*(1), 141–153. doi:10.1016/j.techfore.2008.03.016

Filip, N., & Simu, I. (2011). About the noise energy conversion from agriculture tractor engines. In *Proceedings of the 39thInternational Symposiumon Agricultural Engineering*.

Filote, C., & Ciufudean, C. (2010). Discrete Event Models for Flexible Manufacturing Cells. In Future Manufacturing Systems. Sciyo.

Flynn, B. W., & Fotopolou, K. (2013, March-April). Rectifying loose coils: Wireless power transfer in loosely coupled inductive links with lateral and angular misalignment. *IEEE Microwave Magazine*, *14*(2), 48–54. doi:10.1109/MMM.2012.2234634

Fonseca, B. M., Paquete, C. M., Neto, S. E., Pacheco, I., Soares, C. M., & Louro, R. O. (2013). Mind the gap: Cytochrome interactions reveal electron pathways across the periplasm of Shewanella oneidensis MR-1. *The Biochemical Journal, 449*(1), 101–108. doi:10.1042/BJ20121467 PMID:23067389

Françoise, B. (2008). Does environmental information overcome practice compartmentalization and change consumers' behaviors? *Journal of Cleaner Production, 16*(11), 1170–1180. doi:10.1016/j.jclepro.2007.08.013

Frick, S., Kaltschmitt, M., & Schröder, G. (2010). Life cycle assessment of geothermal binary power plants using enhanced low-temperature reservoirs. *Energy, 35*(5), 2281–2294.

Fridleifsson, I. B., & Freeston, D. H. (1994). Geothermal energy research and development. *Geothermics, 23*(2), 175–214.

Fry, D. N., Holcomb, D. E., Munro, J. K., Oakes, L. C., & Maston, M. J. (1997). Compact Portable Electric Power Sources, Report, Oak Ridge National Laboratory. *ORNL. U. S. Atomic Energy Commission*, TM-13360.

Galperti, C. A. (2008). An Adaptive System for Optimal Solar Energy Harvesting in Wireless Sensor Network Nodes. *IEEE Transactions on Circuits and Systems, 5*(6), 1742–1750.

Ganesan, D., Govindan, R., Shenker, S., & Estrin, D. (2001). Highly-resilient, energy-efficient multipath routing in wireless sensor networks. Paper presented at the ACM SIGMOBILE. *Mobile Computing and Communications Review, 5*(4), 1125. doi:10.1145/509506.509514

Garlan, D. (2001). Software architecture. In J. Marciniak (Ed.), *Wiley Encyclopedia of Software Engineering*. John Wiley & Sons.

George, T. P., & Thomas, T. (2010). Discrete wavelet transform de-noising in eukaryotic gene splicing. *BMC Bioinformatics, 11*(Suppl 1), S50. doi:10.1186/1471-2105-11-S1-S50 PMID:20122225

Gevorkian, P. (2012). *Large Scale Solar Power Systems: Construction and Financials*. Cambridge University Press.

Ghaffari, A., Rezg, N., & Xie, X. (2003). Feedback control logic for forbiden state problems of marked graphs: Application to a real manufacturing system. *IEEE Transactions on Automatic Control, 48*(1), 18–29. doi:10.1109/TAC.2002.806651

Gholamhosseinian, A., & Khalifeh, A. (2012). *Cloud computing and sustainability: Energy efficiency aspects*. Academic Press.

Gilbert, J. M., & Balouchi, F. (2008). Comparison of energy harvesting systems for wireless sensor networks. *International Journal of Automation and Computing, 5*(4), 334–347. doi:10.1007/s11633-008-0334-2

Gipe, P. (1995). *Wind energy comes of age* (Vol. 4). Wiley.

Girianna, M., & Benekohal, R. F. (2002). Dynamic signal coordination for networks with oversaturated intersections. *Transportation Research Record, 1811*, 122–130. doi:10.3141/1811-15

Goiri, Í., Le, K., Haque, M. E., Beauchea, R., Nguyen, T. D., Guitart, J., & Bianchini, R. et al. (2011, November). Greenslot: scheduling energy consumption in green datacenters. In *Proceedings of 2011 International Conference for High Performance Computing, Networking, Storage and Analysis* (p. 20). ACM. doi:10.1145/2063384.2063411

Goiri, Í., Le, K., Nguyen, T. D., Guitart, J., Torres, J., & Bianchini, R. (2012, April). GreenHadoop: leveraging green energy in data-processing frameworks. In *Proceedings of the 7th ACM european conference on Computer Systems* (pp. 57-70). ACM. doi:10.1145/2168836.2168843

Gordon, J. (1987). Optimal sizing of stand-alone photovoltaic solar power systems. *Solar Cells, 20*(4), 295–313.

Grandhi, D. G., & Vijay Kumar, C. (2008). *2-Simplex mapping for identifying the protein coding regions in DNA. In IEEE region conference* (pp. 1–3). TENCON.

Grant, R. (1996). Toward a knowledge-based theory of the firm. *Strategic Management Journal, 17*(S2), 109–122. doi:10.1002/smj.4250171110

Greenberg, A., Hamilton, J., Maltz, D. A., & Patel, P. (2008). The cost of a cloud: Research problems in data center networks. *Computer Communication Review, 39*(1), 68–73. doi:10.1145/1496091.1496103

GSM Based Autonomous Street Illumination System for Efficient Power Management. (2013). *Engineering Trends and Technology*, 54-60.

Gui, C., & Mohapatra, P. (2003). *SHORT: Self-healing and optimizing routing techniques for mobile adhoc networks.* Paper presented at the 4th ACM international symposium on Mobile ad hoc networking & computing.

Gungor, V. C., & Hancke, G. P. (2009). Industrial wireless sensor networks: Challenges, design principles, and technical approaches. *Industrial Electronics. IEEE Transactions on, 56*(10), 4258–4265.

Guo & Zhu. (2009). Prediction of Protein Coding Regions by Support Vector Machine. *International Symposium on Intelligent Ubiquitous Computing and Education.*

Gupta, R., Mittal, A., Singh, K., Bajpai, P., & Prakash, S. (2007). A Time Series Approach for Identification of Exons and Introns. *10th International Conference on Information Technology.* doi:10.1109/ICIT.2007.54

Gurumurthi, S., Sivasubramaniam, A., Kandemir, M., & Franke, H. (2003, June). DRPM: dynamic speed control for power management in server class disks. In *Computer Architecture, 2003. Proceedings. 30th Annual International Symposium on* (pp. 169-179). IEEE. doi:10.1109/ISCA.2003.1206998

Gurumurthi, S., Zhang, J., Sivasubramaniam, A., Kandemir, M., Franke, H., Vijaykrishnan, N., & Irwin, M. J. (2003, March). Interplay of energy and performance for disk arrays running transaction processing workloads. In *Performance Analysis of Systems and Software, 2003. ISPASS. 2003 IEEE International Symposium on* (pp. 123-132). IEEE.

Hande, A., Polk, T., Walker, W., & Bhatia, D. (2007). Indoor solar energyharvesting for sensor network router nodes. *Microprocessors and Microsystems, 31*(6), 420–432. doi:10.1016/j.micpro.2007.02.006

Han, J., Mol, A. P. J., & Lu, Y. (2010). Solar water heaters in China: A new day dawning. *Energy Policy, 38*(1), 383–391. doi:10.1016/j.enpol.2009.09.029

Hanna, M. A., Mutashar, S., Samad, S. A., & Hussain, A. (2014). Energy harvesting for the implantable biomedical devices: Issues and challenges. *Biomedical Engineering Online, 13*(1), 79. doi:10.1186/1475-925X-13-79 PMID:24950601

Harizopoulos, S., Shah, M., Meza, J., & Ranganathan, P. (2009). *Energy efficiency: The new holy grail of data management systems research.* arXiv preprint arXiv:0909.1784.

Harvey, L. D. D. (2006). *A Handbook on Low-Energy Buildings and District-Energy Systems: Fundamentals, Techniques and Examples.* Sterling, VA: Earthscan.

Hawkes, A., & Leach, M. (2007). Cost-effective operating strategy for residential microcombined heat and power. *Energy, 32*(5), 711–723. doi:10.1016/j.energy.2006.06.001

Hazrina & Siti. (2008). Gene prediction system. *International Symposium on Information Technology.*

He, Z., Minteer, S., & Angenent, L. (2005). Electricity generation from artificial wastewater using an upflow microbial fuel cell. *Environmental Science & ...*. Retrieved from http://pubs.acs.org/doi/abs/10.1021/es0502876

Heer, R., Wissenwasser, J., Milnera, M., Farmer, L., Opfner C. H, & Vellekoop, M. (2010). Wireless powered electronic sensors for biological applications. In Proceedings of IEEE EMBC 2010.

Heinzelman, W. R., Chandrakasan, A., & Balakrishnan, H. (2000). *Energy-efficient communication protocol for wireless microsensor networks.* Paper presented at the 33rd annual Hawaii international conference on. doi:10.1109/HICSS.2000.926982

Hi-globe. (n.d.). Retrieved September 15, 2015, from http://hi-globe.com/?page_id=85

Hobbs, W. B., & Hu, D. L. (2012). Tree-inspired piezoelectric energy harvesting.[tree inspired]. *Journal of Fluids and Structures*, *28*, 103–114. doi:10.1016/j.jfluidstructs.2011.08.005

Holloway, L.E., & Krogh, B., (1990). Synthesis of feedback control logic for a class of controlled Petri nets. *IEEE Trans. Autom. Contr.*, (5), 514-523.

Hota, M.K. & Srivastava, V.K. (2008). DSP technique for gene and exon prediction taking complex indicator sequence. *IEEE Region 10 Conference.*

Huang, W. S., Tzeng, K. E., Cheng, M. C., & Huang, R. S., (2003). Design and fabrication of a vibrational micro-generator for wearable MEMS. *Proceedings of Eurosensors XVII.*

Huang, L., Chai, X., Quan, X., Logan, B. E., & Chen, G. (2012). Reductive dechlorination and mineralization of pentachlorophenol in biocathode microbial fuel cells. *Bioresource Technology*, *111*, 167–174. doi:10.1016/j.biortech.2012.01.171 PMID:22357291

Huang, L., Gan, L., Wang, N., Quan, X., Logan, B. E., & Chen, G. (2012). Mineralization of pentachlorophenol with enhanced degradation and power generation from air cathode microbial fuel cells. *Biotechnology and Bioengineering*, *109*(9), 2211–2221. doi:10.1002/bit.24489 PMID:22392229

Huang, L., Pop, V., de Francisco, R., Vullers, R., Dolmans, G., de Groot, H., & Imamura, K. (2010). Ultra low power wireless and energy harvesting technologies — An ideal combination. In *2010 IEEE International Conference on Communication Systems* (pp. 295–300). IEEE. doi:10.1109/ICCS.2010.5686436

Hua, W. S., Li-Dong, D., De-Yi, K., Hao-Yue, P., Zhen, F., & Zhan, Z. (2014). Hybrid device for acoustic noise reduction and energy harvesting based on a silicon micro-perforated panel structure. *Chinese Physics B*, *23*(4), 044302. doi:10.1088/1674-1056/23/4/044302

IATA. (n.d.). *iata.org.* Retrieved from http://www.iata.org/html_email/car1001654/lithium_batteries.pdf

IEA. (2007). *Renewables for Heating and Cooling – Untapped Potential.* Paris, France: International Energy Agency.

Ieropoulos, I. A., Greenman, J., Melhuish, C., & Hart, J. (2005). Comparative study of three types of microbial fuel cell. *Enzyme and Microbial Technology*, *37*(2), 238–245. doi:10.1016/j.enzmictec.2005.03.006

Ieropoulos, I. A., Ledezma, P., Stinchcombe, A., Papaharalabos, G., Melhuish, C., & Greenman, J. (2013). Waste to real energy: The first MFC powered mobile phone. *Physical Chemistry Chemical Physics : PCCP*, *15*(37), 15312–15316. doi:10.1039/c3cp52889h PMID:23939246

Imran, Asifullah, Asad, & Tae-Sun. (2009). Reversible Watermarking Based on Intelligent Coefficient Selection and Integer Wavelet Transform. *International Journal of Innovative Computing, Information and Control*, *5*(12), 4675-4682.

Inoue, K., Qian, X., Morgado, L., Kim, B.-C., Mester, T., Izallalen, M., & Lovley, D. R. et al. (2010). Purification and characterization of OmcZ, an outer-surface, octaheme c-type cytochrome essential for optimal current production by Geobacter sulfurreducens. *Applied and Environmental Microbiology*, *76*(12), 3999–4007. doi:10.1128/AEM.00027-10 PMID:20400562

Intanagonwiwat, C., Govindan, R., & Estrin, D. (2000). *Directed diffusion: a scalable and robust communication paradigm for sensor networks*. Paper presented at the 6th annual international conference on Mobile computing and networking. doi:10.1145/345910.345920

Internet of Thing (IoT). (n.d.). Retrieved from Cisco web site: http://www.cisco.com/web/solutions/trends/iot/overview.html

Internet of Thing. (1994-2014). Retrieved from Tech Target web site: http://whatis.techtarget.com/definition/Internet-of-Things

Internet of Things. (2014). Retrieved from Microsoft web site: http://www.microsoft.com/windowsembedded/en-us/internet-of-things.aspx

Ishaque, K., Salam, Z., & Taheri, H. (2011). Simple, fast and accurate two-diode model for photovoltaic modules. *Solar Energy Materials and Solar Cells*, *95*(2), 586–594. doi:10.1016/j.solmat.2010.09.023

Ishii, S., Logan, B. E., & Sekiguchi, Y. (2012). Enhanced electrode-reducing rate during the enrichment process in an air-cathode microbial fuel cell. *Applied Microbiology and Biotechnology*, *94*(4), 1087–1094. doi:10.1007/s00253-011-3844-8 PMID:22223104

ITU, OECD, & GeSI. (2009). *ICTs and climate change: Official side-event by ITU, OECD, and GeSI*. United Nations Climate Change Talks. Retrieved on February 2, 2010 form: OECD: www.oecd.org/sti/ict/green-ict

Jain, S., & Agarwal, V. (2007). Comparison of the performance of maximum power point tracking schemes applied to single-stage grid-connected photovoltaic systems. *IET Electric Power Applications*, 753–763. doi:10.1049/iet-epa

Jherniak. (2012, September 27). *Reduce Electricity Costs on Street Lights*. Retrieved February 12, 2014, from Lighting Controller: www.egenergy.com

Jia, D., & Liu, J. (2009). Human power-based energy harvesting strategies for mobile electronic devices. *Frontiers of Energy and Power Engineering in China*, *3*(1), 27–46. doi:10.1007/s11708-009-0002-4

Jiang, X., Li, Y., Li, J., Wang, J., & Yao, J. (2014). Piezoelectric energy harvesting from traffic-induced pavement vibrations. *Journal of Renewable and Sustainable Energy, 6*(4). Retrieved from http://scitation.aip.org/content/aip/journal/jrse/6/4/10.1063/1.4891169

Jiang, Y. (2013). Adaptive step size with adaptive-perturbation-frequency digital MPPT controller for a single-sensor photovoltaic solar system. *Power Electronics, IEEE, 28*(7), 3195–3205. Retrieved from http://ieeexplore.ieee.org/xpls/abs_all.jsp?arnumber=6316180

Jiang, X., Polastre, J., & Culler, D. (2005). *Perpetual Environmentally Powered Sensor Networks*. In *Fourth International Symposium on Information Processing in Sensor Networks*.

Jin-qiu, Z., Zhi-zhao, P., Lei, Z., & Yu, Z. (2013). A Review on Energy-Regenerative Suspension.*Proceedings of the World Congress on Engineering 2013*.

Jose, G. L., Antonio, R., & Francesc, M. (2003, March). Human Powered Piezoelectric Batteries to Supply Power to Wearable Electronic Devices. *International Journal of Society of Materials Engineering for Resources*, *10*, 34–40.

Joseph, A. D. (2005). Works in Progress-Energy Harvesting Projects. *IEEE Pervasive Computing / IEEE Computer Society [and] IEEE Communications Society*, 69–71.

Kakumani, R., Devabhaktuni, V., & Ahmad, M. O. (2008). Prediction of protein-coding regions in DNA sequences using a model-based approach. *IEEE International Symposium on Circuits and Systems*. doi:10.1109/ISCAS.2008.4541818

Kamarzaman, N. A., & Tan, C. W. (2014). A comprehensive review of maximum power point tracking algorithms for photovoltaic systems. *Renewable & Sustainable Energy Reviews*, *37*, 585–598. doi:10.1016/j.rser.2014.05.045

Kansal, A., & Srivastava, M. B. (2003). *An environmental energy harvesting framework for sensor networks*. In *Low Power Electronics and Design, 2003. ISLPED'03, Proceedings of the 2003 International Symposium*. doi:10.1145/871506.871624

Kansal, A., & Srivastava, M. B. (2003). An Environmental Energy harvesting framework for Sensor Networks. *Proceedings of International Symposium on Low Power Electronics and Design (ISLPED '03)*. ACM Press. doi:10.1145/871619.871624

Kansal, A., Potter, D., & Srivastava, M. B. (2004). Performance Aware Tasking for Environmentally Powered Sensor Networks. *Proceedings on the Measurement and Modeling of Computer Systems in Joint International Conference*. ACM Press. doi:10.1145/1005686.1005714

Kansal, A., Hsu, J., Zahedi, S., & Srivastava, M. B. (2007). Power management in energy harvesting sensor networks. *ACM Transactions on Embedded Computing Systems*, *6*(4), 32, es. doi:10.1145/1274858.1274870

Kansal, A., & Srivastava, M. B. (2005). Distributed Energy Harvesting for Energy Neutral Sensor Networks. *IEEE Pervasive Computing / IEEE Computer Society [and] IEEE Communications Society*, 4.

Karami, M. A., Bilgen, O., Inman, D. J., & Friswell, M. I. (2011). Experimental and analytical parametric study of single-crystal unimorph beams for vibration energy harvesting. *IEEE Transactions on Ultrasonics, Ferroelectrics, and Frequency Control*, *58*(7), 1508–1520. doi:10.1109/TUFFC.2011.1969 PMID:21768034

Kareem, A., Kijewski, T., & Tamura, Y. (1999). Mitigation of motions of tall buildings with specific examples of recent applications. *Wind and Structures*, *2*(3), 201–251. doi:10.12989/was.1999.2.3.201

Karmakar, S., Kundu, K., & Kundu, S. (2010). Design and Development of Microbial Fuel cells. *Technology and Education Topics in* .... Retrieved from http://www.formatex.info/microbiology2/1029-1034.pdf

Karp, B., & Kung, H. T. (2000). *GPSR: Greedy perimeter stateless routing for wireless networks*. Paper presented at the 6th annual international conference on Mobile computing and networking. doi:10.1145/345910.345953

Karube, I., Matsunaga, T., Tsuru, S., & Suzuki, S. (1977). Biochemical fuel cell utilizing immobilized cells of clostridium butyricum. *Biotechnology and Bioengineering*, *19*(11), 1727–1733. doi:10.1002/bit.260191112

Kautz, R. L. (1978). Picosecond pulses on superconducting striplines. *Journal of Applied Physics*, *49*(1), 308–314. doi:10.1063/1.324387

Kerr, D. (2007), Marine energy. *Philosophical Transactions of the Royal Society London, Series A (Mathematical, Physical and Engineering Sciences)*, *365*(1853), 971-92.

Khripkov, A., Hong, W., & Pavlov, K. (2012). Integrated resonant structure for simultaneous wireless power transfer and data telemetry. *Antennas And Wireless Propagation Letters, IEEE*, *11*, 1659–1662. doi:10.1109/LAWP.2013.2238602

Kim, B., Chang, I., & Gadd, G. (2007). Challenges in microbial fuel cell development and operation. *Applied Microbiology and Biotechnology*. Retrieved from http://link.springer.com/article/10.1007/s00253-007-1027-4

Kim, J. R., Cheng, S., Oh, S.-E., & Logan, B. E. (2007). Power Generation Using Different Cation, Anion, and Ultrafiltration Membranes in Microbial Fuel Cells. *Environmental Science & Technology*, *41*(3), 1004–1009. doi:10.1021/es062202m PMID:17328216

Kim, S.-T., Bang, T.-H., Lee, S.-C., & Park, J.-W. (2014). Real-Time Maximum Power Point Tracking Method Based on Three Points Approximation by Digital Controller for PV System. *Journal of Electrical Engineering and Technology*, *9*(5), 1447–1453. doi:10.5370/JEET.2014.9.5.1447

Kim, Y., Hatzell, M. C., Hutchinson, A. J., & Logan, B. E. (2011). Capturing power at higher voltages from arrays of microbial fuel cells without voltage reversal. *Energy & Environmental Science*, *4*(11), 4662. doi:10.1039/c1ee02451e

Klein, S., & Kromen, J. (1995). IT-enabled co-operations: a resource-based approach. In *Proceedings of the 3rd European Conference on Information Systems.*

Kollimalla, S. K., & Mishra, M. K. (2014). A Novel Adaptive P&O MPPT Algorithm Considering Sudden Changes in the Irradiance. *IEEE Transactions on Energy Conversion*, *29*(3), 602–610. doi:10.1109/TEC.2014.2320930

Koomey, J. (2011). *Growth in data center electricity use 2005 to 2010.* A report by Analytical Press, completed at the request of The New York Times, 9.

Koopman, P. (2003). Elements of the self-healing problem space. In *ICSE Workshop on Architecting Dependable Systems.*

Kraus, J. D. (1986). *Radio Astronomy.* Cynus-Quasar.

Krioukov, A., Goebel, C., Alspaugh, S., Chen, Y., Culler, D. E., & Katz, R. H. (2011).. . *Integrating Renewable Energy Using Data Analytics Systems: Challenges and Opportunities. IEEE Data Eng. Bull.*, *34*(1), 1–9.

Kulah, H., & Najafi, K. (2004). An electromagnetic micro power generator for low-frequency environmental vibrations. *MicroElectro Mechanical Systems—17th IEEE Conference on MEMS.*

Kumar, A., Sharma, A., Kumar, R., Vaish, R., & Chauhan, V. S. (2014). Finite element analysis of vibration energy harvesting using lead-free piezoelectric materials: A comparative study. *Journal of Asian Ceramic Societies*, *2*(2), 138–143. doi:10.1016/j.jascer.2014.02.001

Kumar, R., & Jyotishree. (2012). Novel encoding scheme in genetic algorithms for better fitness. *International Journal of Engineering And Advanced Technology*, *1*(6), 214–218.

Kuphaldt, T. R. (n.d.). *pH measurement: Electrical Instrumentation Signals - Electronics Textbook.* Retrieved January 14, 2014, from http://www.allaboutcircuits.com/vol_1/chpt_9/6.html

Kurs, A., Karalis, A., Moffatt, R., Joannopoulos, J. D., Fisher, P., & Soljacic, M. (2007). Wireless power transfer via strongly coupled magnetic resonances. *Science*, *317*(5834), 83–86. doi:10.1126/science.1143254 PMID:17556549

Laftit, S., Proth, J. M., & Xie, X. L. (1992). Optimization of invariant criteria for event graphs. *IEEE Transactions on Automatic Control*, *37*(12), 547–555. doi:10.1109/9.135488

Lanas, V., & Logan, B. E. (2013). Evaluation of multi-brush anode systems in microbial fuel cells. *Bioresource Technology*, *148*, 379–385. doi:10.1016/j.biortech.2013.08.154 PMID:24063821

Lanthaler, M. (n.d.). *Self-Healing Wireless Sensor Network.* Available at: http://74.125.155.132/scholar?q=cache:qoVzFSeebusJ:scholar.google.com/+selfhealing+wireless+sensor+netowrk+lanthaler&hl=en&as_sdt=2000

Lee, R.-G., Chen, K.-C., Chiang, S.-S., Lai, C.-C., Liu, H.-S., & Wei, M.-S. (2006). A Backup Routing with Wireless Sensor Network for Bridge Monitoring System. In *Annual Conference on Communication Networksand Services Research.* IEEE Computer Society.

Lee, V. C. (2012). *Energy Harvesting for Wireless Sensor Network.* (Doctoral dissertation). University of California, Berkeley, CA.

Leifer, R., McDermott, C., O'Connor, G., Peters, L., & Rice, et al. (2000). *Radical Innovation How Mature Companies Can Outsmart Upstarts*. Boston: Harvard Business School Press.

Leonowicz, F. L. (n.d.). *Intelligent Wireless Street Lighting System.* Electronic Engineering Department, University Roma Tre.

Li, L., Zhang, Y., San, H., Guo, Y. I., & Chen, X. (2008). MEMS for vibration energy harvesting. In *MEMS/MOEMS Technologiesand Applications III, Proceedings of the SPIE.*

Li, X., Guo, M., & Dong, S. (2011). A flex-compressive-mode piezoelectric transducer for mechanical vibration/strain energy harvesting. *IEEE Transactions on Ultrasonics, Ferroelectrics, and Frequency Control*, *58*(4), 698–703.

Liang, P., Wu, W., Wei, J., Yuan, L., Xia, X., & Huang, X. (2011). Alternate charging and discharging of capacitor to enhance the electron production of bioelectrochemical systems. *Environmental Science & Technology*, *45*(15), 6647–6653. doi:10.1021/es200759v PMID:21710972

Li, J., & Wonham, W. M. (1993). Control of vector discrete event systems i – The base model. *IEEE Transactions on Automatic Control*, *38*(8), 1214–1227. doi:10.1109/9.233154

Liu, Y. C. (2009). Distributed Intelligent City Street Lamp Monitoring and Control System Based on Wireless Communication Chip nR401. In *Network Security, Wireless Communications and Trusted Computing* (pp. 278-281). NSWCTC.

Liu, H., & Logan, B. E. (2004). Electricity Generation Using an Air-Cathode Single Chamber Microbial Fuel Cell in the Presence and Absence of a Proton Exchange Membrane. *Environmental Science & Technology*, *38*(14), 4040–4046. doi:10.1021/es0499344 PMID:15298217

Liu, H., Ramnarayanan, R., & Logan, B. E. (2004). Production of Electricity during Wastewater Treatment Using a Single Chamber Microbial Fuel Cell. *Environmental Science & Technology*, *38*(7), 2281–2285. doi:10.1021/es034923g PMID:15112835

Liu, J.-Q., Fang, H.-B., Xu, Z.-Y., Mao, X.-H., Shen, X.-C., Chen, D., & Cai, B.-C. et al. (2008). A MEMS-based piezoelectric power generator array for vibration energy harvesting. *Microelectronics Journal*, *39*(5), 802–806. doi:10.1016/j.mejo.2007.12.017

Liu, L., Wang, H., Liu, X., Jin, X., He, W. B., Wang, Q. B., & Chen, Y. (2009, June). GreenCloud: a new architecture for green data center. In *Proceedings of the 6th international conference industry session on Autonomic computing and communications industry session* (pp. 29-38). ACM. doi:10.1145/1555312.1555319

Liu, X., Du, X., Wang, X., Li, N., Xu, P., & Ding, Y. (2013). Improved microbial fuel cell performance by encapsulating microbial cells with a nickel-coated sponge. *Biosensors & Bioelectronics*, *41*, 848–851. doi:10.1016/j.bios.2012.08.014 PMID:22939511

Liu, Y., Li, M., Ji, X., Luo, X., Wang, M., & Zhang, Y. (2014). A comparative study of the maximum power point tracking methods for PV systems. *Energy Conversion and Management*, *85*, 809–816. doi:10.1016/j.enconman.2014.01.049

Liu, Y., Liu, H., Wang, C., Hou, S.-X., & Yang, N. (2013). Sustainable energy recovery in wastewater treatment by microbial fuel cells: Stable power generation with nitrogen-doped graphene cathode. *Environmental Science & Technology*, *47*(23), 13889–13895. doi:10.1021/es4032216 PMID:24219223

Liu, Y., Ren, K. L., Hofmann, H. F., & Zhang, Q. (2005). Investigation of Electrostrictive Polymers for Energy Harvesting. *IEEE Transactions on Ultrasonics, Ferroelectrics, and Frequency Control*, *52*(12), 2411–2417. doi:10.1109/TUFFC.2005.1563285 PMID:16463508

Liu, Z., Lin, M., Wierman, A., Low, S. H., & Andrew, L. L. (2011, June). Greening geographical load balancing. In *Proceedings of the ACM SIGMETRICS joint international conference on Measurement and modeling of computer systems* (pp. 233-244). ACM. doi:10.1145/1993744.1993767

Li, W. J., Chan, G. M. H., Ching, N. N. H., Leong, P. H. W., & Wong, H. Y. (2000). Dynamical modeling and simulation of a lasermicromachined vibration-based micro power generator. *International Journal of Nonlinear Science Simulation*, *1*, 345–353.

Li, Y., Yu, H., Su, B., & Shang, Y. (2008). Hybrid micropower source for wireless sensor network. *IEEE Sensors Journal*, *8*(6), 678–681. doi:10.1109/JSEN.2008.922692

Loewe, P., & Dominiquini, J. (2006). Overcoming the barriers to effective innovation. *Journal of Strategy and Leadership*, *34*(1), 24–31. doi:10.1108/10878570610637858

Logan, B. E. (2008). *Microbial Fuel Cells*. Retrieved from http://books.google.pt/books/about/Microbial_Fuel_Cells.html?id=cXnc2wmrE9gC&pgis=1

Logan, B. E. (n.d.). *Patents Related to Microbial Fuel Cells and Microbial Electrolysis Cells*. Retrieved February 11, 2014, from http://www.engr.psu.edu/ce/enve/logan/bioenergy/mfc_patents.htm

Logan, B. E. (2009). Exoelectrogenic bacteria that power microbial fuel cells. *Nature Reviews. Microbiology*, *7*(5), 375–381. doi:10.1038/nrmicro2113 PMID:19330018

Logan, B. E., Hamelers, B., Rozendal, R., Schröder, U., Keller, J., Freguia, S., & Rabaey, K. et al. (2006). Microbial Fuel Cells: Methodology and Technology †. *Environmental Science & Technology*, *40*(17), 5181–5192. doi:10.1021/es0605016 PMID:16999087

Logan, B. E., & Regan, J. M. (2006). Microbial fuel cells-challenges and applications. *Environmental Science & Technology*, *40*(17), 5172–5180. doi:10.1021/es0627592 PMID:16999086

Lopez-Lapena, O., Penella, M. T., & Gasulla, M. (2010). A New MPPT Method for Low-Power Solar Energy Harvesting. *IEEE Transactions on Industrial Electronics*, *57*(9), 3129–3138. doi:10.1109/TIE.2009.2037653

Louro, R. O., & Paquete, C. M. (2012). The quest to achieve the detailed structural and functional characterization of CymA. *Biochemical Society Transactions*, *40*(6), 1291–1294. doi:10.1042/BST20120114 PMID:23176470

Lovley, D. R. (2006). Bug juice: Harvesting electricity with microorganisms. *Nature Reviews. Microbiology*, *4*(7), 497–508. doi:10.1038/nrmicro1442 PMID:16778836

Lu, C., Park, S. P., Raghunathan, V., & Roy, K. (2010). Analysis and design of ultra low power thermoelectric energy harvesting systems. In *Proceedings of the ACM/IEEE International Symposiumon Low-Power Electronics and Design (ISLPED '10)*. doi:10.1145/1840845.1840882

Lucy, E.-A. (2010). *Innowattech: Harvesting Energy and Data A stand alone technology*. Retrieved March 2015, 2, from Innowattech: http://www.iroads.co.il/sites/default/files/mtsgt_1_innowattech_presentation_-lucy_edery-azulay.pdf

Lu, F., Lee, H. P., & Lim, S. P. (2004). Modeling and analysis of micro piezoelectric power generators for micro-electromechanical systems applications. *Smart Materials and Structures*, *13*(1), 57–63. doi:10.1088/0964-1726/13/1/007

Lund, J. W. (1995). Onion dehydration. *Transactions - Geothermal Resources Council*, *19*, 69–74.

Luthra, J., & Pal, S. K. (2011). A hybrid firefly algorithm using genetic operators for the cryptanalysis of a monoalphabetic substitution cipher. In *Information and Communication Technologies (WICT), World Congress I* (pp. 202-206). Mumbai, India: IEEE.

Mahmood, E., & Ambikairajah. (2007), On DNA numerical representations for period-3 based exon prediction. *IEEE International Workshop on Genomic Signal Processing and Statistics.*

Maidment, G., & Tozer, R. (2002). Combined cooling heat and power in supermarkets. *Applied Thermal Engineering, 22*(6), 653–665. doi:10.1016/S1359-4311(01)00117-X

Mainwaring, A., Culler, D., Polastre, J., Szewczyk, R., & Anderson, J. (2002). Wireless Sensor Networks for Habitat Monitoring. In *Proc. 1st ACM International Workshop on Wireless Sensor Networks and Applications*. ACM.

Ma, J. (1995). Properties of high-T_c superconducting circular waveguides with Meissner boundary. *International Journal of Infrared and Millimeter Waves, 16*(1), 147–157. doi:10.1007/BF02085853

Ma, J. (1998). TM-properties of HTS's rectangular waveguides with Meissner boundary condition. *International Journal of Infrared and Millimeter Waves, 19*(3), 399–408. doi:10.1023/A:1022599324690

Ma, J. (1999). Wave propagation properties in high-temperature superconducting parallel-plate waveguides. *IEEE Microwave and Guided Wave Letters, 9*(5), 183–185. doi:10.1109/75.766758

Ma, J., Wang, Z., Li, X., Wang, Y., & Wu, Z. (2014). Bioelectricity generation through microbial fuel cell using organic matters recovered from municipal wastewater. *Environmental Progress & Sustainable Energy, 33*(1), 290–297. doi:10.1002/ep.11745

Mancelos, N., Correia, J., Pires, L. M., Oliveira, L. B., & Oliveira, J. P. (2014). Piezoelectric Energy Harvester for a CMOS Wireless Sensor. In *Technological Innovation for Collective Awareness Systems* (pp. 470–477). Springer Berlin Heidelberg. doi:10.1007/978-3-642-54734-8_52

Mandal S., Turicchia L., &Sarpeshkar R. (2010). A low-power, battery-free tag for body sensor networks. *IEEE Pervasive Computing, 9*(1), 71-77.

Manyika, J. (2013). *Disruptive technologies: Advance that will transform life, business and the global economy*. Retrieved from McKinsey & Company web site: http://www.mckinsey.com/insights/business_technology/disruptive_technologies

Mardamutu, K., Joon, R. A. L. A., Jegatheesan, J., Ponnusamy, V., & Mei, Y. (2008). A Critical Analysis of Green Database System for Energy Efficiency and Green Computing. *Computer Communication Review, 39*(1), 68.

Mareca, H., Darryl, G., & Chi, C. (2013). *Mobile Sensing Health & Wellness.* ON World Survey Report.

Martin, M. (2012). *Cardiac Pacemakers Market: Global industry size, share, trends, analysis and forecasts 2012-2018.* Retrieved March 9, 2015, from http://www.allvoices.com/contributed-news/12404227-cardiac-pacemakers-market-global-industry-size-share-treds-analysis-and-forecasts-2012-2018

Martins, G., Peixoto, L., Ribeiro, D. C., Parpot, P., Brito, A. G., & Nogueira, R. (2010). Towards implementation of a benthic microbial fuel cell in lake Furnas (Azores): Phylogenetic affiliation and electrochemical activity of sediment bacteria. *Bioelectrochemistry (Amsterdam, Netherlands), 78*(1), 67–71. doi:10.1016/j.bioelechem.2009.07.003 PMID:19716775

Mascarenas, D. L. (2006). *Development of an impedance-based wireless sensor node for monitoring of bolted joint preload.* (M.S. Thesis). Dept. of Structural Engineering, University of California, San Diego, CA

Mathkour, H., Ahmad, M., & Khan, H. M. (2009). An integrated statistical comparative analysis between variant genetic datasets of Mus musculus. *International Journal of Computational Intelligence in Bioinformatics and Systems Biology*, *1*(2), 163–176. doi:10.1504/IJCIBSB.2009.030647

Matos, C. T., & Lopes da Silva, T. (2013). Using multi-parameter flow cytometry as a novel approach for physiological characterization of bacteria in microbial fuel cells. *Process Biochemistry*, *48*(1), 49–57. doi:10.1016/j.procbio.2012.11.003

Matova S. P., Elfrink R., Vullers R. J. M., & Schaijk R. V. (2011). Harvesting energy from airow with a michromachined piezoelectric harvester inside a Helmholtzresonator. *Journal of Micromechanics and Microengineering, 21*(10), 1-6.

Mattis, D. C., & Bardeen, J. (1958). Theory of the anomalous skin effect in normal and superconducting metals. *Physical Review*, *111*(2), 412–417. doi:10.1103/PhysRev.111.412

Maulik, U., & Bandyopadhyay, S. (2000). Genetic algorithm-based clustering technique. *Pattern Recognition*, *33*(9), 1455–1465. doi:10.1016/S0031-3203(99)00137-5

Mbed. (n.d.). Platforms. Retrieved November 02, 2013, from http://developer.mbed.org/platforms/

McGowin, C. (2008). *Renewable Energy Technical Assessment Guide. TAG-RE: 2007, Electric Power Research Institute*. Palo Alto, CA: EPRI.

Meehan, A., Gao, H., & Lewandowski, Z. (2011). Energy Harvesting With Microbial Fuel Cell and Power Management System. *IEEE Transactions on Power Electronics*, *26*(1), 176–181. doi:10.1109/TPEL.2010.2054114

Mei, Q., Shan, M., Liu, L., & Guerrero, J. M. (2011). A Novel Improved Variable Step-Size Incremental-Resistance MPPT Method for PV Systems. *IEEE Transactions on Industrial Electronics*, *58*(6), 2427–2434. doi:10.1109/TIE.2010.2064275

Membranes International Inc. (n.d.). *CMI-7000 Cation Exchange Membranes Technical Specifications.* Retrieved June 16, 2014, from http://www.membranesinternational.com/tech-cmi.htm

Mena-Chalco, J.P., Carrer, H., Zana, Y. & Cesar, R.M. (2008). Identification of Protein Coding Regions Using the Modified Gabor-Wavelet Transform. *IEEE/ACM Transactions on Computational Biology and Bioinformatics, 5*(2), 198 – 207.

Meninger, S. (1999). *A low power controller for a MEMS based energy converter.* M Sc Massachusetts Institute of Technology.

MIT Technology Review. (2003). *10 Emerging Technologies That Will Change the World.* Retrieved September 15, 2015 from http://www2.technologyreview.com/featured-story/401775/10-emerging-technologies-that-will-change-the/2/

Mitcheson, P. D., Green, T. C., Yeatman, E. M., & Holmes, A. S. (2004). Architectures for Vibration-Driven Micro-Power Generators, IEEE. *Journal of Microelectromechanical Systems*, *13*(3), 429–440. doi:10.1109/JMEMS.2004.830151

Mitcheson, P. D., Yeatman, E. M., Rao, G. K., Holmes, A. S., & Green, T. C. (2008). Energy harvesting from human and machine motion for wireless electronic devices. *Proceedings of the IEEE*, *96*(9), 1457–1486. doi:10.1109/JPROC.2008.927494

Mitrovic, B., & Rozema, L. A. (2008). On the correct formula for the lifetime broadened superconducting density of states. *Journal of Physics Condensed Matter*, *20*(1), 015215. doi:10.1088/0953-8984/20/01/015215

Mizuno, M., & Chetwynd, D. (2003). Investigation of a resonance microgenerator, Micromech. *Microeng*, *13*(2), 209–216. doi:10.1088/0960-1317/13/2/307

Mohamaddoust, R., Haghighat, A. T., Motahari Sharif, M. J., & Capanni, N. (2011). A Novel Design of An Automatic Lighting Control System for a Wireless Sensor Network with Increased Sensor Lifetime and Reduced Sensor Numbers. *Sensors (Basel, Switzerland)*, *11*(12), 8933–8952. doi:10.3390/s110908933 PMID:22164114

Mohamed, M. I., Wu, W. Y., & Moniri, M. (2011). Power harvesting for smart sensor networks in monitoring water distribution system. In *Proceedings of the International Conference onNetworking, Sensing and Control (ICNSC '11)*. doi:10.1109/ICNSC.2011.5874897

Mohammadi, E., Karimi, M., & Heikalabad, S. R. (2011). A novel virtual machine placement in cloud computing. *Australian Journal of Basic and Applied Sciences*, *5*(10), 1549–1555.

Molognoni, D., Puig, S., Balaguer, M. D., Liberale, A., Capodaglio, A. G., Callegari, A., & Colprim, J. (2014). Reducing start-up time and minimizing energy losses of Microbial Fuel Cells using Maximum Power Point Tracking strategy. *Journal of Power Sources*, *269*, 403–411. doi:10.1016/j.jpowsour.2014.07.033

Morais, R., Matos, S. G., Fernandes, M. A., Valente, A. L. G., Soares, S. F. S. P., Ferreirac, P. J. S. G., & Reis, M. J. C. S. (2008). Sun, wind and water flow as energy supply for small stationary data acquisition platforms. *Computers and Electronics in Agriculture*, *64*(2), 120–132. doi:10.1016/j.compag.2008.04.005

Morgado, L., Dantas, J. M., Bruix, M., Londer, Y. Y., & Salgueiro, C. A. (2012). Fine Tuning of Redox Networks on Multiheme Cytochromes from Geobacter sulfurreducens Drives Physiological Electron/Proton Energy Transduction. *Bioinorganic Chemistry and Applications*, *298739*. doi:10.1155/2012/298739 PMID:22899897

Morton, S., Burns, D., & Dainty, A. (2007). Beyond Lean: Overcoming *Resistance* to Innovation to Improve Productivity. *POMS 18th Annual Conference.*

Mousavi, A., Vyatkin, V., Berezovskaya, Y., & Zhang, X. (2015, September). Towards energy smart data centers: Simulation of server room cooling system. In *Emerging Technologies & Factory Automation (ETFA), 2015 IEEE 20th Conference on* (pp. 1-6). IEEE.

Murphy, H., & Niitsuma, H. (1999). Strategies for compensating for higher costs of geothermal electricity with environmental benefits. *Geothermics*, *28*(6), 693–711. doi:10.1016/S0375-6505(99)00018-8

Muthukumar, M., & Sangeetha, T. (2014). The Harnessing Of Bioenergy From A Dual Chambered Microbial Fuel Cell (Mfc) Employing Sago-Processing Wastewater As Catholyte. *International Journal of Green Energy*, *11*(2), 161–172. doi:10.1080/15435075.2013.771581

Muthuramalingam, M., & Manoharan, P. S. (2014). Comparative analysis of distributed MPPT controllers for partially shaded stand alone photovoltaic systems. *Energy Conversion and Management*, *86*, 286–299. doi:10.1016/j.enconman.2014.05.044

Myron, L. Company. (2012). *Application Bulletin: Oxidation Reduction Potential (ORP)/Redox and Free Chlorine.* Retrieved from http://www.myronl.com/PDF/application_bulletins/orp_ab.pdf

Nacef, A. B., Senouci, S.-M., Doudane, Y. G., & Beylot, A.-L. (2011). A cooperative low power Mac protocol for wireless sensor networks. In *Proceedings of the IEEE InternationalConference on Communications (ICC '11)*. doi:10.1109/icc.2011.5962416

Nadhirah, A., Mohd, A. O., Mohd, N. H., & Mohd, H. M. (2014). A review of firefly algorithm. *Journal of Engineering and Applied Sciences, 9*(10).

Nadler, D., & Gerstein, M. (1992). *What is organizational architecture?* Harvard Business Review September–October.

Nanda, B., & Das, G. (2014). Ant colony optimization. A computational intelligence technique. *International Journal of Computer & Communication Technology*, *2*(6), 105–110.

Nawaz, I., & Tiwari, G. (2006). Embodied energy analysis of photovoltaic (PV) system based on macro-and micro-level. *Energy Policy*, *34*(17), 3144–3152. doi:10.1016/j.enpol.2005.06.018

Ngô, C., & Natowitz, J. (2009). *Our energy future: resources, alternatives and the environment*. Wiley. doi:10.1002/9780470473795

Nguyen, T.-H., Yu, Y.-Y., Wang, X., Wang, J.-Y., & Song, H. (2013). A 3D mesoporous polysulfone-carbon nanotube anode for enhanced bioelectricity output in microbial fuel cells. *Chemical Communications (Cambridge, England)*, *49*(91), 10754–10756. doi:10.1039/c3cc45775c PMID:24108240

Ni, T., Zuo, L., & Kareem, A. (2011). Assessment of Energy Potential and Vibration Mitigation of Regenerative Tuned Mass Dampers on Wind Excited Tall Buildings. In Proceedings of 2011 ASME design engineering technical conferences, (vol. 1, pp. 333-342). doi:10.1115/DETC2011-48728

Nilsson, J. W., & Riedel, S. A. (2001). *Electric Circuits*. Retrieved from http://books.google.pt/books/about/Electric_Circuits.html?id=sehSAAAAMAAJ&pgis=1

Noguchi, T., Naruse, M., & Sekimoto, Y. (2012). RF conductivity and surface impedance of a superconductor taking into account the complex superconducting gap energy. *Physics Procedia*, *36*, 318–323. doi:10.1016/j.phpro.2012.06.166

Noguchi, T., Naruse, M., & Sekimoto, Y. (2013). Contribution of quasiparticles in the subgap states to the surface impedance of superconductors. *IEEE Transactions on Applied Superconductivity*, *23*(3), 1501404. doi:10.1109/TASC.2013.2240761

Noguchi, T., Suzuki, T., Endo, A., & Tamura, T. (2009). Contribution of the imaginary part of the superconducting gap energy on the SIS tunneling current. *Physica. C, Superconductivity*, *469*(15-20), 1585–1588. doi:10.1016/j.physc.2009.05.218

Noguchi, T., Suzuki, T., & Tamura, T. (2011). Subgap tunneling current at low temperature in Nb/Al-AlN/Nb SIS junctions. *IEEE Transactions on Applied Superconductivity*, *21*(3), 756–759. doi:10.1109/TASC.2010.2089033

OECD. (2009A). *Towards Green ICT Strategies. Assessing Policies and Programmes on ICT and the Environment*. DSTI/ICCP/IE (2008)3/FINAL, May 2009. Retrieved February 5, 2010 from www.oecd.org/sti/ict/green-ict

Ohara, Zhang,, & Toda, Miyake, & Imamura. (2007). De-noising Method by Combining Adaptive Line Enhancer and Complex Discrete Wavelet Transform. *ICIC Express Letters*, *1*(2), 145–151.

Olivo, J., Carrara, S., & De Micheli, G. (2011). Energy harvesting and remote powering for implantable biosensors. *IEEE Sensors Journal*, *11*, 1573–1586.

Ooi, B. L. (2010). *Optimisation and Frequency Tuning Concepts for a Vibration Energy Harvester.* (PhD Thesis). University of Hull.

Ooi, B. L. A., Aziz, R. A., & Gilbert, J. M. (2014). Analytical and numerical modelling for dual-resonator vibration-based electromagnetic generator.*5th International Conference on Intelligent and Advanced Systems (ICIAS)* (pp. 1-5). Kuala Lumpur: IEEE. doi:10.1109/ICIAS.2014.6869443

Ooi, B. L., & Gilbert, J. M. (2014). Design of wideband vibration-based electromagnetic generator by means of dual-resonator. *Sensors and Actuators. A, Physical*, *213*(0), 9–18. doi:10.1016/j.sna.2014.03.037

Oyama, A., Obayashi, S., Nakahashi, K., & Nakamura, T. (1997). Transonic wing optimization using genetic algorithm. *AIAA Paper*, *97*, 1854.

Padhy, P., Martinez, K., Riddoch, A., Hart, J., & Ong, H. (2005). Glacialenvironmentmonitoring using sensor networks. In *Proceedings of the Conference on Real-World Wireless Sensor Networks.*

Paine, S., Papa, D. C., Leombruno, R. L., Zhang, X., & Blundell, R. (1994). Beam waveguide and receiver optics for the SMA.*Proceedings of the 5th International Symposium on Space Terahertz Technology*. University of Michigan.

Palosaari, J., Leinonen, M., Hannu, J., Juuti, J., & Jantunen, H. (2012). Energy harvesting with a cymbal type piezoelectric transducer from low frequency compression. *Journal of Electroceramics*, *28*(4), 214–219. doi:10.1007/s10832-012-9713-8

Pant, D., Van Bogaert, G., Diels, L., & Vanbroekhoven, K. (2010). A review of the substrates used in microbial fuel cells (MFCs) for sustainable energy production. *Bioresource Technology*, *101*(6), 1533–1543. doi:10.1016/j.biortech.2009.10.017 PMID:19892549

Paradiso, J. A., & Starner, T. (2005). Energy Scanvenging for Mobile and Wireless Electronics. *IEEE Pervasive Computing / IEEE Computer Society [and] IEEE Communications Society*, *4*(1), 18–27. doi:10.1109/MPRV.2005.9

Paradiso, J. A., & Starner, T. (2005). Energy scavenging for mobile and wireless electronics. *Pervasive Computing, IEEE*, *4*(1), 18–27.

Parisio, A., Molinari, M., Varagnolo, D., & Johansson, K. H. (n.d.). Control of HVAC Systems in Sweden. *Current Status and Future Directions. Academic Press.*

Park, C., & Chou, P.H. (2007). AmbiMax: Autonomous Energy Harvesting Platform for Multi-Supply Wireless Sensor Nodes. *3rd Annual IEEE Communications Society on Sensor and Ad Hoc Communications and Networks.*

Park, J.-D., & Ren, Z. (2011). Efficient energy harvester for microbial fuel cells using DC/DC converters. In 2011 IEEE Energy Conversion Congress and Exposition (pp. 3852–3858). IEEE. doi:10.1109/ECCE.2011.6064292

Park, C., & Chou, P. H. (2004). Power utility maximization for multiple-supply systems by a load-matching switch. *Proceedings of ACM/IEEE International Symposium on Low Power Electronics and Design.* doi:10.1145/1013235.1013281

Park, G., Rosing, T., Todd, M. D., Farrar, C. R., & Hodgkiss, W. (2008). Energy harvesting for structural health monitoring sensor networks. *Journal of Infrastructure Systems*, *14*(1), 64–79. doi:10.1061/(ASCE)1076-0342(2008)14:1(64)

Park, J.-D., & Ren, Z. (2012a). High efficiency energy harvesting from microbial fuel cells using a synchronous boost converter. *Journal of Power Sources*, *208*, 322–327. doi:10.1016/j.jpowsour.2012.02.035

Park, J.-D., & Ren, Z. (2012b). Hysteresis controller based maximum power point tracking energy harvesting system for microbial fuel cells. *Journal of Power Sources*, *205*, 151–156. doi:10.1016/j.jpowsour.2012.01.053

Patil, R. U., & Gawade, S. S. (2012). Design and Static Magnetic Analysis of Electromagnetic Regenerative Shock Absorber. *International Journal of Advanced Engineering Technology*, *III*(II), 54–59.

Paul, C., Ramamurthy, K., & Alla, L. (2002). Organizational context, climate and innovativeness: Adoption of imaging technology. *Journal of Engineering and Technology Management*, *19*(3-4), 221–247. doi:10.1016/S0923-4748(02)00019-X

Pawlish, M., Varde, A. S., Robila, S. A., & Ranganathan, A. (2014). A call for energy efficiency in data centers. *SIGMOD Record*, *43*(1), 45–51. doi:10.1145/2627692.2627703

Peng, X., Yu, H., Yu, H., & Wang, X. (2013). Lack of anodic capacitance causes power overshoot in microbial fuel cells. *Bioresource Technology*, *138*, 353–358. doi:10.1016/j.biortech.2013.03.187 PMID:23624054

Petrone, G., Spagnuolo, G., & Vitelli, M. (2012). An analog technique for distributed MPPT PV applications. *Industrial Electronics, IEEE*, *59*(12), 4713–4722. Retrieved from http://ieeexplore.ieee.org/xpls/abs_all.jsp?arnumber=6092483

Pfeffer, J., & Salancik, G. (1978). *The External Control of Organizations: A Resource Dependence Perspective*. New York: Harper and Row Publishers.

Philip, G., & Booth, M. E. (2001). A new six 'S' framework on the relationship between the role of information systems (IS) and competencies in 'IS' management. *Journal of Business Research*, *51*(3), 233–247. doi:10.1016/S0148-2963(99)00051-X

Pobering, S., & Schwesinger, N. (2004). A novel hydropower harvesting device. In *Proceedings of the International Conference onMEMS, NANO and Smart Systems (ICMENS '04).*

Pobering, S., & Schwesinger, N. (2009). Power supply for wireless sensor systems. In *Proceedings of the IEEE Sensors Conference(SENSORS '08).*

Potter, M. (1911). Electrical effects accompanying the decomposition of organic compounds. *Proceedings of the Royal Society of London. Series B*. Retrieved from http://www.jstor.org/stable/80609

Powers, C. (2011). *Leadership in Green IT (Brochure) (No. DOE/GO-102011-3295). National Renewable Energy Laboratory*. Golden, CO: NREL.

Prajapati, N. B., Agravat, R. R., & Hasan, M. I. (2009). *Comparative study of various cooling schedules for location area planning in cellular networks using simulated annealing. In Networks and Communications NETCOM'09* (pp. 146–150). Chennai, India: IEEE.

Proth, J. M., & Xie, X. L. (1994). Cycle time of stochastic event graphs: Evaluation and marking optimization. *IEEE Transactions on Automatic Control*, *39*(7), 1482–1486. doi:10.1109/9.299640

Purkus, A., & Barth, V. (2011). Geothermal power production in future electricity markets— A scenario analysis for Germany. *Energy Policy*, *39*(1), 349–357. doi:10.1016/j.enpol.2010.10.003

Putha, R., Quadrifoglio, L., & Zechman, E. (2012). Comparing ant colony optimization and genetic algorithm approaches for solving traffic signal coordination under oversaturation conditions. *Computer-Aided Civil and Infrastructure Engineering*, *27*(1), 14–28. doi:10.1111/j.1467-8667.2010.00715.x

Rabaey, K., & Verstraete, W. (2005). Microbial fuel cells: novel biotechnology for energy generation. *TRENDS in Biotechnology*. Retrieved from http://www.sciencedirect.com/science/article/pii/S0167779905000922

Rabaey, J. M., Ammer, M. J., da Silva, J. L. Jr, Patel, D., & Roundy, S. (2000). PicoRadio supports ad hoc ultra-low power wireless networking. *Computer*, *33*(7), 42–48. doi:10.1109/2.869369

Rabaey, K., Rodríguez, J., Blackall, L. L., Keller, J., Gross, P., Batstone, D., & Nealson, K. H. et al. (2007). Microbial ecology meets electrochemistry: Electricity-driven and driving communities. *The ISME Journal*, *1*(1), 9–18. doi:10.1038/ismej.2007.4 PMID:18043609

Rachinski, S., Carubelli, A., Mangoni, A. P., & Mangrich, A. S. (2010). Pilhas de combustíveis microbianas utilizadas na produção de eletricidade a partir de rejeitos orgânicos: Uma perspectiva de futuro. *Quimica Nova*, *33*(8), 1773–1778. doi:10.1590/S0100-40422010000800026

Radeckas, B., & Lukosevicius, V. (2000). Klaipeda Geothermal demonstration project. In *Proceedings World Geothermal Congress 2000.*

Raghunathan, V., Kansal, A., Hsu, J., Friedman, J., & Srivastava, M. (2005). Design considerations for solar energy harvesting wireless embedded systems. In *Proceedings of the 4th InternationalSymposium on Information Processing in Sensor Networks(IPSN '05)*. doi:10.1109/IPSN.2005.1440973

Rahimi, M., Shah, H., Sukhatme, G. S., & Estrin, D. (2003). Studying the feasibility of Energy harvesting in a mobile sensor network. *Proceedings of IEEE International Conference on Robotics and Automation*. doi:10.1109/ROBOT.2003.1241567

Rahman, & Leong, K. (2011). Investigation of useful ambient vibration sources for the application of energy harvesting. *IEEE Student Conference on Research and Development (SCORed)* (pp. 391-396). Cyberjaya, Malaysia: IEEE.

Rahmani, R., & Seyedmahmoudian, M. (2013). Implementation of Fuzzy Logic Maximum Power Point Tracking Controller for Photovoltaic System. *American Journal of Applied Sciences*, *10*(3), 209–218. doi:10.3844/ajassp.2013.209.218

Raju, M., & Grazier, M. (2008). *Energy harvesting. ULP meets energy harvesting: a game-changing combination for design engineers*. Retrieved from http://scholar.google.pt/scholar?hl=pt-PT&q=ULP+Meets+Energy+Harvesting&btnG=&lr=#0

Ramachandran, T., Jung, L. T., Ponnusamy, V., & Hudaya, A. (2014, June). ZigBee wireless communication for monitoring renewable street light system. In *Computer and Information Sciences (ICCOINS), 2014 International Conference on* (pp. 1-6). IEEE.

Ramasur, D., & Hancke, G. P. (2012). A wind energy harvester for low power wireless sensor networks. In *Proceedings of the IEEE International Instrumentation and Measurement Technology Conference (I2MTC '12)*. doi:10.1109/I2MTC.2012.6229698

Ranganathan, P., Rivoire, S., & Moore, J. (2009). Models and Metrics for Energy-Efficient Computing. *Advances in Computers*, *75*, 159–233. doi:10.1016/S0065-2458(08)00803-6

Rao, Y., McEachern, K. M., & Arnold, P. D. (2013). A compact human-powered energy harvesting system. *Journal of Physics: Conference Series*, *476*(1). doi:10.1088/1742-6596/476/1/012011

Ravindran, S., Kroener, M., & Woias, P. (2012). *A standalone Pyroelectric Harvester for Thermal Energy Harvesting* (pp. 80–83). Atlanta, GA: PowerMEMS.

Rawat, R., & Chandel, S. S. (2013). Hill climbing techniques for tracking maximum power point in solar photovoltaic systems - a review. *Special Issue of International Journal of Sustainable Development and Green Economics, 2*(1-1), 90–95. Retrieved from http://irnet.sg/irnet_journal/IJSDGE/IJSDGE_doc/IJSDGE_V2I1,2_papers/17.pdf

Reif, T. (2008). Profitability analysis and risk management of geothermal projects. *Geo-Heat Center Quarterly Bulletin*, *28*(4), 1–4.

Reinisch H., Gruber S., Unterassinger H., Wiessecker M., Hofer G., Pribyl W., & Holweg G. (2011). An electro-magnetic energy harvesting system with 190 nW idle mode power consumption for a BAW based wireless sensor node. *IEEE Journal of Solid-State Circuits, 46*(7), 1728-1741.

Ren, S., Xia, X., Yuan, L., Liang, P., & Huang, X. (2013). Enhancing charge harvest from microbial fuel cells by controlling the charging and discharging frequency of capacitors. *Bioresource Technology*, *146*, 812–815. doi:10.1016/j.biortech.2013.08.055 PMID:23993288

Riener, R. M. (2011). An energy efficient pedestrian aware Smart Street Lighting. *International Journal of Pervasive*, 147-161.

Rocha, J. G., Goncalves, L. M., Rocha, P. F., Silva, M. P., & Lanceros-Mendez, S. (2008). Energy harvesting from piezoelectric materials fully integrated in footwear. *Industrial Electronics. IEEE Transactions on*, *57*(3), 813–819.

Rodrigo, M. A., Cañizares, P., García, H., Linares, J. J., & Lobato, J. (2009). Study of the acclimation stage and of the effect of the biodegradability on the performance of a microbial fuel cell. *Bioresource Technology*, *100*(20), 4704–4710. doi:10.1016/j.biortech.2009.04.073 PMID:19487121

Rodrigues, E., & Melício, R. (2011). Simulation of a solar cell considering single-diode equivalent circuit model. *... on Renewable Energies ...*. Retrieved from http://icrepq.com/icrepq'11/339-rodrigues.pdf

Roundy, S. (2003). *Energy Scavenging fro wireless sensor nodes with a focus on Vibration to electricity conversion.* (PhD Thesis). University of California Berkeley.

Roundy, S. J. (2003). *Energy scavenging for wireless sensor nodes with a focus on vibration to electricity conversion.* (Doctoral dissertation). University of California, Berkeley, CA.

Roundy, S., Steingart, D., Frechette, L., Wright, P. K., & Rabaey, J. (2004). Power Sources for Wireless Sensor Networks. In H. Karl, A. Wolisz, & A. Willig (Eds.), Wireless Sensor Networks (Vol. 2920, pp. 1-17). Springer Berlin Heidelberg. doi:10.1007/978-3-540-24606-0_1

Roundy, S., Wright, P. K., & Rabaey, J. (2003, July1). A study of low level vibrations as a power source for wireless sensor nodes. *Computer Communications*, *26*(11), 1131–1144. doi:10.1016/S0140-3664(02)00248-7

Rowe, D. (2012). *Modules, Systems, and Applications in Thermoelectrics.* CRC Press. Retrieved from https://books.google.com.my/books?id=CVLSBQAAQBAJ

Roy, M., Biswas, S., & Barman, S. (2009). Identification and Analysis of Coding and Noncoding Regions of a DNA Sequence by Positional Frequency Distribution of Nucleotides (PFDN) Algorithm. *4th International Conference on Computers and Devices for Communication.*

Rozendal, R. A., Hamelers, H. V. M., Rabaey, K., Keller, J., & Buisman, C. J. N. (2008). Towards practical implementation of bioelectrochemical wastewater treatment. *Trends in Biotechnology*, *26*(8), 450–459. doi:10.1016/j.tibtech.2008.04.008 PMID:18585807

Saadon, S., & Sidek, O. (2011). A review of vibration-based MEMS piezoelectric energy harvesters. *Energy Conversion and Management*, *52*(1), 500–504. doi:10.1016/j.enconman.2010.07.024

Saadon, S., & Sidek, O. (2011). Vibration-based MEMS piezoelectric energy harvester (VMPEH) modeling and analysis for green energy source. In *Proceedings of the 4th International Conferenceon Developments in eSystems Engineering (DeSE '11)*. doi:10.1109/DeSE.2011.85

Sabherwal, R., & Tsoumpas, P. (1993). The development of strategic information systems: Some case studies and research proposals. *European Journal of Information Systems*, *2*(24), 240–259. doi:10.1057/ejis.1993.36

Salas, V., Olías, E., Barrado, A., & Lázaro, A. (2006). Review of the maximum power point tracking algorithms for stand-alone photovoltaic systems. *Solar Energy Materials and Solar Cells*, *90*(11), 1555–1578. doi:10.1016/j.solmat.2005.10.023

Saleem, M., Di Caro, G. A., & Farooq, M. (2011). Swarm intelligence based routing protocol for wireless sensor networks: Survey and future directions. *Information Sciences*, *181*(20), 4597–4624. doi:10.1016/j.ins.2010.07.005

Sander & Saulny. (2007). *Bridge collapse in Minneapolis Kills at least 7.* Retrieved from New York Times web site: http://www.nytimes.com/2007/08/02/us/02bridge.html?_r=0

Santoro, C., Ieropoulos, I., Greenman, J., Cristiani, P., Vadas, T., Mackay, A., & Li, B. (2013). Current generation in membraneless single chamber microbial fuel cells (MFCs) treating urine. *Journal of Power Sources*, *238*, 190–196. doi:10.1016/j.jpowsour.2013.03.095

Santoro, C., Ieropoulos, I., Greenman, J., Cristiani, P., Vadas, T., Mackay, A., & Li, B. (2013). Power generation and contaminant removal in single chamber microbial fuel cells (SCMFCs) treating human urine. *International Journal of Hydrogen Energy*, *38*(26), 11543–11551. doi:10.1016/j.ijhydene.2013.02.070

Sardini, E., & Serpelloni, M. (2011). Self-powered wireless sensor for air temperature and velocity measurements with energy harvesting capability. *IEEE Transactions on Instrumentation and Measurement, 60*(5), 1838-1844.

Sari, I., Balkan, T., & Kulah, H. (2008). An electromagnetic micro power generator for wideband environmental vibrations. *Sensors and Actuators A: Physical, 145-146*(0), 405-413. Retrieved from http://www.sciencedirect.com/science/article/pii/S0924424707008564

Scott, S. (1904). *History of the Moorish Empire in Europe*. Academic Press.

SDF Solar Aware Distributed Flow in Wireless Sensor Networks. (n.d.). Darmstadt: Technische Universitat Darmstadt.

Seah, W. K., Eu, Z. A., & Tan, H. P. (2009). Wireless sensor networks powered by ambient energy harvesting (WSN-HEAP)-Survey and challenges. In *Wireless Communication, Vehicular Technology, Information Theory and Aerospace & Electronic Systems Technology, 2009. Wireless VITAE 2009. 1st International Conference on.*

Seah, W. K. G., Zhi, A. E., & Tan, H.-P. (2009). Wireless Sensor Networks Powered by Ambient Energy Harvesting (WSNHEAP)—survey and challenges. In *Proceedings of the 1stInternational Conference on Wireless Communication, Vehicular Technology, Information Theory and Aerospace and Electronic Systems Technology, Wireless (VITAE '09).*

Search Mobile Computing. (2006, April 17). Retrieved Jun 3, 2013, from Definition of Wireless: http://searchmobilecomputing.techtarget.com/definition/wireless

Sell, D., Kramer, P., & Kreysa, G. (1989). Use of an oxygen gas diffusion cathode and a three-dimensional packed bed anode in a bioelectrochemical fuel cell. *Applied Microbiology and Biotechnology*, *31*(2), 211–213. doi:10.1007/BF00262465

Sera, D., Mathe, L., & Kerekes, T. (2013). On the perturb-and-observe and incremental conductance MPPT methods for PV systems. *IEEE Journal of, 3*(3), 1070–1078. Retrieved from http://ieeexplore.ieee.org/xpls/abs_all.jsp?arnumber=6517453

Seyedi, A., & Sikdar, B. (2008). Energy Efficient Transmission Strategies for Body Sensor Networks with Energy Harvesting. In *Proc. of the42nd Annual Conference on Information Sciences and Systems (CISS)*. doi:10.1109/CISS.2008.4558613

Sha, K., Du, J., & Shi, W. (2005). *Wear: A balanced, fault-tolerant, energy-aware routing protocol for wireless sensor networks.* Technical Report MIST-TR-2005-001, Wayne State University.

Shah, R. C., & Rabaey, J. (2002). *Energy Aware Routing for Low Energy Ad Hoc Sensor Networks.* Paper presented at the IEEE Wireless Communications and Networking Conference (WCNC), Orlando, FL. doi:10.1109/WCNC.2002.993520

Shang, Y., Li, Y., Yu, H., Sun, H., & Su, B. (2007). Experimental investigation on thermoelectric generator of micro hybrid power source. In *Proceedings of the 3rd International Symposiumon Advanced Optical Manufacturing and Testing Technologies: Design, Manufacturing, and Testing of Micro- and Nano-OpticalDevices and Systems*, *Proceedings of the SPIE.*

Shankar, N. U. (1986). *Application of digital techniques to radio astronomy measurements.* (Doctoral dissertation). Available from Raman Research Institute Digital Repository. (2289/3391)

Shearwood, C., & Yates, R. B. (1997). Development of a resonant electromagnetic micro-generator. *Electronic Letters, 33*(22), 1883-1884.

Shifeng, Z., & Steve, G. (2007). A software architecture and framework for Web-based distributed Decision Support Systems. *Decision Support Systems*, (43): 1133–1150.

Shmilovitz, D. (2005). On the control of photovoltaic maximum power point tracker via output parameters. *IEE Proceedings. Electric Power Applications*, *152*(2), 239. doi:10.1049/ip-epa:20040978

Shuo & Yi-Sheng. (2008). An integrative algorithm for predicting protein coding regions. *IEEE Asia Pacific Conference on Circuits and Systems.*

Silva, R., Farinholt, K., & Park, G. (2011). Hybrid energy sources for embedded sensor nodes. In *Proceedings of the Industrialand Commercial Applications of Smart Structures Technologies.*

Simedics. (2014, Aug). Harvesting Energy to Power Medical Devices. *The Journal of mHealth, 1*(4), 38-39.

Simjee, F. I., & Chou, P. H. (2008). Efficient Charging of Supercapacitors forExtended Lifetime of Wireless Sensor Nodes. *IEEE Transactions on Power Electronics*, *23*(3), 1526–1536. doi:10.1109/TPEL.2008.921078

Simjee, F., Pai, H., & Chou, P. H. (2006). Everlast: Long-life, Supercapacitor operated Wireless Sensor Node. In *Proc 2006 International Symposium on Low Power Electronics and Design*. ACM. doi:10.1145/1165573.1165619

Smith, P. (2011). *Comparing Low-Power Wireless Technologies | DigiKey*. Retrieved December 03, 2014, from http://www.digikey.com/en/articles/techzone/2011/aug/comparing-low-power-wireless-technologies

Smoker, J., Nouh, M., Aldraihem, O., & Baz, A. (2012). Energy harvesting from a standing wave thermoacoustic-piezoelectric resonator. *Journal of Applied Physics*, *111*(10), 104901. doi:10.1063/1.4712630

Snarski, S. R., Kasper, R. G., & Bruno, A. B. (2004). Device for electromagnetohydrodynamic (EMHD) energy harvesting. *Proceedings of the Society for Photo-Instrumentation Engineers*, *5417*, 147–161. doi:10.1117/12.545170

Sobha, G., Chittal, R. R., & Kumar, K. (2007). Medical applications of wireless networks. In *Systems and Networks Communications. ICSNC 2007.Second International Conference*. Cap Esterel, France: IEEE.

Sodano, H. A., Inman, D. J., & Park, G. (2005). Generation and storage of electricity from power harvesting devices. *Journal of Intelligent Material Systems and Structures*, *16*(1), 67–75. doi:10.1177/1045389X05047210

Soga, K. (2011, May 4). *eh-network.org*. Retrieved from 2nd Energy Harvesting Research Theme Workshop - Structural Monitoring: http://eh-network.org/events/workshop2/slides1.pdf

Sohrabi, K., Gao, J., Ailawadhi, V., & Pottie, G. J. (2000). Protocols for Self-Organization of a Wireless Sensor Network. *IEEE Personal Communications*, *7*(5), 16–27. doi:10.1109/98.878532

Solar-Aware Clustering in Wireless Sensor Networks. (n.d.). Swedish Institute of Computer Science (SICS).

Soo, C., Devinne, Y., Midgley, D., & Deering, A. (2002). Knowledge management: Philosophy, processes, and pitfalls. *California Management Review*, *44*(4), 129–150. doi:10.2307/41166146

Special Interest Group. (2013). *Report on the UK Energy Harvesting Mission to Japan*. Energy Harvesting.

Starner, T. (1996). Human-powered wearable. *IBM Systems Journal, 35*(3-4), 618-629. doi:10.1147/sj.353.0618

Stojcev, M. K., Kosanovic, M. R., & Golubovic, L. R. (2009). Power management and energy harvesting techniques for wireless sensor nodes. In *Proceedings of the 9th International Conference onTelecommunication in Modern Satellite, Cable, and Broadcasting Services (TELSIKS '09)*. doi:10.1109/TELSKS.2009.5339410

Stokes, C. D., Duff, A., Mantini, M. J., Grant, B. A., & Venkatasubramanian, R. (2010). Nanostructured thermoelectric material and device technology for energy harvesting applications. In *Proceedings of the 4th IEEE Nanotechnology Materials and Devices Conference (NMDC '10)*. doi:10.1109/NMDC.2010.5652061

Strassner, B., & Chang, K. (2013). Microwave power transmission: Historical milestones and system components. *Proceedings of the IEEE*, *101*(6), 1379–1396. doi:10.1109/JPROC.2013.2246132

Stratton, J. A. (1941). *Electromagnetic Theory*. New York: McGraw-Hill.

Stützle, T., & Hoos, H. H. (2000). MAX--MIN ant system. *Future Generation Computer Systems*, *16*(8), 889–914. doi:10.1016/S0167-739X(00)00043-1

Subramanian, M. (2005). The influence of intellectual capital on the types of innovative capabilities. *Academy of Management Journal*, *48*(3), 450–463. doi:10.5465/AMJ.2005.17407911

Tabbakh, S. R. K., Maarefdoust, R., Kyun, N. C., & Ali, B. M. (2010). Environmental taxonomy of power scavenging techniques for autonomous self powered wireless sensors. In *Proceedings of theAsia Pacific Conference on Circuit and System (APCCAS '10)*.

Talbi, E. G., Roux, O., Fonlupt, C., & Robillard, D. (1999). Parallel ant colonies for combinatorial optimization problems. In J. Rolim (Eds.), Parallel and Distributed Processing, 11 IPPS/SPDP'99 Workshops, (LNCS), (Vol. 1586, pp. 239–247). Springer. doi:10.1007/BFb0097905

Tan, H. P., Eu, Z. A., & Seah, W. K. (2009). Impact of Power Control in Wireless Sensor Networks Powered by Ambient Energy Harvesting for Railroad Health Monitoring. In *Proc. of the 2nd Intl Workshop on Applications of Ad hoc and Sensor Networks* (AASNET). doi:10.1109/WAINA.2009.113

Tan, Y. K., & Panda, S. K. (2011). Energy harvesting from hybrid indoor ambient light and thermal energy sources for enhanced performance of wireless sensor nodes. *IEEE Transactions on Industrial Electronics, 58*(9), 4424-4435.

Tan, Y. K., & Panda, S. K. (2011). Self-autonomous wireless sensor nodes with wind energy harvesting for remote sensing of wind-driven wild re spread. *IEEE Transactions on Instrumentation and Measurement, 60*(4), 1367-1377.

Tang, L., Yang, Y., & Soh, C. (2010, December). Toward Broadband Vibration-based Energy Harvesting. *Journal of Intelligent Material Systems and Structures*, *21*(18), 1867–1897. doi:10.1177/1045389X10390249

Tang, X., & Zuo, L. (2011). Enhanced vibration energy harvesting using dual-mass systems. *Journal of Sound and Vibration*, *330*(21), 5199–5209. doi:10.1016/j.jsv.2011.05.019

Tang, X., & Zuo, L. (2012). Simultaneous energy harvesting and vibration control of structures with tuned mass dampers. *Journal of Intelligent Material Systems and Structures*, *23*(18), 2117–2127. doi:10.1177/1045389X12462644

Tan, Y. K., & Panda, S. K. (2007). A novel piezoelectric based wind energy harvester for low-power autonomous wind speedsensor. In *Proceedings of the 33rd Annual Conference of the IEEE Industrial Electronics Society (IECON '07).*

Technavio. (2015, January 09). *Technavio*. Retrieved March 13, 2015, from Discover market opportunities: http://www.technavio.com/pressrelease/increased-demand-from-consumer-electronics-segment-driving-growth-in-the-global-mems

TECTEG MFR. (n.d.). Retrieved February 3, 2015, from TECTEG MFR Home: http://thermoelectric-generator.com/

Terry Street. (2007). Green ICT –taking the strategic approach. *Socitm Consulting*. Retrieved on February 2, 2010 from socitm.gov.uk/consulting

Terziovski, M., Fitzpatrick, P., & O'Neill, P. (2003). Successful predictors of business process reengineering (BPR) in financial services. *International Journal of Production Economics*, *84*(1), 35–50. doi:10.1016/S0925-5273(02)00378-X

Thad, S. E. (2003). Powerful Change Part 1: Batteries and possible alternatives for the mobile market. *IEEE Pervasive Computing/IEEE Computer Society [and] IEEE Communications Society*, *2*(4), 86–88. doi:10.1109/MPRV.2003.1251172

Tham, C. Y., McCowen, A., & Towers, M. S. (2003). Modelling of PCB transients with boundary elements/method of moments in the frequency domain. *Engineering Analysis with Boundary Elements*, *27*(4), 315–323. doi:10.1016/S0955-7997(02)00119-4

Thomson Reuters. (n.d.). *ISI Web of Knowledge*. Retrieved January 01, 2014, from http://wokinfo.com/

Todar, K. (2008). *Online Textbook of Bacteriology*. Retrieved from http://textbookofbacteriology.net/index.html

Tong, L., Zhao, Q., & Adireddy, S. (2003). *Sensor networks with mobile agents*. Paper presented at the Military Communications Conference.

Tsai, H., Tu, C., & Su, Y. (2008). Development of generalized photovoltaic model using MATLAB/SIMULINK. *Proceedings of the World Congress on …*. Retrieved from http://s2i.bordeaux.free.fr/Espace Terminale/Ressources/Projet/Projet 2 2013-2014/2.DevelopmentofGeneralizedPhotovoltaicModelusingMatlab.pdf

Tuovenin, T., Berg, M., Yazdandoost, K. Y., Salonen, E., & Iinatti, J. (2012). Reactive near-field region radiation of planar UWB antennas close to a dispersive tissue model.*Antennas and Propagation Conference (LAPC)* (pp. 1-4). Loughborough, UK: IEEE. doi:10.1109/LAPC.2012.6403037

Ucar, A., & Balo, F. (2009). Investigation of wind characteristics and assessment of wind generation potentiality in Uludaˇg-Bursa, Turkey. *Applied Energy*, *86*(3), 333–339. doi:10.1016/j.apenergy.2008.05.001

Ultra low power wireless connectivity technology backgrounder. (n.d.). Retrieved October 18, 2014, from http://www.nordicsemi.com/eng/News/Press-Center/Press-Backgrounders/ULP-wireless-connectivity-backgrounder

Umeda, M., Nakamura, K., & Ueha, S. (1997). Energy Storage characteristics of a piezo-generator using impact induced vibration. *Journal of Applied Physics*, *36*(Part 1, No. 5B), 3146–3151. doi:10.1143/JJAP.36.3146

US Department of Energy, Pacific Northwest National Laboratory. (n.d.). Retrieved June, 28, 2013, from http://eioc.pnnl.gov/ research/sitawareness.stm

US Environment Protection Agency. Energy star. (2013). *Report to Congress on Server and Data Center Efficiency: Public Law 109-431*. Retrieved June, 21, 2013, from http://www.energystar.gov/ia/ partners/prod_development/downloads/EPA_Datacenter_Report_Congress_Final1.pdf

Vaidyanathan & Byung-jun. (2002). *Digital filters for gene prediction applications*. IEEE Asilomar on Signals, Systems, and Computers.

Vaidyanathan & Byung-jun. (2002). Digital filters for gene prediction applications. *Thirty-Sixth Asilomar Conference on Signals, Systems and Computers*. doi:10.1109/ACSSC.2002.1197196

Vaidyanathan & Byung-jun. (2002). Gene and exon prediction using allpass-based filters. *IEEE International Workshop on Genomic Signal Processing and Statistics GENSiPS.*

van Breusegem, V., & Boel, R. K. (1992). Maximally permissive feedback controls for controlled state machine. In *Proc. WODES'92*.

Vance, A. (2013). *Inside the Arctic Circle, Where Your Facebook Data Lives*. Retrieved June, 21, 2013, from http://www.bloomberg.com/bw/articles/2013-10-03/facebooks-new-data-center-in-sweden-puts-the-heat-on-hardware-makers

Virginie, G., & Francois, G. (2006). Design of information system architectures using a key-problem framework. *Computers in Industry*, *57*(2), 189–200. doi:10.1016/j.compind.2005.09.001

Vlajic, N., & Moniz, N. (2007). *Self-healing wireless sensor networks: Results that may surprise*. Paper presented at the Globecom Workshops, 2007 IEEE.

Vullers, R. J. M., Schaijk, R. V., Doms, I., Hoof, C. V., & Mertens, R. (2009). Micropower energy harvesting. *Solid-State Electronics*, *53*(7), 684–693. doi:10.1016/j.sse.2008.12.011

Walker, D., & Bier, A. W. (2009). *Oxidation Reduction Potential: Understanding a Challenging Measurement*. Retrieved from http://www.hach.com/asset-get.download.jsa?id=7639984589

Walker, C. K., Kooi, J. W., Chan, M., Leduc, H. G., Schaffer, P. L., Carlstrom, J. E., & Phillips, T. G. (1992). A low-noise 492 GHz SIS waveguide receiver. *International Journal of Infrared and Millimeter Waves*, *13*(6), 785–798. doi:10.1007/BF01011595

Wang, D.-A., & Chang, K.-H. (2010). Electromagnetic energy harvesting from flow induced vibration. *Microelectronics Journal*, *41*(6), 356–364. doi:10.1016/j.mejo.2010.04.005

Wang, D.-A., & Ko, H.-H. (2010). Piezoelectric energy harvesting from flow-induced vibration. *Journal of Micromechanics and Microengineering*, *20*(2), ID025019. doi:10.1088/0960-1317/20/2/025019

Wang, D.-A., & Liu, N.-Z. (2011). A shear mode piezoelectric energy harvester based on a pressurized water flow. *Sensors andActuators, A. Physical*, *167*(2), 449–458.

Wang, H., Park, J.-D., & Ren, Z. (2012). Active energy harvesting from microbial fuel cells at the maximum power point without using resistors. *Environmental Science & Technology*, *46*(9), 5247–5252. doi:10.1021/es300313d PMID:22486712

Wang, H., & Ren, Z. J. (2013). A comprehensive review of microbial electrochemical systems as a platform technology. *Biotechnology Advances*, *31*(8), 1796–1807. doi:10.1016/j.biotechadv.2013.10.001 PMID:24113213

Wang, X. (2012). Piezoelectric nanogenerators-Harvesting ambient mechanical energy at the nanometer scale. *Nano Energy*, *1*(1), 13–24. doi:10.1016/j.nanoen.2011.09.001

Wang, Y., Qiu, Z. A., & Yalamanchili, R. (1994). Meissner model of superconducting rectangular waveguides. *International Journal of Electronics*, *76*(6), 1151–1171. doi:10.1080/00207219408926024

Wan, Z. G., Tan, Y. K., & Yuen, C. (2011). Review on energy harvesting and energy management for sustainable wireless-sensor networks. In *Proceedings of the IEEE 13th International Conference on Communication Technology (ICCT '11).*

Watson, V. J., Nieto Delgado, C., & Logan, B. E. (2013a). Improvement of activated carbons as oxygen reduction catalysts in neutral solutions by ammonia gas treatment and their performance in microbial fuel cells. *Journal of Power Sources*, *242*, 756–761. doi:10.1016/j.jpowsour.2013.05.135

Watson, V. J., Nieto Delgado, C., & Logan, B. E. (2013b). Influence of chemical and physical properties of activated carbon powders on oxygen reduction and microbial fuel cell performance. *Environmental Science & Technology*, *47*(12), 6704–6710. doi:10.1021/es401722j PMID:23692057

Weddell, A. (2010). *Notes from "Energy Harvesting for Wireless Automation" Conference*. Munich: Hanson Wade.

Weddell, A. (2010). *Notes from "Energy Harvesting for Wireless Automation"*. Munich: Conference.

Wei, X., & Zhi, L. (2010). The multi-objective routing optimization of WSNs based on an improved ant colony algorithm. In *Wireless Communications Networking and Mobile Computing (WiCOM), 2010 6th International Conference* (pp. 1-4). Cheng Du, China: IEEE. doi:10.1109/WICOM.2010.5601072

Wei, B., Tokash, J. C., Chen, G., Hickner, M. A., & Logan, B. E. (2012). Development and evaluation of carbon and binder loading in low-cost activated carbon cathodes for air-cathode microbial fuel cells. *RSC Advances*, *2*(33), 12751–12758. doi:10.1039/C2RA21572A

Wengler, M. J. (1992). Submillimeter waves detection with superconducting tunnel junctions. *Proceedings of the IEEE*, *80*(11), 1810–1826. doi:10.1109/5.175257

Werner-Allen, G., Lorincz, K., Ruiz, M., Marcillo, O., Johnson, J., Lees, J., & Welsh, M. (2006, March-April). Deploying a Wireless Sensor Network on an Active Volcano. *IEEE Internet Computing*, *10*(2), 18–25. doi:10.1109/MIC.2006.26

White, N. M., Glynne-Jones, P., & Beeby, S. (2001). A novel thickfilm piezoelectric micro-generator. *Smart Materials and Structures*, *10*(4), 850–852. doi:10.1088/0964-1726/10/4/403

Williams, C. B., Pavic, A., Crouch, R. S., & Wood, R. C. (1998). Feasibility study of vibration-electric generator for bridge vibration sensors. *IMAC-Proceedings 16th International Modal Analysis Conference.*

Williams, C., & Yates, R. (1996). Analysis of a micro-electric generator for microsystems. *Sensors and Actuators A: Physical*, *52*(1-3), 8-11. Retrieved from http://www.sciencedirect.com/science/article/pii/092442479680118X

Williams, C. B., & Yates, R. B. (1995). Analysis of a micro-electric generator for microsystems," in Proceedings of Solid-StateSensors and Actuator and in Eurosensors IX. *Transducers*, *1*, 369–372.

Winfield, J., Greenman, J., Huson, D., & Ieropoulos, I. (2013). Comparing terracotta and earthenware for multiple functionalities in microbial fuel cells. *Bioprocess and Biosystems Engineering*, *36*(12), 1913–1921. doi:10.1007/s00449-013-0967-6 PMID:23728836

Winters, J. H., & Rose, C. (1991). High-T_c superconductors waveguides: Theory and applications. *IEEE Transactions on Microwave Theory and Techniques*, *39*(4), 617–623. doi:10.1109/22.76423

Wired, U. K. (2012). *NASA Wants to Power Robots With Microbes*. Retrieved from http://www.wired.com/2012/01/navy-microbial-fuel-cells/

Wireless Streetlight Control System. (2013). *International Journal of Computer Applications*, 1-7.

Wischke, M., Masur, M., Goldschmidtboeing, F., & Woias, P. (2010). Electromagnetic vibration harvester with piezoelectrically tunable resonance frequency. *Journal of Micromechanics and Microengineering*, *20*(3), 035025. doi:10.1088/0960-1317/20/3/035025

Wischke, M., Masur, M., Kroner, M., & Woias, P. (2011). Vibration harvesting in traffic tunnels to power wireless sensor nodes. *Smart Materials and Structures*, *20*(8), 085014. doi:10.1088/0964-1726/20/8/085014

Woodward, L., Perrier, M., Srinivasan, B., Pinto, R. P., & Tartakovsky, B. (2010). Comparison of real-time methods for maximizing power output in microbial fuel cells. *AIChE Journal. American Institute of Chemical Engineers*, *56*(10), 2742–2750. doi:10.1002/aic.12157

WordPress.com. (2011, March 21). *Cool Formula For Calculating Skyscraper Sway*. Retrieved March 5, 2015, from https://mathspig.wordpress.com/tag/taipei-101-tower/

Xen Users' Manual v 3.3. (2008). Retrieved June, 3, 2013, from http://bits.xensource.com/Xen/docs/user.pdf

Xia, L., Cheng, J., Glover, N. E., & Chiang, P. (2014, June6). 0.56 V, –20 dBm RF-Powered, Multi-Node Wireless Body Area Network System-on-a-Chip With Harvesting-Efficiency Tracking Loop. *IEEE Journal of Solid-State Circuits*, *49*(6), 1345–1355. doi:10.1109/JSSC.2014.2305074

Xiao, W., & Elnosh, A. (2011). Overview of maximum power point tracking technologies for photovoltaic power systems. *IECON 2011-37th*. Retrieved from http://ieeexplore.ieee.org/xpls/abs_all.jsp?arnumber=6119946

Xia, X., Zhang, F., Zhang, X., Liang, P., Huang, X., & Logan, B. E. (2013). Use of pyrolyzed iron ethylenediaminetetraacetic acid modified activated carbon as air-cathode catalyst in microbial fuel cells. *ACS Applied Materials & Interfaces*, *5*(16), 7862–7866. doi:10.1021/am4018225 PMID:23902951

Xie, X. (1998). Perturbation analysis of stochastic Petri nets. *IEEE Transactions on Automatic Control*, *43*(1), 76–80. doi:10.1109/9.654889

Xu, W., Mu, C., & Jin, J. (2014). Novel Linear Iteration Maximum Power Point Tracking Algorithm for Photovoltaic Power Generation. *IEEE Transactions on Applied Superconductivity*, *24*(5), 1–6. doi:10.1109/TASC.2014.2333534

Xu, Z. (2010, June). Building a power-aware database management system. In *Proceedings of the Fourth SIGMOD PhD Workshop on Innovative Database Research* (pp. 1-6). ACM. doi:10.1145/1811136.1811137

Yalamanchili, R., Qiu, Z. A., & Wang, Y. (1995). Rectangular waveguides with two conventional and two superconducting walls. *International Journal of Electronics*, *78*(4), 715–727. doi:10.1080/00207219508926205

Yang, X. S. (2013). Firefly algorithm: Recent advances and applications. *Int J. Swarm Intelligence.* DOI: 10.1504/IJSI.2013.055801

Yang, Q., Feng, Y., & Logan, B. E. (2012). Using cathode spacers to minimize reactor size in air cathode microbial fuel cells. *Bioresource Technology*, *110*, 273–277. doi:10.1016/j.biortech.2012.01.121 PMID:22342583

Yang, X. S. (2010). *Engineering optimization: An introduction with metaheuristic applications.* Wiley & Sons. doi:10.1002/9780470640425

Yassin, G., Tham, C. Y., & Withington, S. (2003). Propagation in lossy and superconducting cylindrical waveguides. *Proceedings of* the *14th International Symposium on Space Terahertz Technology*.

Yassin, G., Jung, G., Dikovsky, V., Barboy, I., Kambara, M., Cardwell, D. A., & Withington, S. (2001). Investigation of microwave propagation in high-temperature superconducting waveguides. *IEEE Transactions on Microwave Theory and Techniques*, *11*, 413–415.

Yeap, K. H., Tham, C. Y., Nisar, H., & Loh, S. H. (2013). Analysis of Probes in a Rectangular Waveguide. In J. Rolf (Ed.), *Frequenz* (pp. 145–154). Berlin: Schiele und Schön. doi:10.1515/freq-2012-0117

Yeap, K. H., Tham, C. Y., Nisar, H., & Loh, S. H. (2013). Analysis of probes in a rectangular waveguide. *Frequenz J. of RF-Engineering and Telecommunications*, *67*, 145–154.

Yeap, K. H., Tham, C. Y., Yassin, G., & Yeong, K. C. (2011a). Attenuation in rectangular waveguides with finite conductivity walls. *Radioengineering J.*, *20*, 472–478.

Yeap, K. H., Tham, C. Y., Yassin, G., & Yeong, K. C. (2011b). Propagation in lossy rectangular waveguides. In A. A. Kishk (Ed.), *Electromagnetic waves propagation in complex matter* (pp. 255–272). Intech.

Yeap, K. H., Tham, C. Y., Yeong, K. C., & Yeap, K. H. (2009). A simple method for calculating attenuation in waveguides. *Frequenz J. of RF-Engineering and Telecommunications*, *63*, 236–240.

Yee, S. T., & Ventura, J. A. (2000). Phase-type approximation of Petri nets for analysis of manufacturing systems. *IEEE Trans. on Rob. and Autom.*, (3), 318-322.

Ye, G., & Soga, K. (2012). Energy harvesting from water distribution systems. *Journal of Energy Engineering*, *138*(1), 7–17. doi:10.1061/(ASCE)EY.1943-7897.0000057

Yick, J., Mukherjee, B., & Ghosal, D. (2008). Wireless sensor network survey. *Computer Networks*, *52*(12), 2292–2330. doi:10.1016/j.comnet.2008.04.002

Yin, C., & Yue, S. S.-T. (2007). Prediction of protein coding regions by the 3-base periodicity analysis of a DNA sequence. *Journal of Theoretical Biology*, *247*(4), 687–694. doi:10.1016/j.jtbi.2007.03.038 PMID:17509616

Yong, Y.-T., Chow, C.-O., Kanesan, J., & Ishii, H. (2009). A survey on design of self-powered wireless sensor network. In *Proceedings of the International Conference for Technical Postgraduates(TECHPOS '09).* doi:10.1109/TECHPOS.2009.5412044

Youssef, M., Younis, M. F., & Arisha, K. (2002). *A constrained shortest-path energy-aware routing algorithm for wireless sensor networks.* Paper presented at the Wireless Communications and Networking Conference, 2002. doi:10.1109/WCNC.2002.993370

YSI. (2009). *The Dissolved Oxygen Handbook.* Retrieved from https://www.fondriest.com/pdf/ysi_do_handbook.pdf

Yu, Y., Govindan, R., & Estrin, D. (2001). *Geographical and energy aware routing: A recursive data dissemination protocol for wireless sensor networks. Technical report ucla/csd-tr-01-0023.* UCLA Computer Science Department.

Zhang, F., Chen, G., Hickner, M., & Logan, B. (2012). Novel anti-flooding poly (dimethylsiloxane)(PDMS) catalyst binder for microbial fuel cell cathodes. *Journal of Power Sources*. Retrieved from http://www.sciencedirect.com/science/article/pii/S0378775312010981

Zhang, L., Hurley, W., & W[UNKNOWN ENTITY &odie;]lfle, W. (2011). A new approach to achieve maximum power point tracking for PV system with a variable inductor. *Power Electronics, IEEE, 1*, 948–952. Retrieved from http://ieeexplore.ieee.org/xpls/abs_all.jsp?arnumber=5609212

Zhang, X., Ruoshui, W., Molin, H., & Martinot, E. (2010). A study of the role played by renewable energies in China's sustainable energy supply. *Energy, 35*(11), 4392–4399. doi:10.1016/j.energy.2009.05.030

Zhao, C., Gai, P., Liu, C., Wang, X., Xu, H., Zhang, J., & Zhu, J.-J. (2013). Polyaniline networks grown on graphene nanoribbons-coated carbon paper with a synergistic effect for high-performance microbial fuel cells. *Journal of Materials Chemistry A, 1*(40), 12587. doi:10.1039/c3ta12947k

Zhou, G., Huang, L., Li, W., & Zhu, Z. (2014). Harvesting Ambient Environmental Energy for Wireless Sensor Networks: A Survey. *Journal of Sensors*, 815467.

Zhou, H.-Y., Wu, F., & Hou, K.-M. (2008). An event-driven multithreadingreal-time operating system dedicated to wirelesssensor networks. In *Proceedings of the International Conferenceon Embedded Software and Systems (ICESS '08).*

Zhou, Z., Holland, P. M., & Igic, P. (2014). MPPT algorithm test on a photovoltaic emulating system constructed by a DC power supply and an indoor solar panel. *Energy Conversion and Management, 85*, 460–469. doi:10.1016/j.enconman.2014.06.007

Zhu, Q., David, F. M., Devaraj, C. F., Li, Z., Zhou, Y., & Cao, P. (2004, February). Reducing energy consumption of disk storage using power-aware cache management. In *Software, IEEE Proceedings* (pp. 118-118). IEEE.

Zhuang, L., Zheng, Y., Zhou, S., Yuan, Y., Yuan, H., & Chen, Y. (2012). Scalable microbial fuel cell (MFC) stack for continuous real wastewater treatment. *Bioresource Technology, 106*, 82–88. doi:10.1016/j.biortech.2011.11.019 PMID:22197329

Zhu, D., Beeby, S., Tudor, J., White, N., & Harris, N. (2010). A novel miniature wind generator for wireless sensing applications. In *Proceedings of the 9th IEEE Sensors Conference (SENSORS '10).*

Zhu, D., Tudor, M. J., & Beeby, S. P. (2010, February). Strategies for increasing the operating frequency range of vibration energy harvester: A review. *Measurement Science & Technology, 21*(2), 1–29. doi:10.1088/0957-0233/21/2/022001

Zhu, X., Tokash, J. C., Hong, Y., & Logan, B. E. (2013). Controlling the occurrence of power overshoot by adapting microbial fuel cells to high anode potentials. *Bioelectrochemistry (Amsterdam, Netherlands), 90*, 30–35. doi:10.1016/j.bioelechem.2012.10.004 PMID:23178374

Zorlu, O., Topal, E. T., & Kulah, H. (2011, February). A vibration-based electromagnetic energy harvester using mechnical frequency up-conversion method. *Sensors Journal, IEEE, 11*(2), 481–488. doi:10.1109/JSEN.2010.2059007

Zuo, L., & Cui, W. (2013). Dual-Functional Energy-Harvesting and Vibration Control: Electromagnetic Resonant Shunt Series Tuned Mass Dampers. *Journal of Vibration and Acoustics, 135*(5), 510181–510189. doi:10.1115/1.4024095 PMID:23918165

Zuo, L., & Tang, X. (2013). Large-scale vibration energy harvesting. *Journal of Intelligent Material Systems and Structures, 24*(11), 1405–1430. doi:10.1177/1045389X13486707

Zurawski, R., & Zhon, M. C. (1994). Petri nets and industrial applications: A tutorial. *IEEE Transactions on Industrial Electronics, 41*(6), 567–583. doi:10.1109/41.334574

About the Contributors

Vasaki Ponnusamy is an Assistant Professor at Universiti Tunku Abdul Rahman, Malaysia. She obtained her Bachelor of Computer Science and MSc (Computer Science) from Science University of Malaysia and her PhD in IT from Universiti Teknologi PETRONAS (UTP), Malaysia (2013). She is currently working on biologically-inspired computing, wireless sensor network and energy harvesting.

Noor Zaman acquired his degree in Engineering in 1998, and Master's in Computer Science at the University of Agriculture at Faisalabad in 2000. His academic achievements further extended with a PhD in Information Technology at University Technology Malaysia (UTP). He has vast experience of 16 years in the field of teaching and research. He is currently working as an Assistant Professor at the College of Computer Science and Information Technology, King Faisal University, in Saudi Arabia. He has contributed well in King Faisal University for achieving ABET Accreditation, by working as a member and Secretary for Accreditation and Quality cell for more than 08 years. He takes care of versatile operations including teaching, research activities, leading ERP projects, IT consultancy and IT management. He headed the department of IT, and administered the prometric center in the Institute of Business and Technology (BIZTEK), in Karachi Pakistan. He has worked as a consultant for Network and Server Management remotely in Apex Canada USA base Software house and call center. Dr. Noor Zaman has authored several research papers, and edited five international reputed Computer Science area books, has many publications to his credit. He is an associate Editor, Regional Editor and reviewer for reputed international journals and conferences around the world. He has completed several research grants and currently involved with funded projects in different courtiers. His areas of interest include Wireless Sensor Network (WSN), Internet of Things IoT, Mobile Application Programming, Ad hoc Networks, Cloud Computing, Big Data, Mobile Computing, Software Engineering, Unix, and Linux.

Tang Jung Low obtained his Bachelor of Engineering in Computer Technology from Teesside University, UK (1989), MSc IT from National University of Malaysia (2001), and his PhD in IT from Universiti Teknologi PETRONAS (UTP), Malaysia (2012). He is currently a Senior Lecturer in the Computer and Information Sciences Department of UTP. Dr. Low has been in the academic line for two decades already as lecturer cum researcher in various public and private institutes of higher learning. He teaches various engineering and ICT courses. His research interest include wireless/mobile technology, embedded systems, wireless sensor networking, robotics, and Green IT. His current R&D include: bio-inspired energy efficient routing in WSN, sensors location optimization, and green by IT.

Anang Hudaya Muhamad Amin is a senior lecturer in the Faculty of Information Science and Technology at Multimedia University in Malaysia. He received a BTech (Hons.) in information technology from Universiti Teknologi PETRONAS and a masters in network computing and PhD from Monash University. His research interests include artificial intelligence with specialization in distributed pattern recognition and bio-inspired computational intelligence, wireless sensor networks, and distributed computing.

* * *

Muneer Ahmad has completed his Bachelors and Masters in Computer Sciences, he further extended his qualification with PhD in Information Technology from University Technology PETRONAS UTP, Malaysia. He has credit for several research publications including world recognized conference and journals. He has more than 12 years of University level teaching experience.

Shakeel Ahmed received his B.Sc. (Computer. Science in 1997) from Kakatiya University and M.C.A. (Master of Computer Applications in 2000) from M. K. University, India. He earned his PhD degree in Computer Science from Devi Ahilya University, Indore, India, in 2013. He has served at various universities and scientific organizations in India and Saudi Arabia. Currently, he is working at King Faisal University as Assistant Professor in the College of Computer Sciences and Information Technology. His current research interests include Mobile Ad hoc networks, Software Engineering, Modeling of Systems using Formal Approaches, Big data and data mining.

A. Rashid A. Aziz received his Bachelor of Science in 1988, the Master of Science degree in 1990 and a Ph.D. degree in Mechanical Engineering from the University of Miami, USA, in 1995. He is a Professor in the Mechanical Engineering Dept. and heads the Centre for Automotive Research and Electric Mobility (CAREM), Universiti Teknologi PETRONAS. His research interests include internal combustion engines, fluid mechanics, computational fluid dynamics, heat transfer and alternative energy. He is currently holding the position of the Deputy Vice Chancellor responsible for research & innovation.

Agnes Ruey Chyi Cheah received her Bachelor of Engineering (HONS) Electronic Engineering from Universiti Tunku Abdul Rahman in 2014, currently pursuing her Master of Engineering Science in Universiti Tunku Abdul Rahman as well. Besides, she is a graduate engineer of BEM.

Calin Horatiu Ciufudean is an Associate Professor Ph.D. Eng. "Stefan cel Mare" University of Suceava, Faculty of Electrical Engineering and Computer Science, Department of Computers, Control Systems, and Electronics. - Courses and laboratories: Industrial Control Systems, Reliability and Diagnosis, Discrete Event Systems, Data Acquisition Systems; – Researcher in control systems field, environment and sustainable development; – Researcher in the teams and director of over 34 research projects; – Trainer of student groups for national and international student competitions; – Editor of the Romanian Journal of Ecology; – Author of over 170 scientific papers published in scientific journals and conferences proceedings abroad and in Romania; – Author of 12 books published in Romania and abroad; – Author or co-author of 32 inventions. - Technical Expert of the Romanian Ministry of Justice. - President of the Romanian Society of Electrical & Control Engineering, Suceava Branch. - He is a member of the editorial boards of several international scientific journals and conferences of control systems and electric engineering science. He was designated chairmen and/or key note speaker at 35 international conferences.

António Espírito-Santo started in 1996 his professional activity at the Electrical Energy Management Laboratory of the University of Coimbra. At the same time, he cooperated with a private enterprise that offers services in the energy management field. Since 1997, he works at University of Beira Interior (UBI), developing the following activities: teaching theoretical and practical classes of Electronics, Embedded Systems and Signal Analysis; supervise and advice the work of graduation, master, and PhD students; research and publication of findings. He obtained a M.Sc. in Electrical Engineering in 2001. He concluded a Ph.D. in Electrical Engineering in 2008.

Lam Hong Lee received a Bachelor of Computer Science degree from Universiti Putra Malaysia in 2004, and a Ph.D. degree in Computer Science from the University of Nottingham in 2009. He joined Universiti Tunku Abdul Rahman, Malaysia since March 2009 as an Assistant Professor. He is currently an Associate Professor at Quest International University Perak, Malaysia since September 2013. His current research interest lies in improving text categorization using AI techniques, specifically Support Vector Machines. Besides this, he is also investigating on the implementation of data mining, pattern recognition and machine learning techniques in various kinds of intelligent systems.

Kanahavalli Mardamutu works with the Global Top Fortune 100 company namely PETRONAS since August 2015. She is currently engaged with the Technology enabler division - SAP doing strategic and planning to optimise resources and also researching on integration of Internet of Things to reduce cost and maximise data management. She received her Bachelor of Computer Science from Quest International University Perak, Malaysia. Her research interest includes analysis and designing new formula to optimise energy utilisation.

Saira Muzafar acquired her First Masters degree in Applied Physics with specialization in Electronics from Pakistan and Second Masters Degree in Telecommunications from Pakistan. Her areas of interest include Wireless Sensor Network (WSN), Cloud Computing, Big Data, and Mobile Computing She is currently working as a Faculty member in the College of Computer Science and Information Technology, King Faisal University, in Saudi Arabia. She takes care of versatile operations including teaching and research activities. Contact Information College of Computer Science & Information Technology, ABET Accredited King Faisal University, Al Ahsa, Al Hofuf, Saudi Arabia.

Humaira Nisar received the BE (Honors) in Electrical Engineering from University of Engineering and Technology, Lahore, Pakistan. She received MS degree in Nuclear Engineering from Quaid-e-Azam University, Islamabad, Pakistan. She received MS degree in Mechatronics and Ph.D. in Information and Mechatronics from Gwangju Institute of Science and Technology, Republic of Korea. Currently, she is Assistant Professor at Department of Electronic Engineering, Universiti Tunku Abdul Rahman, Malaysia. She is also senior member of IEEE. Her research interests include signal and image processing, bio medical imaging, brain signal analysis, motion estimation and video compression.

Beng Lee Ooi obtained a master of engineering degree from the University of Hull,UK, in 2007 and with Ph.D. in energy harvesting in 2011. He was a postdoctoral researcher with Centre for Automotive Research and Electric Mobility (CAREM), Universiti Teknologi PETRONAS (Malaysia) between 2013 and 2014. In 2014, he joined Quest International University Perak (Malaysia) as a Lecturer. His research interest includes vibration energy harvesting devices for environmental sensing purposes.

Pedro Serra has an undergraduation in Biomedical Sciences and a master degree in Electrotechnical and Computer Engineering (Bionics), and has always strived with his deepest commitment. Pursuing a PhD in Electrotechnical Engineering, he's also a freelance translator and a gymnastics coach. His current academic work is related with low power devices and wastewater energy harvesting.

Yen Pei Tay has 13 years of industry experience at top Global Fortune 500 companies i.e. Motorola, Nokia and IBM. During his tenure, he gained valuable professional experience by holding various job roles in software development, systems integration, product management, pre-sales & consulting. With strong telecommunications background, he has in-depth knowledge in 3GPP standards development in 4G LTE and Access Network Discovery and Selection Function (ANDSF) technologies. With strong affection in continuing his post-graduate studies, Yen Pei current research largely focuses on new disciplines of innovation namely Disruptive Innovation and Big Data. He is currently a lecturer of Faculty of Business, Management and Social Sciences and the Advisor of Innovation Society at Quest International University Perak, Malaysia.

Choy Yoong Tham retired as professor and dean of the School of Science and Technology, Wawasan Open University, Penang, Malaysia in 2013.

Chung Ket Thein obtained his Bachelors and PhD in Mechanical Engineering from University of Hull, United Kingdom. He joined Taylor's University, Malaysia in 2013 as a lecturer. His research interests are the development of reliability assessment and the engineering design optimisation using finite element method.

Kim Ho Yeap received his Bachelor of Engineering (Hons) Electrical and Electronic Engineering from Petronas University of Technology (UTP) in 2004, Master of Science in Microelectronics from National University of Malaysia (UKM) in 2005, and PhD in Engineering from Tunku Abdul Rahman University (UTAR) in 2011. In 2008, He was funded by both Tunku Abdul Rahman University and University of Oxford to undergo 2 months of research attachment in University of Oxford. Kim Ho is a member of the IET, IEEE, and IAENG, a graduate engineer of the BEM, as well as, a graduate member of the IEM. He is also a Chartered Engineer (CEng) registered with the UK Engineering Council. From July 2005 to December 2007, he was a lecturer with Tunku Abdul Rahman University (Malaysia) in the KL campus. In early December 2007, he was deployed to the Perak main campus. He is currently an assistant professor in the Department of Electronic Engineering (DEE) in the Faculty of Engineering and Green Technology (FEGT). Throughout his career, Kim Ho has served in various administrative capacities, including the Head of Programme of Master of Engineering Science, the Head of Department of Electronic Engineering, and the chairperson of the Self Assessment Committee (SAC). Kim Ho has been actively involved in teachings, as well as research. Since 2005, he has taught more than 10 different subjects. Kim Ho's research areas of interests are in electromagnetics and microelectronics. Since 2006, he has published about 19 international refereed journal papers, 4 national refereed journal papers, 24 international conference proceedings, 2 books, and 2 book chapters. He also serves as the reviewer for 1 book chapter, 10 journals and 36 conferences / symposiums / workshops. Kim Ho is a member of the advisory board of the International Journal of Electronics, Communication and, Computer Technology (IJECCT), Advances in Computational Research, and the World Research Journal of Telecommunications Systems. He is also a member of the editorial board of the International Journal of Engineering and Manufacturing (IJEM) and the i-manager's Journal on Digital Signal Processing (JDP). Kim Ho has been given various awards which include 4 Kudos awards from Intel Microelectronics and 13 research grants. In 2010 and 2011, Kim Ho was invited by the Perak Entrepreneur and Skills Development Centre (PESDC) and Penang Skills Development Centre (PSDC) to conduct training on electronic sensors and RF / microwave, respectively. Kim Ho's biography is also listed in the 2012, 2013, and 2014 editions of Marquis Who's Who in the World and published in the Jan/Feb 2012 edition of the MBA Edge magazine.

Kee-Choon Yeong works as an Associate Professor in the Department of Electronic Engineering, UTAR Malaysia. He received his BSc degree from NUS, Singapore, MSc from BGSU, Ohio, US and PhD from RPI, Troy, US. His research interests include analysis and design of electromagnetic structures and photonics.

Chun Keat Yew is a final year PhD candidate of School of Engineering, University of Hull. His research interest includes energy harvesting, MEMS, human center electronics and renewable energy.

Cheah Wai Zhao completed his degree in Bachelor of Computer Science (Hons) majoring in distributed system at Quest International University Perak. His research mainly focus on cloud computing, Internet of Thing (IoT) and mobile ad hoc network. Previously, he received an award as IPV6 Silver certified engineer.

Index

M

N

O

P

Q

R

S

V

W

www.ingramcontent.com/pod-product-compliance
Ingram Content Group UK Ltd.
Pitfield, Milton Keynes, MK11 3LW, UK
UKHW050951290926
15470UKWH00005B/1763